Current Topics in Bioenergetics

Volume 7
Photosynthesis: Part A

Current Topics in Bioenergetics

Edited by
D. RAO SANADI

Boston Biomedical Research Institute
Boston, Massachusetts

and
LEO P. VERNON

Brigham Young University
Provo, Utah

VOLUME 7
PHOTOSYNTHESIS: PART A

1978

ACADEMIC PRESS
NEW YORK   SAN FRANCISCO   LONDON
A Subsidiary of Harcourt Brace Jovanovich, Publishers

ACADEMIC PRESS, INC.
111 Fifth Avenue, New York, New York 10003

United Kingdom Edition published by
ACADEMIC PRESS, INC. (LONDON) LTD.
24/28 Oval Road, London NW1 7DX

LIBRARY OF CONGRESS CATALOG CARD NUMBER: 66–28678

ISBN 0–12–152507–4

PRINTED IN THE UNITED STATES OF AMERICA

Contents

PHOTOCHEMICAL ASPECTS

Photochemistry of Chlorophyll in Solution: Modeling Photosystem II

G. R. SEELY

Picosecond Events and Their Measurement

MICHAEL SEIBERT

The Primary Electron Acceptors in Green-Plant Photosystem I and Photosynthetic Bacteria

BACON KE

The Primary Reaction of Chloroplast Photosystem II

DAVID B. KNAFF AND RICHARD MALKIN

ELECTRON TRANSPORT AND PHOTOPHOSPHORYLATION

Photosynthetic Electron-Transport Chains of Plants and Bacteria and Their Role as Proton Pumps

A. R. CROFTS AND P. M. WOOD

The ATPase Complex of Chloroplasts and Chromatophores

RICHARD E. MCCARTY

[The third article in this section will be in Volume 8.]

List of Contributors

Numbers in parentheses indicate the pages on which the authors' contributions begin.

A. R. CROFTS* (175), *Department of Biochemistry, University of Bristol, Medical School, Bristol, England*

BACON KE (75), *Charles F. Kettering Research Laboratory, Yellow Springs, Ohio*

DAVID B. KNAFF† (139), *Department of Cell Physiology, University of California, Berkeley, California*

RICHARD E. MCCARTY (245), *Section of Biochemistry, Molecular and Cell Biology, Cornell University, Ithaca, New York*

RICHARD MALKIN (139), *Department of Cell Physiology, University of California, Berkeley, California*

G. R. SEELY (3), *Charles F. Kettering Research Laboratory, Yellow Springs, Ohio*

MICHAEL SEIBERT‡ (39), *GTE Laboratories, Inc., Waltham, Massachusetts*

P. M. WOOD (175), *Department of Biochemistry, University of Cambridge, Cambridge, England*

* Present address: Department of Physiology and Biophysics, University of Illinois, Urbana, Illinois.
† Present address: Department of Chemistry, Texas Tech University, Lubbock, Texas.
‡ Present address: Solar Energy Research Institute, 1536 Cole Blvd., Golden, Colorado 80401.

Preface

Photosynthesis has always been a fruitful field for investigation, not only because of its inherent importance, but also because it is an integrated biological process which has traditionally attracted the interest of the physicist, the chemist, and the biologist. The major advances in our understanding of the process include contributions from each of the disciplines: the fundamental study by Priestley of the gases involved, the quantum requirement experiments of Warburg and Emerson, Arnold's research into the size of the photosynthetic unit, the pathway of carbon dioxide fixation by Calvin's group, Van Niel's elegant work on comparative photosynthesis, and so on.

The articles contained in this volume and the forthcoming Volume 8 reflect this same broad approach to the modern study of photosynthesis. The experimental sophistication has extended markedly to permit picosecond measurements of the primary photophysical and photochemical reactions, the study of individual polypeptide components of the chlorophyll–protein reaction center complexes, detection of the individual complexes and their arrangement in the photosynthetic membrane, and the measurement of proton and other ion movement across the membrane as a function of the photosynthetic electron transfer reactions.

In reviewing the articles for Volumes 7 and 8, it became evident that we are no longer dealing with black boxes with alphabets in them, but with specific molecules, their function, and their interaction with other specific molecules. One gains the impression that the major areas of investigation of photosynthesis have been defined and we are in a "mopping up" phase in which we supply those final, critical pieces of the puzzle which tie everything together. It is obvious that we are moving quickly on all fronts, just as in most areas of biology, and the pieces are quickly falling into place. We feel, however, that there are still a few surprises left before the picture is complete.

Contents of Previous Volumes

xi

Photochemical Aspects

Photochemistry of Chlorophyll in Solution: Modeling Photosystem II[1]

G. R. SEELY

Charles F. Kettering Research Laboratory
Yellow Springs, Ohio

I. Photophysical and Photochemical Processes of Chlorophyll in Solution

A. INTRODUCTION: THE MODEL SYSTEM APPROACH

The investigator who wishes to understand how photosynthesis works is confronted with systems that have undergone a billion years or more of evolution. The steps by which the photosynthetic system of green plants evolved, with its two photosystems and their ability to oxidize water, lie perhaps forever beyond the reach of paleontology. For example, it is not easy to tell which components of the photosynthetic unit (PSU) are really necessary for it to function, which assist the process in some way, and which merely occupy space or are involved in other reactions.

The essence of the model system approach is to take a small number of presumably necessary components, and seek conditions under which they operate effectively to produce the desired result. In photosynthesis,

---

[1] Contribution No. 583 from C. F. Kettering Research Laboratory.

a desired result is the efficient conversion of sunlight into chemical free energy. The investigation of models of the photosynthetic energy conversion system is already well advanced. It is hoped through these studies not only to learn how photosynthesis works in the plant, but also how to construct entirely artificial systems for conversion of solar energy into chemical or electrochemical energy.

It should be remembered that the primary purpose of studying model systems is to understand the operation of the systems they are modeled after, not to duplicate them. We are bound to rediscover the mistakes of the remote evolutionary past, and with luck, to obviate them eventually. Failure may be much more instructive than success, and is certainly much easier to come by.

After a critical review of chlorophyll photoprocesses, we shall examine in detail one intensively studied system, the photochemical reaction between chlorophyll and quinones in homogeneous systems as a model of the primary reactions of photosystem II (PS II). The emphasis will be on the kinetics and energetics of reactions of chlorophyll a (chlorophyll, chl), with other photosynthetic pigments introduced sparingly for comparison only.

B. THE SINGLET EXCITED STATE

1. Absorption of Light

Chlorophyll absorbs light throughout the visible spectrum. Weiss (1972) discerned seven electronic transitions between 300 and 700 nm. Absorption in the region of 660–670 nm raises chlorophyll to the lowest vibrational level of the first singlet excited state (chl*), process (1). The

$$\text{chl} \xrightarrow{h\nu} \text{chl*}, \quad \Delta E = 180 \text{ kJ/mole} \tag{1}$$

position of the band peak is slightly affected by solvent and temperature, being shifted to the red by solvents of high polarizability (refractive index) and by lowering the temperature (Seely and Jensen, 1965; Broyde and Brody, 1967; Singhal et al., 1968). These shifts are small, and in most organic solvents the peak is located within 1% of 665 nm. Since absorption of light takes place much faster than molecular or atomic motions can conform, this corresponds to an internal energy increase (ΔE) at constant volume of 180 kJ/mole. Energy absorbed in electronic transitions at higher frequencies is so rapidly degraded by the process of *internal conversion* into vibrational energy (heat), that the absorption of sunlight of any frequency may be described simply by process (1).

The natural or radiative lifetime τ_0 can be calculated by a weighted integration of the part of the absorption spectrum that corresponds to the first electronic excited state, provided the transition is strongly allowed, and the fluorescence spectrum bears an approximately mirror-image relationship to the absorption spectrum (Strickler and Berg, 1962). Fetisova and Borisov (1973–1974) have carried out the integration for chlorophyll in various media and concluded a value of $\tau_0 = 15.6 \times 10^{-9}$ second in ether, or a rate constant of $k_f^* = 6.4 \times 10^7$ sec^{-1} for process (2).

$$\text{chl}^* \xrightarrow{\;k_f^{\,\circ}\;} \text{chl} + \text{fluorescence} \tag{2}$$

2. Singlet State Decay

The peak of the fluorescence spectrum lies at slightly lower frequencies than the peak of absorption, because a small amount of work is done against the medium after excitation of the molecule. The difference, or Stokes shift, is small, usually from 6 to 10 nm (Singhal and Hevesi, 1971), and the enthalpy (ΔH) retained in the first excited state is about 99% of ΔE, e.g., 178 kJ/mole or 1.84 eV in ethanol. Because this is an energy difference between two definite, thermally equilibrated electronic states of the system, it is also equals (or very nearly equals) the free energy (ΔG) stored in the excited state.

Chl* may decay not only by process (2) but also by internal conversion to the ground state (3), and by intersystem crossing to the triplet

$$\text{chl}^* \xrightarrow{\;k_{ic}^*\;} \text{chl} \tag{3}$$

$$\text{chl}^* \xrightarrow{\;k_{isc}^*\;} \text{chl}^T \tag{4}$$

state of chlorophyll, chl^T (4). Process (3) requires dissipation of electronic energy into vibrational motions, and its rate decreases exponentially with the amount of energy that must be redistributed (Gouterman and Holten, 1977). Process (4) also requires redistribution of energy, and in addition a change of spin multiplicity. However, if higher triplet states exist that can receive energy from the singlet state, the energy difference may be small and the rate constant large. Dreeskamp et al. (1974) have demonstrated this for a number of hydrocarbons and heterocyclics.

Bowers and Porter (1967) showed that for chlorophylls a and b in ether, chl* decays only by processes (2) and (4), within experimental error, and (3) is too slow to detect. Their fluorescence yields (ϕ_f) for chl a and chl b were 0.32 and 0.12, and their corresponding triplet state yields (ϕ_T) were 0.64 and 0.88. It is not known whether (3) is important in some other solvent, but the chances are that it is not.

The observed lifetime τ of the singlet state is then given by (5). The lifetime of chl* reported by Kelly and Porter (1970) in ethanol (5.3 $\times$

$$\tau \equiv 1/k_d{}^* = 1/(k_f{}^* + k_{isc}^*) = \phi_f\tau_0 \tag{5}$$

10^{-9} sec) is close to values tabulated by Livingston (1960) for both ethanol and ether, and compares well with the calculated value of $\phi_f\tau_0$ (5.0 $\times$ 10^{-9} sec) in the latter solvent.

3. Fluorescence Quenching

When foreign substances are in solution together with chlorophyll, other modes of energy loss from the singlet excited state become possible. In the presence of a quenching substance Q, the fluorescence of chlorophyll is diminished, usually in accord with the Stern–Volmer equation (6), where $\phi_f{}^0$ is the yield in the absence of quencher, and

$$\phi_f{}^0/\phi_f = 1 + K_{sv}[Q] \tag{6}$$

$K_{sv} = k_Q{}^*\tau$, where $k_Q{}^*$ is the bimolecular rate constant of reaction between chl* and Q.

Quenching mechanisms vary. Those for which substantial evidence exists for chlorophyll are the following.

a. Energy Transfer. Singlet-state energy can be transferred directly to an acceptor molecule of equal or lower excited state energy by interaction with the short-range radiation field of the donor. The molecules need not be in contact, and transfer can span distances of 50–60 Å efficiently (Knox, 1975). Few examples are known of transfer from chl* *a*, probably because there are few stable molecules that absorb at lower frequencies than chl *a*. Transfer to bacteriochlorophyll has been observed, by excitation of acceptor fluorescence, for closely associated pairs of molecules (Zen'kevich *et al.*, 1975) and in solution when the acceptor and donor molecules are separated, but bound to the same high polymer (Seely, 1976). Of course, transfer *to* chl *a* from chl *b* and other light-gathering pigments is a normal part of the operation of the photosynthetic unit.

b. Action of Oxidants and Reductants. An early tabulation of Stern–Volmer constants for organic quenchers of chlorophyll fluorescence showed clearly that the better quenchers were known oxidants and reductants, that oxidants were generally much more effective than reductants, and that aromatic compounds with similar form but without marked redox properties were inert (Livingston and Ke, 1950). There exists a definite correlation between K_{sv} and the polarographic reduction potential of a series of aromatic nitro compounds (Seely, 1969b). These facts strongly suggest that the quenching process is transfer of an electron from chlorophyll to the quencher, or vice versa. However, it is also possible that quenching proceeds through a molecular complex (exciplex) between chl* and Q, not involving complete electron transfer, the strength of which depends on the redox properties of Q. This question is not yet resolved and will be brought up again in Section II,G.

The rate constants k_Q^*, calculated for good oxidants from K_{sv} and τ, are often very large, being comparable to, and sometimes larger than, calculated diffusion-limited rate constants in the same solvent. Although this by itself is no cause for alarm, they are often about ten times as large as the corresponding constants for electron transfer from chl^T to the same oxidants (cf. Table I), which are often regarded also as "diffusion limited." Zamaraev and Khairutdinov (1974) have emphasized the importance of "tunneling" of electrons through potential barriers of up to 30 Å in length to electron transfer processes in condensed media including biological systems. This, and the suspected retardation of the mutual diffusion of two molecules at close range (Emeis and Fehder, 1970), may cooperate to explain this apparent paradox, but the proper development of this theme lies far beyond the scope of this review.

    c. Enhanced Intersystem Crossing. Addition of solutes containing atoms of large atomic number increases the rate of crossing from the excited singlet state to a triplet state by increasing the spin-orbit coupling factor (McGlynn *et al.*, 1963). Holten *et al.* (1976) have employed 8 M CH$_3$I to reduce the lifetime of singlet excited bacteriopheophytin from 2×10^{-9} to 0.35×10^{-9} sec. Dzhagarov (1970) has reported a similar effect of CH$_3$I on chlorophyll. Song (cited in Weiss, 1972) calculated that the third triplet state of chlorophyll should lie close in energy to the first singlet excited state, so intersystem crossing might proceed rapidly through that.

    d. Self-quenching. The fluorescence of chlorophyll in concentrated solutions in ether is quenched (Watson and Livingston, 1950). Beddard and Porter (1976) have interpreted these data in terms of transfer of energy to weakly associated pairs of chlorophyll molecules, where it is quenched by an unknown mechanism. The phenomenon is rather general in condensed chlorophyll systems where the concentration is high enough, and has been measured, for example, in micelles (Zen'kevich *et al.*, 1972), polymer solutions (Seely, 1967), monolayers (Costa *et al.*, 1972), and lecithin films (Kelly and Patterson, 1971). The triplet state of the chlorophyll is not produced by this kind of quenching, so far as can be determined, but the weakness of the association of the chlorophyll pairs responsible for it is a poor basis for arguing a much increased rate of internal conversion. It is energetically possible, and logically plausible, that quenching is the result of electron transfer from one chlorophyll molecule to another, with subsequent immediate degradation to the ground state by the mechanism proposed in Section II,G.

The possibility of all these forms of quenching exists within the chloroplast.

C. THE TRIPLET EXCITED STATE

1. Spectrum and Energy

The first direct evidence for the existence of this state consisted of changes in the spectrum at short times after flash photolysis of solutions of chlorophyll (Livingston, 1955). The spectrum of chl^T shows indistinct broad bands throughout the visible region and into the infrared (Linschitz and Sarkanen, 1958). Later, the electron spin resonance (ESR) spectrum of chl^T was detected at low temperature (Rikhireva et al., 1968). Norris et al. (1975) have recorded ESR spectra of the triplet states of several chlorophylls in various states of aggregation and have tabulated their zero-field splitting (ZFS) parameters, which determine the separation of the triplet sublevels. Electrons in the triplet states are well delocalized, and the ZFS parameters are not much affected by aggregation.

The phosphorescence of chlorophyll a, that part of the luminescence emanating directly from the triplet state, long eluded detection because of its weakness. However, Krasnovskii et al. (1974) have recorded phosphorescence spectra of chlorophylls and pheophytins a and b at 77°K in several solvents. Although the spectra are rather broad, and are sensitive to the solvent and the state of aggregation, it is possible to estimate from them a value of 930 nm (10750 cm^{-1}) for the vibrationless triplet-to-ground state transition. This corresponds to a free energy of 128 kJ/mole or 1.33 eV. The triplet state therefore retains 72% of the free energy of the singlet, or at most 71% of the energy absorbed by the ground state of chlorophyll.

2. Triplet State Decay

The rate constant for intersystem crossing to the ground state (7) or reciprocal of the lifetime at low triplet concentration, ranges from 440 sec^{-1} in benzene (Linschitz and Sarkanen, 1958) to 2300 sec^{-1} in methanol (Livingston and McCartin, 1963).

$$chl^T \xrightarrow{\ k_d^T\ } chl \qquad\qquad (7)$$

The natural lifetime of chl^T is unknown because the phosphorescence yield is unknown, but if the latter is $\leq 0.1\%$, the former would be at least 1 second.

The triplet state consists of three sublevels, differing only slightly in energy, but differing markedly in their interaction with the ground state and the singlet excited state. In consequence, the triplet sublevels are populated unequally from the singlet excited state and decay at unequal rates to the ground state. The observed lifetime at low temperatures is therefore a composite of the separate sublevel lifetimes, weighted by their probabilities of population from the singlet excited state. Clarke *et al.* (1976) have summarized their determinations of the several rate constants for chlorophylls *a* and *b*, and Zn pheophytins *a* and *b*, in a few solvents at 2°K and 95°K. The solvent affects both the population probabilities and the depopulation rates, i.e., both (4) and (7), and suggests that variations in fluorescence yield with solvent may be attributed to this cause.

Some other processes for the decay of chl^T in the absence of foreign quenchers exist. Parker and Joyce (1967) measured E-type delayed fluorescence (8) as a function of temperature to estimate the singlet–

$$chl^T \xrightarrow{+\Delta_{st}} chl^* \xrightarrow{-h\nu} chl \tag{8}$$

triplet energy difference (Δ_{st}). At high light-flash intensities, bimolecular decay (9) is important and may be dominant, as it proceeds with "dif-

$$chl^T + chl^T \rightarrow chl + chl^T \tag{9}$$
$$(or\ 2\ chl)$$

fusion controlled" rates (ca. 2×10^9 M^{-1} sec^{-1}) (Linschitz and Sarkanen, 1958). This reaction may also lead to measurable production of ions in acetonitrile (Imura *et al.*, 1975). A self-quenching reaction (10) was reported (Linschitz and Sarkanen, 1958), but it is slow and its existence has been questioned (Livingston and McCartin, 1963).

$$chl^T + chl \rightarrow 2\ chl \tag{10}$$

Finally, photoejection of an electron into the solvent is energetically possible in visible or near-ultraviolet light (11), since the ionization potential of chlorophyll in tetramethylsilane is 4.50 eV (Nakato *et al.*,

$$chl^T \xrightarrow{h\nu \geq 3.17\ eV} chl^{\cdot+} + e^-(Solv) \tag{11}$$

1974). This process may account for oxidation of chlorophyll at low temperatures in the absence of known oxidants.

3. Quenching of the Triplet State

Similar opportunities exist for quenching the triplet state as for quenching the singlet.

a. *Energy Transfer.* Quenching by transfer (12) to carotenoids is well known *in vitro* and is believed to be responsible for elimination of chlorophyll triplet states in bacterial and green plant photosynthetic

$$chl^T + Q \rightarrow chl + Q^T \tag{12}$$

systems. Transfer leads to isomerization of the carotenoid, and a minimum of 5 conjugated double bonds seems necessary (Claes, 1961; Krasnovskii and Drozdova, 1961). A transient absorption of the carotene triplet state has been detected (Chessin *et al.*, 1966).

It is necessary that the acceptor have a lowest triplet energy level not greater than that of chlorophyll (128 kJ/mole). Thus, transfer to tetracene (123 kJ/mole) is possible (Chibisov *et al.*, 1969a). Transfer to O_2 is also possible, exciting it to the $^1\Delta_g$ state (94 kJ/mole), and initiating sensitized photooxidations such as the well-studied one of allylthiourea (Livingston and Owens, 1956; Krasnovskii and Sapozhnikova, 1966).

b. *Quenching by Electron Transfer.* Unlike quenching of the singlet state, quenching of the triplet state by reductants and oxidants is known to proceed through electron transfer. The compounds that react with chl^T are mostly the same ones that quench chl*, and oxidants are generally much more active than reductants. These reactions, which lead to the production of $chl \cdot^+$ or $chlH \cdot$, have been reviewed (Seely, 1966; Chibisov, 1969). Among oxidants, quinones and aromatic nitro compounds have received a large share of attention.

c. *Enhanced Intersystem Crossing.* There has been little investigation of quenching by paramagnetic metal ions or by heavy atoms. Linschitz and Sarkanen (1958) mentioned quenching of chl^T by salts of Ni, Co, and Cu.

D. ENERGETICS OF ELECTRON TRANSFER REACTIONS

1. Excited-State Oxidation-Reduction Potentials

Successful modeling of photosynthetic reactions requires not only a knowledge of reaction rate constants, but also an accounting of energy transactions beginning with reactions of chlorophyll excited states. Redox processes are best discussed in terms of electrochemical potential. These are rather hard to measure directly for excited states, but fortunately they can be estimated reliably from excitation free energies and ground state one-electron oxidation and reduction potentials, at least for chlorophyll and similarly "well-behaved" compounds. The manner in which this can be done was described earlier (Seely, 1966); recently Gouterman and Holten (1977) have applied this approach to a discussion of the energetics of bacteriopheophytin reactions. Here we update the calculations for chlorophyll using best available data.

The potential of the reaction (13) for oxidation of chlorophyll has been

$$chl^{\cdot+} + e^- \rightarrow chl \qquad E_{13} = +0.76 \text{ V} \qquad (13)$$

determined by polarography in propionitrile, methylene chloride, and aqueous acetone (Stanienda, 1965; Borg et $al.$, 1970; Kutyurin et $al.$, 1966). We neglect possible small junction potentials and refer all potentials to the normal hydrogen electrode. When this is added to the free energy of process (14), the oxidation potential of chl* results [Eq. (15)].

$$chl \rightarrow chl^* \qquad E_{14} = -1.84 \text{ eV} \qquad (14)$$

$$chl^{\cdot+} + e^- \rightarrow chl^* \qquad E_{15} = -1.08 \text{ V} \qquad (15)$$

If Q and Q$^-$ form a redox couple with potential E_Q [reaction (16)], the photoreduction of Q by chl* (17) is exothermic if $E_{17} > 0$, that is, if

$$Q + e^- \rightarrow Q^- \qquad E_{16} = E_Q \qquad (16)$$

$$chl^* + Q \rightarrow chl^{\cdot+} + Q^- \qquad E_{17} = +1.08 + E_Q \qquad (17)$$

$E_Q > -1.08$ V. Put another way, chl* should be able to reduce an oxidant if its polarographic reduction potential is not less than -1.08 V.

Similarly, since the energy of chl^T is estimated to be 1.33 eV [Eq. (18)], the energy of reaction (19) is $+0.76 - 1.33 = -0.57$ V and reac-

$$chl \rightarrow chl^T \qquad E_{18} = -1.33 \text{ eV} \qquad (18)$$

$$chl^{\cdot+} + e^- \rightarrow chl^T \qquad E_{19} = -0.57 \text{ V} \qquad (19)$$

tion (20) should be exothermic if $E_Q > -0.57$ V. The prediction of Eq. (20) is substantiated by rates of reaction of pyrochlorophyll with

$$chl^T + Q \rightarrow chl^{\cdot+} + Q^- \qquad E_{20} = +0.57 + E_Q \qquad (20)$$

a series of aromatic nitro compounds covering the potential range of interest (Seely, 1969a).

Although one-electron reductions of chlorophyll are probably not important to photosynthesis, the potentials for these reactions may also be estimated with the aid of the polarographic reduction potential in 40% aqueous ethanol, as measured by Kiselev *et al.* (1974) [Eq. (21)]. From the potentials E_{22} and E_{23}, it follows that chl* and chl^T should be

$$\text{chl} + \text{e}^- \rightarrow \text{chl} \cdot^- \qquad E_{21} = -0.99 \text{ V} \qquad (21)$$

$$\text{chl*} + \text{e}^- \rightarrow \text{chl} \cdot^- \qquad E_{22} = +0.85 \text{ V} \qquad (22)$$

$$\text{chl}^T + \text{e}^- \rightarrow \text{chl} \cdot^- \qquad E_{23} = +0.34 \text{ V} \qquad (23)$$

reduced by substances whose anodic halfwave potentials are $< +0.85$ V and $< +0.34$ V, respectively. Reductants so easily oxidized are not usually stable in air, which explains why so few photoreductants of chl^T or chl* are known. It is quite possible that photoreduction by ascorbic acid and phenylhydrazine, two well-studied reductants, proceeds through H transfer or concerted e$^-$ and H$^+$ transfer, which of course would modify the above argument.

Also of interest are the electron transfer reactions (24)–(26). Electron transfer is possible between chl* and chl (Stanienda, 1968), but not

$$\text{chl*} + \text{chl} \rightarrow \text{chl} \cdot^+ + \text{chl} \cdot^- \qquad E_{24} = +0.09 \text{ V} \qquad (24)$$

$$\text{chl}^T + \text{chl} \rightarrow \text{chl} \cdot^+ + \text{chl} \cdot^- \qquad E_{25} = -0.42 \text{ V} \qquad (25)$$

$$\text{chl}^T + \text{chl}^T \rightarrow \text{chl} \cdot^+ + \text{chl} \cdot^- \qquad E_{26} = +0.91 \text{ V} \qquad (26)$$

likely between chl^T and chl. Reaction (26) apparently occurs (Imura *et al.*, 1975) but must compete with reaction (9).

2. Ion Pair Formation

The energetics of the preceding reactions between chlorophyll species and quenchers pertain to isolated and solvated ionic species at equilibrium with the reaction solvent. Since electron transfer reactions require collision between the reagents, or at least a close approach, the immediate reaction product is a pair of ions very close together, the energy of which differs from that of the separated ions. This energy difference is discussed in the next section; here we wish to point out, with the aid of Fig. 1, a situation involving singlet-state reactions that may limit the efficiency of photosynthesis.

Reaction of the chl^T with a quencher Q_1 can only give, in significant yield, an ion pair $^3[\text{chl} \cdot^+ Q_1 \cdot^-]$ of lower total energy than that of chl^T, as illustrated on the right side of Fig. 1. Reaction of chl* with the same quencher gives an ion pair $^1[\text{chl} \cdot^+ Q_1 \cdot^-]$ which has almost the same

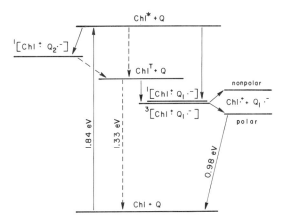

FIG. 1. Ion-pair formation from chlorophyll excited states with oxidants (Q) of different redox potentials. Energy differences drawn to scale for chl*, chl^T, and chl$^{\cdot+}$ + Q$_1^{\cdot-}$, with Q$_1$ = benzoquinone in ethanol. Intersystem crossings are marked with dashed lines.

energy as the triplet ion pair. However, chl* is also capable of transferring electrons to quenchers Q$_2$, which form ion pairs of energy intermediate between that of chl* and that of chl^T, as shown on the left side of Fig. 1. Because the energy difference between ^{1}[chl$^{\cdot+}$ Q$_2^{\cdot-}$] and chl^T + Q is small, the rate of crossing to this state may be very fast, limited only by the rate of spin reversal within the ion pair. This rate should be at least comparable to the spin-lattice relaxation time, ca. 10^6 sec^{-1}, and may approach the rate of intersystem crossing in chlorophyll, ca. 10^8 sec^{-1} (Gouterman and Holten, 1977). If the ion pair does not separate within this time, it may decay by back transfer of an electron to chl^T and Q$_2$.

It may be that there is no way of avoiding this sort of outcome in the PSU, and only oxidants like Q$_1$ give rise to separation of charge. In effect, the energy of the triplet state could impose an upper limit to the efficiency of conversion of light into electrochemical energy, even though the reaction proceeds through the excited singlet, and the triplet is never directly involved. Most of what is known of the energetics of the PS I reaction center suggests that this limitation exists.

E. DISSOCIATION OF ION PAIRS

1. Criteria for Separation of Ions

We have noted that the energetic predictions of Section I,D,1 may not apply to real photochemical reactions without modification, when in the real process ion pairs are formed that differ in energy from separated ions. This is the case with the chlorophyll–quinone reaction discussed in Section II; in case there is no creation of charge, as when chlorophyll reduces a cationic dye molecule, the modification would be different, but still important. We confine ourselves here to an interpretation of the case of ion-pair formation.

When the energy difference (U) between separated ions and the ion pair is considered at all, it is usually evaluated by Eq. (27), where the ions C^{q_i} and A^{q_j}, of radii r_i and r_j, form a pair with charges separated by $r_i + r_j$ in a medium of dielectric constant ϵ (for example, Seely, 1966;

$$U = (-q_i q_j)/[\epsilon (r_i + r_j)] \tag{27}$$

Gudkov et al., 1975a; Gouterman and Holten, 1977). Here and in the sequel we accept the pleasant fiction of replacing an ion by a point charge within a sphere the radius of which is comparable to its geometric dimensions.

This expression is good enough if the ions are well separated, but it is not valid if they are so close together that each perturbs the solvent shell polarized about the other ion. Indeed, there may be no solvent molecules at all separating the ions in a pair formed photochemically.

A better approach is to equate the ion pair to a dipole of moment $p = q(r_i + r_j)$, where $q = |q_i| = |q_j|$, in a sphere of radius a embedded in a dielectric. The energy $U^{(\epsilon)}$ of separation is that the separated ions minus that of the dipole (Frenkel, 1955). The calculation is assisted by Fig. 2.

The energy $W_{CA}^{(\epsilon)}$ of stabilization of a dipole in a dielectric is that of its interaction with the reaction field (Onsager, 1936) set up by it in the dielectric [Eq. (28)]. The free-energy difference between the separated ions in the medium and in vacuum is given by Eq. (29).

$$W_{CA}^{(\epsilon)} = \left(\frac{\epsilon - 1}{2\epsilon + 1}\right) \frac{p^2}{a^3} \tag{28}$$

$$-(W_A^{(\epsilon)} + W_C^{(\epsilon)}) = \left(1 - \frac{1}{\epsilon}\right) \left(\frac{q_i^2}{2r_i} + \frac{q_j^2}{2r_j}\right) \tag{29}$$

Combining (27)–(29) for the cycle of Fig. 2 gives

$$U^{(\epsilon)} = \left(\frac{\epsilon - 1}{2\epsilon + 1}\right) \frac{p^2}{a^3} - \frac{q_i q_j}{(r_i + r_j)} - \left(\frac{\epsilon - 1}{\epsilon}\right) \left(\frac{q_i^2}{2r_i} + \frac{q_j^2}{2r_j}\right) \tag{30}$$

In order to draw any conclusions from (30), it is necessary to propose some relation between the sum of radii $R = r_i + r_j$ and the dipole radius

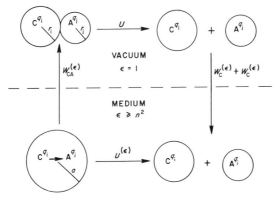

FIG. 2. Calculation of energy of stabilization of an ion pair in a medium of dielectric constant ϵ, following Frenkel (1955).

a, which in Onsager's derivation is quite arbitrary. If $a = fR$, $r_i = r_j$, and $q_i = -q_j$, Eq. (30) can be put into a form, Eq. (31), comparable to Eq. (27).

$$U^{(\epsilon)} = \frac{q^2}{\epsilon R} \left(\frac{\epsilon(\epsilon - 1) - f^3(2\epsilon + 1)(\epsilon - 2)}{f^3(2\epsilon + 1)} \right) \qquad (31)$$

Geometric arguments constrain the value of f^3 to the range 0.25 to 1. If $f^3 = 1$, $U^{(\epsilon)} > 0$ and the ion pair is stable with respect to the separated ions only when $\epsilon < 2.7$. When $f^3 \leq 0.5$, the ion pair is stable ($U^{(\epsilon)} > 0$) for all values of $\epsilon > 1$. When $f^3 > 0.5$, there is a value of ϵ, which rapidly decreases as f increases, above which the ion pair exothermically dissociates into separated ions. The existence of a "threshold value" of ϵ for appearance of separated ions better describes the experimental situation than Eq. (27), according to which ion pairs are stable for all values of ϵ.

Equation (30) describes the energy difference between initial and final states but says nothing about its dependence on ionic separation $r > r_i + r_j$, and therefore gives no hint of possible activation energy barriers or rates of ion movement.

A completely satisfactory description of the ion-pair potential in a dielectric as a function of separation is not available. Friedman and Ramanathan (1970) employed an ion pair potential of the form (32). In (32), the second term is the core repulsion term that limits the mutual

$$U_{FR}^{(S)}(r) = \frac{q_i q_j}{\epsilon r} + \frac{B_{ij}}{r^9} + \left(\frac{\epsilon - \epsilon_c}{2\epsilon + \epsilon_c} \right) \frac{(q_i^2 r_j^3 + q_j^2 r_i^3)}{2\epsilon r^4} + A_{ij} N(i, j, r) \qquad (32)$$

approach of ions, the third is a repulsive term which arises in the fact that each ion is surrounded by a polarized solvent shell of lower dielectric constant ϵ_c, and the fourth is a term proportional to the average number $N(i, j, r)$ of solvent molecules excluded from the overlapping solvation shells about ions a short distance r apart.

Bahe (1972) considered explicitly the gradient of the dielectric constant, $d\epsilon/dr$, over the volume V of polarized solvent about an ion and derived Eq. (33), valid for distances at which the solvation shells do

$$U_B^{(S)}(r) = \frac{q_i q_j}{\epsilon r} + \frac{\pi q^2}{8\epsilon_c^2} \left| \frac{d\epsilon}{dr} \right| \frac{V}{v} \frac{r^3} \tag{33}$$

not overlap. The arbitrariness in some of the terms in these equations adumbrates the difficulties that would be encountered in a serious attempt to match theoretical prediction with experimental observations of separated ion formation.

2. Criteria for Creation of Ion Pairs

In the preceding section we discussed the separation of an ion pair, equilibrated with its solvent shell, into constituent ions. However, the ion pair as initially formed is not in equilibrium with the solvent. The time required for electron transfer after collision, ca. 10^{-15} second, is much shorter than the time required for rotation of solvent molecules into a low-energy orientation. The time required for this should be comparable to dielectric relaxation times, which in the usual organic solvents vary from ca. 10^{-10} to 10^{-9} second for ethanol (Smyth, 1955).

The energy relations involved in the creation of ion pairs from the singlet state are detailed in Fig. 3. The energy of dissociation of the instantaneously formed ion pair, $^1[\text{chl}^{\cdot+} \; Q^{\cdot-}]_n$, is of the same form as that of the stabilized ion pair $^1[\text{chl}^{\cdot+}Q^{\cdot-}]_\epsilon$, except that the dipole energy is calculated with the dielectric constant at optical frequencies, n^2, instead of the static value ϵ [Eq. (34)]. The photochemical process occurs only if $U_\phi = E_Q - E_{15} - U^{(n)} > 0$. Separated ions are formed exothermically only if $U^{(\epsilon)} > 0$.

$$U^{(n)} = \left(\frac{n^2 - 1}{2n^2 + 1} \right) \frac{p^2}{a^3} + \frac{q^2}{R} - \left(\frac{\epsilon - 1}{\epsilon} \right) \left(\frac{q_i^2}{2r_i} + \frac{q_j^2}{2r_i} \right) \tag{34}$$

If these conditions are combined, a necessary, but *not* a sufficient, condition for both the occurrence of electron transfer and the appear-

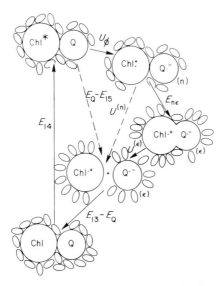

FIG. 3. Energy relations involving in creation and stabilization of ion pairs from the singlet excited state. Dipolar solvent molecules are indicated by ellipses. The accessible reaction path is drawn with solid lines. Subscripts (n) and (ϵ) mark ion pairs stabilized by the optical and low-frequency values of the dielectric constant.

ance in quantity of separated ions is Eq. (35), in which compatible energy units (eV or kJ) must be used.

$$E_Q - E_{15} > U^{(n)} - U^{(\epsilon)} = \frac{p^2}{a^3}\left(\frac{\epsilon - 1}{2\epsilon + 1} - \frac{n^2 - 1}{2n^2 + 1}\right) \qquad (35)$$

By chance some arrangements of solvent molecules around the collision complex [chl*Q] should permit exothermic electron transfer. The quenching of pyrochlorophyll fluorescence by nitro compounds was correlated successfully by assuming that a favorable solvent distribution was required, and that the fraction of distributions that were favorable increased with the value of E_Q (Seely, 1969b). Electron transfer from the triplet state may not be so severely constrained because the lifetime of chl^T is much longer than the orientation time of solvent molecules, and many different solvent arrangements may be experienced.

II. The Chlorophyll–Quinone Reaction as a Model of PS II

A. THE REACTIONS OF PS II

The participants in the reactions immediately surrounding the PS II reaction center are still not well known, but it was recognized soon after the discovery of the two photosystems that plastoquinone was reduced here (Weikard et al., 1963). The apparent primary electron acceptor was characterized by its ability in an oxidized state to quench fluorescence of chlorophyll in PS II, and was called "Q" by Duysens and Sweers (1963). More recently, absorption transients attributed to the primary electron acceptor have been reported (Stiehl and Witt, 1969; Knaff and Arnon, 1969), and the evidence appears to be accumulating that these derive from a bound form of plastoquinone (van Gorkom, 1974). The composition of PS II is discussed in detail in the chapter by Knaff and Malkin in this volume.

Knowledge of the involvement of quinones in photosynthesis stimulated a great deal of research on the photochemical interaction between chlorophyll and quinones in model systems, usually dilute solutions. Some of this work has been directed toward demonstrating not only photoreduction of the quinone, but also oxidation of the solvent, which may contain water. The implied goal of much PS II model system work is to define the conditions under which water can be oxidized, after photooxidation of chlorophyll or an analogous pigment by a quinone.

The ability of benzoquinone (BQ) to quench the fluorescence of chlorophylls in organic solvents was demonstrated by Evstigneev and Krasnovskii in 1948. Livingston and Ke (1950) noted that compounds which were effective quenchers of chlorophyll fluorescence, including BQ, chloranil, and duroquinone (DQ), were oxidants, but stopped short of suggesting electron transfer as a mechanism of quenching. It has been only very recently, with the development of picosecond flash techniques, that any direct evidence has appeared concerning the nature of the quenching act.

Linschitz and Rennert (1952) noted that quinones stabilized the photobleaching of chlorophyll in EPA (8:3:5 ether:isopentane:alcohol) glass at 77°K. Theirs may have been a case where electrons photoejected from chlorophyll were trapped by quinone.

Fujimori and Livingston (1957) noted that BQ quenched the triplet state of chlorophyll about as well as carotenoids, but the first definite evidence of charge transfer was provided by Tollin and Green's (1962) observation of the unmistakable ESR signal of semiquinones during irradiation of several quinones and chlorophyll in EPA. This observation was followed by a vigorous investigation of the reaction by a number of techniques, including flash photolysis, low-temperature spectrometry, photovoltaic potential, pH transients, and much more ESR spectroscopy.

In recent years, especially from Porter's group, there have appeared quantitative studies of interaction between chlorophylls and quinones in more ordered systems, such as lecithin films, monolayers, and multilayers (Costa et al., 1972; Costa and Porter, 1974; Beddard et al., 1975). We shall not attempt to cover model systems of this nature here, but we do wish to affirm their importance to the further investigation of these reactions.

B. Flash Photolysis

In 1965 Okayama and Chiba demonstrated the photoreduction of plastoquinone (PQ) by ascorbic acid in the presence of chlorophyll, further establishing the relevance of the model system to PS II. In the same year, Ke *et al.* (1965) employed flash photolysis to show that ubiquinone (UQ) and trimethylquinone (TMQ), which have redox potentials similar to that of PQ, delayed the recovery of the ground-state chlorophyll spectrum. The delay was attributed to oxidation of chl^T by the quinones, and some sensitized redox reactions were explained as proceeding from this initial step.

Chibisov *et al.* (1967) applied flash photolysis and rapid transient spectrophotometry to the investigation of the reaction between BQ and chlorophyll and related compounds in ethanol. They measured the rapid rate of electron return from $BQ^{\cdot-}$ to $chl^{\cdot+}$, but also noted a pseudo-first-order component to the rate of restoration of chlorophyll, which they attributed to oxidation of the solvent. Continuing this investigation, Kuz'min *et al.* (1971) found that the oxidation of alcohols by $chl^{\cdot+}$ was much slower than was supposed earlier, and that BQ (or an impurity in it) accounted for most of the first-order component. They also could not detect oxidation of water by the cation radicals of chlorophyllins *a* and *b*, chlorin e_6, or rhodin g_7. Pheophytins *a* and *b*, however, showed only the transient of $BQ^{\cdot-}$ at 430 nm, from which it was inferred that their cation radicals oxidized alcohols more rapidly than the apparatus could follow.

In the more viscous solvent isobutanol, Seifert and Witt (1969) were able to follow the oxidation of chl^T by BQ, and the reverse transfer from $BQ^{\cdot-}$ to $chl^{\cdot+}$. They distinguished the difference spectrum for the formation of chl^T from that for formation of $chl^{\cdot+} + BQ^{\cdot-}$ (Fig. 4). White and Tollin (1971a,b) interpreted similar transients in a mixed alcohol solvent as transfer of an electron from an alcohol (ROH) to the quinone, mediated within a complex by the energy of chl^T [Eq. (36)].

$$(ROH - - - chl^T - - - BQ) \rightarrow RO^{\cdot} + chl + BQ^{\cdot-} + H^+ \qquad (36)$$

Kelly and Porter (1970) investigated the reaction of photoexcited chlorophyll with DQ, α-tocopherylquinone, and vitamin K_1 in detail. These quinones, having lower redox potentials than BQ, were more stable in ethanol solution. The reaction sequence below is based on their

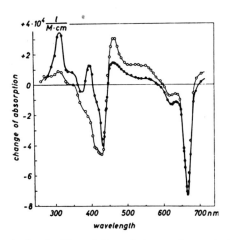

FIG. 4. Flash photolysis of chlorophyll and benzoquinone in isobutanol at 25°C, according to Seifert and Witt (1969). Open circles: chlorophyll alone, difference spectrum of formation of triplet state. Filled circles: BQ present, difference spectrum of formation of chl·$^+$ and BQ·$^-$. From Seifert and Witt (1969).

proposed mechanism, with certain modifications suggested by later work.

$$\text{chl} \xrightarrow{h\nu} \text{chl*} \tag{1}$$

$$\text{chl*} \xrightarrow{k_f^*} \text{chl} \tag{2}$$

$$\text{chl*} \xrightarrow{k_{isc}^*} \text{chl}^T \tag{4}$$

$$\text{chl*} + Q \xrightarrow{k_Q^*} {}^1[\text{chl·}^+Q·^-] \tag{37}$$

$${}^1[\text{chl·}^+Q·^-] \xrightarrow{k_d^1} \text{chl} + Q \tag{38}$$

Reaction (37) leads to quenching of the fluorescence of chl*, perhaps through a charge transfer complex or ion pair $^1[\text{chl·}^+Q·^-]$.

$$\text{chl}^T \xrightarrow{k_d^T} \text{chl} \tag{7}$$

$$\text{chl}^T + Q \xrightarrow{k_Q^T} {}^3[\text{chl·}^+Q·^-] \tag{39}$$

$${}^3[\text{chl·}^+Q·^-] \xrightarrow{k_d^3} \text{chl} + Q \tag{40}$$

$${}^3[\text{chl·}^+Q·^-] \xrightarrow{k_s} \text{chl·}^+ + Q·^- \tag{41}$$

$${}^3[\text{chl·}^+Q·^-] \xrightarrow{k_x} {}^1[\text{chl·}^+Q·^-] \tag{42}$$

Ion-pair formation (39) was first suggested by Tollin and Green (1962) to account for certain low-temperature ESR observations. Reaction (42) involves spin reversal of one of the electrons and is followed immediately by (38); this sequence is kinetically indistinguishable from (40).

$$\text{chl·}^+ + \text{Q·}^- \xrightarrow{k_{43}} \text{chl} + \text{Q} \tag{43}$$

$$\text{chl·}^+ + \text{HQ·} \xrightarrow{k_{44}} \text{chl} + \text{Q} + \text{H}^+ \tag{44}$$

$$\text{chl·}^+ + \text{H}_2\text{Q} \xrightarrow{k_{45}} \text{chl} + \text{HQ·} + \text{H}^+ \tag{45}$$

$$\text{chl·}^+ + \text{S} \xrightarrow{k_{46}} \text{chl} + \text{S·}^+ \tag{46}$$

In the last reaction, S is a molecule of the solvent, or perhaps an impurity in it or the quinone.

$$(\text{S} = \text{C}_2\text{H}_5\text{OH}): \text{C}_2\text{H}_4\text{OH·} + \text{Q} \xrightarrow{k_{47}} \text{CH}_3\text{CHO} + \text{HQ·} \tag{47}$$

$$2\text{HQ·} \xrightarrow{k_{48}} \text{Q} + \text{H}_2\text{Q} \tag{48}$$

$$2\text{Q·}^- \xrightarrow{k_{49}} \text{Q} + \text{Q}^{2-} \tag{49}$$

$$\text{HQ·} + \text{Q·}^- \xrightarrow{k_{50}} \text{HQ}^- + \text{Q} \tag{50}$$

$$\text{Q·}^- + \text{H}^+ \xrightarrow{k_{51}} \text{HQ·} \tag{51}$$

$$\text{HQ·} \xrightarrow{k_{52}} \text{Q·}^- + \text{H}^+ \tag{52}$$

$$\text{HQ}^- + \text{H}^+ \xrightarrow{k_{53}} \text{H}_2\text{Q} \tag{53}$$

Reaction (47) has been established as the principal fate of alcohol radicals in pulse radiolysis experiments (Willson, 1971). The remaining reactions clear away the quinone radicals. The whole sequence would be completely reversible were it not for Eqs. (46) and (47).

Gathered into Table I are many of the available rate-constant data for these reactions.

Because reaction (43) is at least 10 times as fast as (44), and because (48) or (50) is ca. 10^2 times as fast as reaction (49), the outcome of the follow-up reactions will depend materially on the availability of H^+ for reaction (51). The consequences of this are better examined after a summary of the ESR observations on the system.

C. Electron Spin Resonance Observations

Tollin and Green (1962) observed the characteristic spectra of semi-quinone radicals on irradiation of solutions of chlorophyll with BQ,

TABLE I

PUBLISHED VALUES OF RATE CONSTANTS IN CHLOROPHYLL–QUINONE PHOTOCHEMICAL REACTION SCHEME

Rate constant	Value[a]	Quinone[b]	Solvent	Reference[c]
k_d^*	$1.9 \times 10^8 \text{ s}^{-1}$	—	Ethanol	(1)
	$2.0 \times 10^8 \text{ s}^{-1}$	—	Ether, acetone	(2)
chl b:	$2.9 \times 10^8 \text{ s}^{-1}$	—	Ethanol	(2)
k_{isc}^*	$1.3 \times 10^8 \text{ s}^{-1}$	—	Ether	(3)
k_Q^*	$2 \times 10^{10} \text{ M}^{-1}\text{s}^{-1}$	BQ	Methanol	(4)[d]
	$9 \times 10^9 \text{ M}^{-1}\text{s}^{-1}$	DQ	Ethanol	(1)
	$6 \times 10^9 \text{ M}^{-1}\text{s}^{-1}$	TOCO	Ethanol	(1)
	$1 \times 10^{10} \text{ M}^{-1}\text{s}^{-1}$	VK$_1$	Ethanol	(1)
	$5 \times 10^{10} \text{ M}^{-1}\text{s}^{-1}$	2,6-DMQ	Ethanol	(5)
chl b:	$1.3 \times 10^{10} \text{ M}^{-1}\text{s}^{-1}$	2,5-DMQ	Ethanol	(6)[e]
k_d^1	$> 10^{11} \text{ s}^{-1}$	2,6-DMQ	Ethanol	(5)
k_d^T	1300 s^{-1}	—	Ethanol	(1)
	1300 s^{-1}	—	Isobutanol	(7)
k_Q^T	$2 \times 10^9 \text{ M}^{-1}\text{s}^{-1}$	BQ	Isobutanol	(7)
	$1.4 \times 10^9 \text{ M}^{-1}\text{s}^{-1}$	DQ	Ethanol	(1)
	$0.7 \times 10^9 \text{ M}^{-1}\text{s}^{-1}$	TOCO	Ethanol	(1)
	$1.2 \times 10^9 \text{ M}^{-1}\text{s}^{-1}$	VK$_1$	Ethanol	(1)
chl b:	$1 \times 10^9 \text{ M}^{-1}\text{s}^{-1}$	2,5-DMQ	Ethanol	(6)
phe:	$2.5 \times 10^9 \text{ M}^{-1}\text{s}^{-1}$	BQ	Ethanol	(8)
k_d^3	$1.2 \times 10^3 \text{ s}^{-1}$	BQ	Ethanol	(9)
k_s	$> 10^6 \text{ s}^{-1}$	BQ	Alcohols	(10)
k_x	probably $\geq 10^6 \text{ s}^{-1}$	—	—	—
k_{43}	$2 \times 10^9 \text{ M}^{-1}\text{s}^{-1}$	BQ	Ethanol	(11)
	$4 \times 10^9 \text{ M}^{-1}\text{s}^{-1}$	BQ	Isobutanol	(7)
	$3.2 \times 10^9 \text{ M}^{-1}\text{s}^{-1}$	DQ	Ethanol	(1)
	$3.3 \times 10^9 \text{ M}^{-1}\text{s}^{-1}$	TOCO	Ethanol	ʼ(1)
	$4.5 \times 10^9 \text{ M}^{-1}\text{s}^{-1}$	VK$_1$	Ethanol	(1)
chl b:	$4.7 \times 10^9 \text{ M}^{-1}\text{s}^{-1}$	BQ	Ethanol	(11)
chln:	$2.8 \times 10^9 \text{ M}^{-1}\text{s}^{-1}$	BQ	Water	(12)
k_{44}	$1.9 \times 10^8 \text{ M}^{-1}\text{s}^{-1}$	DQ	Ethanol	(1)
	$1.4 \times 10^8 \text{ M}^{-1}\text{s}^{-1}$	TOCO	Ethanol	(1)
	$3.3 \times 10^8 \text{ M}^{-1}\text{s}^{-1}$	VK$_1$	Ethanol	(1)
k_{45}	$5.5 \times 10^5 \text{ M}^{-1}\text{s}^{-1}$	DQ	Ethanol	(1)
	$1.8 \times 10^5 \text{ M}^{-1}\text{s}^{-1}$	TOCO	Ethanol	(1)
k_{46}	$\geq 1000 \text{ M}^{-1}\text{s}^{-1}$	(BQ)	Ethanol	(9)
	$4.0 \text{ M}^{-1}\text{s}^{-1}$	(BQ)	Methanol	(12)
	$15 \text{ M}^{-1}\text{s}^{-1}$	(BQ)	Ethanol	(12)
	$27 \text{ M}^{-1}\text{s}^{-1}$	(BQ)	Isopropanol	(12)
	0.6 s^{-1}	(DQ)	Ethanol	(1)
	1.7 s^{-1}	(TOCO)	Ethanol	(1)
phe:	$> 1000 \text{ M}^{-1}\text{s}^{-1}$	(BQ)	Alcohols	(12)
phe:	$1.5 \times 10^4 \text{ M}^{-1}\text{s}^{-1}$	(BQ)	Ethanol	(8)
k_{47}	$4.5 \times 10^9 \text{ M}^{-1}\text{s}^{-1}$	BQ	Ethanol	(13)

TABLE I (Continued)

Rate constant	Value[a]	Quinone[b]	Solvent	Reference[c]
k_{48}	1.7×10^7 M^{-1}s^{-1}	BQ	Ethanol	(9)
	2.2×10^7 M^{-1}s^{-1}	BQ	sec-butanol	(9)
	6×10^8 M^{-1}s^{-1}	BQ	Water–isopropanol	(14)
	1.1×10^9 M^{-1}s^{-1}	BQ	Water	(15)
	3.6×10^8 M^{-1}s^{-1}	DQ	Water–isopropanol	(14)
	1.1×10^8 M^{-1}s^{-1}	DQ	Ethanol	(1)
	8×10^8 M^{-1}s^{-1}	DQ	1 : Water:ethanol	(16)
	0.4×10^8 M^{-1}s^{-1}	TOCO	Ethanol	(1)
	0.7×10^8 M^{-1}s^{-1}	VK$_1$	Ethanol	(1)
k_{49}	2.8×10^7 M^{-1}s^{-1}	BQ	Water–isopropanol	(14)
	55 M^{-1}s^{-1}	BQ	Methanol (abs.)	(17)
	4.6×10^6 M^{-1}s^{-1}	DQ	1 : 1 Water:ethanol	(16)
	1.5×10^7 M^{-1}s^{-1}	DQ	Water–isopropanol	(14)
	3.8 M^{-1}s^{-1}	TQ	Methanol (abs.)	(17)
k_{50}	4×10^9 M^{-1}s^{-1}	DQ	1 : 1 Water:ethanol	(16)[f]
k_{51}	1.5×10^{10} M^{-1}s^{-1}	BQ	Water	(18)
	7×10^9 M^{-1}s^{-1}	DQ	Ethanol	(1)
k_{52}	1.5×10^6 M^{-1}s^{-1}	BQ	Water	(18)[g]
	7.4×10^3 s^{-1}	DQ	1 : 1 Water:ethanol	(16)
k_{53}	probably $\cong k_{51}$			

[a] Many of the values are reported with probable errors of 10–30%, which have been omitted here in the interest of clarity. Values refer to reactions of chlorophyll a, except as noted for chlorophyll b (chl b), pheophytin a (phe), and chlorophyllin (chln). s^{-1} = seconds^{-1}.

[b] Abbreviations: BQ, benzoquinone; DQ, duroquinone, TOCO, tocopherylquinone; VK$_1$, vitamin K$_1$; DMQ, dimethylquinone; TQ, toluquinone.

[c] References: (1) Kelly and Porter (1970); (2) Livingston (1960); (3) Bowers and Porter (1967); (4) Livingston and Ke (1950); (5) Huppert et al. (1976); (6) Lamola et al. (1975); (7) Seifert and Witt (1969); (8) Tollin (1976); (9) Hales and Bolton (1972); (10) Gudkov et al. (1975a); (11) Chibisov et al. (1967); (12) Kuz'min et al. (1971); (13) Willson (1971); (14) Rao and Hayon (1973); (15) Adams and Michael (1967); (16) Bridge and Porter (1958); (17) Sullivan and Reynolds (1976); (18) Yamazaki and Piette (1965).

[d] Calculated from Stern–Volmer constant with singlet-state lifetime reported in (1).

[e] Calculated from Stern–Volmer constant with chl b lifetime from (2).

[f] Recalculated from data at pH 6.9 assuming reaction (50) and using pK_a of Rao and Hayon (1973).

[g] Calculated from k_{51} and pK_a = 4.0 for HBQ· (Adams and Michael, 1967).

o- and p-chloranil in EPA. The photosignals decayed completely in the dark. No signal of chl·$^+$ was observed, which was tentatively attributed to extreme broadening of the spectrum. On decreasing the temperature, the 5-line spectrum of BQ·$^-$ became accompanied by a broad signal, and below $-130°$C only the broad signal (peak-to-peak width ΔH_{pp} = 7 G) remained. This was attributed to a chlorophyll–quinone complex, [chl·$^+$

BQ·$^-$], unable to dissociate in the viscous solvent. Below $-170°C$ no signal appeared at all.

Tollin et al. (1965) found that the quantum yield of signal production in ethanol with 0.03 M BQ increased from 0.10 at $-10°C$ to 0.70 at $-110°C$, then decreased to 0 at $-196°C$. Mukherjee et al. (1969) determined rate constants for reaction (39) on the assumption of competition between that reaction and reaction (7). Their values, e.g., $k_Q{}^T = 5 \times 10^4 \, M^{-1} \, sec^{-1}$ for Q = BQ in ethanol, are far below the range of values determined by flash photolysis (Table I). At their quinone concentrations ($>10^{-2} \, M$) it is not conceivable that (39) could be the rate-controlling reaction, and in line with the mechanism presented above it is more likely that they were following the competition between (43) and (46), with [S] proportional to [BQ] (Kuz'min et al., 1971), or between (40) and a charge-transfer reaction such as (54). The rate constant for (54) should be at least $5 \times 10^7 \, M^{-1} \, sec^{-1}$ (Meisel and Fessenden, 1976).

$$^3[chl·^+ \, Q·^-] + Q \rightarrow [chl·^+ \, Q] + Q·^- \tag{54}$$

The nature of the low-temperature signal was investigated more closely by Harbour and Tollin (1972, 1974b). With pheophytin as sensitizer in ethanol, the BQ·$^-$ signal was accompanied at $-135°C$ by a narrow signal, identified by the narrowing effect of deuteration as of the radical $CH_3\dot{C}HOH$. With chlorophyll as sensitizer, the solvent radical was not seen; the broad signal seen instead, along with that of BQ·$^-$, was identified by its parameters ($g = 2.0025$, $\Delta H_{pp} = 8$ G) as that of chl·$^+$. However, its formation was attributed to a direct photoejection of an electron from chl*, rather than to reactions (39) and (41) (Harbour and Tollin, 1974a). Bobrovskii and Kholmogorov (1973), by irradiating chlorophyll ($a + b$) with BQ in ethanol while cooling to $-196°C$, distinguished two broad signals in the frozen solution, one of chl·$^+$, the other ($\Delta H_{pp} = 12$ G) of BQ·$^-$ with fine structure suppressed. At low temperature ($-140°C$) and high [BQ] ($10^{-2} \, M$), Hales and Bolton (1972) observed a broad signal like that reported by Tollin and Green (1962), but at $g = 2.0046$. They suggested that the signal belongs to microcrystals of BQ, because deuteration of BQ narrowed it.

Evstigneev et al. (1971) found that in weakly acidic ethanol at $-90°C$ the signal of BQ·$^-$ was replaced by that of chl·$^+$. The appearance of the ESR signal at pH < 4.2 was accompanied by that of a positive photopotential and of the visible absorption spectrum of chl·$^+$ (Voznyak et al., 1974). It seems clear, then, that in neutral or slightly basic solution, the BQ·$^-$ signal is seen because chl·$^+$ is consumed by reaction (46). In weakly acidic solution, BQ·$^-$ is protonated [reaction (51)] and the rapid disproportionation (48) leaves chl·$^+$ unreacted at low temperatures (Kelly and Porter, 1970; Hales and Bolton, 1972; Voznyak et al., 1974). This interpretation is consistent with the known pK_a's of dissociation of semiquinone radicals, which are 4.0 for HBQ· (Adams and Michael, 1967) and 5.1 for HDQ· in aqueous solution (Rao and Hayon, 1973).

Direct evidence for the intervention of the ion pair ^{3}[chl·$^+$ Q·$^-$] from the steady-state photosignal is slight, contrary to first impressions (Tollin and Green, 1962). However, Hales and Bolton (1972) detected a short-lived transient ESR signal, which they assigned to this ion pair. Indirect evidence for the ion pair arises from the solvent dependence of flash photolysis products. By varying the solvent, Chibisov and Barashkov (1972), Kutyurin et al. (1973a), and Gudkov et al. (1975a) found that the yield of observable radicals from the chlorophyll–quinone reaction rose rapidly between dielectric constants of 10 and 20, implying ion pairs that dissociate in media sufficiently polar. These observations suggest the existence of a "threshold" value of ϵ above which ion pairs dissociate spontaneously, as implied by Eq. (31).

There is evidence that water or other H-bonding solvents stabilize the separated ions. Harbour and Tollin (1974c) detected no BQ·$^-$ ESR signal in acetone unless at least 5% water was present. Kostikov et al. (1974) found an unstructured signal in dry ethanol in place of the normal structured signal of BQ·$^-$ observed when 1% or more of the water was present. The signal resembled that of chl·$^+$ and was assigned to the [chl·$^+$ BQ·$^-$] ion pair, but its slow decay time (30 seconds at room temperature) seems inconsistent with the strong electronic interaction presumed to account for the absence of hyperfine structure. Finally, Lamola et al. (1975) interpreted their CIDNP observations of the 2,5-dimethylquinone (DMQ) chlorophyll system to imply the existence of triplet-state ion pairs which either separate or decay according to reactions (40) and (41).

D. VISIBLE SPECTRA OF OXIDIZED CHLOROPHYLLS

The spectrum of chl·$^+$ generated electrochemically in CH_2Cl_2 at 25°C has a strong band at 399 nm, with weaker bands visible at 570, 662, 740, and 820 nm (Borg et al., 1970). The spectrum in acetone at -70°C has corresponding bands near 405, 600, 660, 750, and 855 nm (Kutyurin et al., 1973b). In sec-butyl chloride at -196°C the infrared band is at 830 nm; and in benzonitrile at 25°C, at 810 nm (Seki et al., 1973). These variations imply that the spectrum of chl·$^+$ is considerably influenced by its solvent environment.

Generally speaking, the spectra of chl·$^+$, generated as a transient by flash photolysis at room temperature, or in viscous solution by irradiation at low temperature, resemble the electrochemically generated spectra. Bobrovskii and Kholmogorov (1973) recorded bands of chl$\cdot^+$ at 410, 660, and 830 nm in ethanol at -196°C. The chl$\cdot^+$ band that appears at 470 nm in some difference spectra (Krasnovskii and Drozdova, 1964; Chibisov et al., 1967; Seifert and Witt, 1969) may be real or may be an artifact of the more pronounced difference spectrum of the chlorophyll Soret band (Fig. 4). There is a distinct band at 470 nm in the spectrum of Evstigneev and Gavrilova (1966), however, in ethanol at -70°C at pH 3.6. The appearance of the visible spectrum of chl$\cdot^+$ at low temperature coincides with the appearance of the characteristic ESR signal (Evstigneev et al., 1971; Bobrovskii and Kholmogorov, 1973), just as it does with electrochemical generation (Borg et al., 1970).

Gordeyev et al. (1973) have detected changes in the 470 nm region, depending on the pH, which they interpret as short- and long-lived forms of the radical products. But the absorption changes due to quinone, e.g., reaction (51), strongly overlap those that might be attributed to chl$\cdot^+$.

The evidence is still vague that there is more than one species of chl$\cdot^+$ a, the spectrum of which varies with the solvent environment. However, with chl b and chlorobium chlorophylls, the situation is quite different.

Irradiation of chl b with BQ in ethanol at -100°C and pH 4 gave rise to a product with a distinct absorption band at 440 nm (Fig. 5) (Evstigneev and Gavrilova, 1967; Evstigneev et al., 1969a). After 12 minutes at -100°C, this product was converted into another with a band

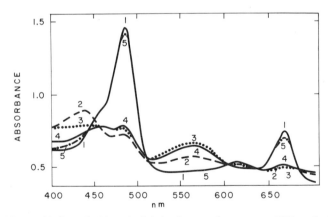

FIG. 5. Photooxidation of chlorophyll b by benzoquinone at -100°C and pH ~4. (1) Absorption spectrum before irradiation; (2) after 3 minutes of red light; (3) after standing in the dark for 12 minutes; (4) after addition of ascorbic acid; (5) after warming to room temperature and recording to -100°C. From Evstigneev et al. (1969a), reproduced by permission of Plenum Publishing Corp.

at 560 nm. On warming to room temperature, the latter reverted quantitatively to chl b. The 560 nm band appeared directly on irradiation at $-70°C$, but was accompanied by some of the 440 nm species. Ascorbic acid reduced the 440 nm species to chl b at $-100°C$, but reduced the 560 nm form only at higher temperatures. The chl b flash photolysis difference spectra of Chibisov $et\ al.$ (1967) and of Seifert and Witt (1969) do not indicate a 560 nm band, but could be compatible with an oxidized product band at 440 nm.

The chlorobium chlorophylls (Cchl, "bacterioviridin") are more easily oxidized than chl a, and their oxidation products are relatively stable at room temperature in the presence of BQ. At 25°C or $-70°C$ and pH 4.3, the primary product or photoreaction with BQ in ethanol has a rather featureless spectrum except for a band near 400 nm, like the spectrum of chl$^{.+}$ a (Evstigneev $et\ al.$, 1967). This reverts slowly to Cchl in the dark. On longer irradiation, an oxidized product with a band at 770 nm is formed, which even more slowly reverts to Cchl in the dark. The first material is reduced by ascorbic acid, the second is not. Reversion of the second product is accelerated by irradiation at wavelengths greater than 720 nm.

These results suggest that chl$^{.+}$ b and Cchl$^{.+}$ are stabilized by some sort of reaction with the solvent, which does not lead immediately to restoration of the pigment. However, they could also represent an internal rearrangement or perhaps an aggregation of the oxidized pigment. Whether these reactions have any relation to the postulated reduction of chl$^{.+}$ a by alcohols is unknown.

E.   Photopotentials and Related Phenomena

1. Photopotentials

Evstigneev and Gavrilova (1966) found that irradiation of an ethanol solution of chlorophyll and BQ at room temperature and pH < 6 produced an electrode-active substance that increased the potential of a Pt electrode by as much as 0.3 V. The photopotential maximized in about 3 minutes and decayed in about the same time in the dark. In the dark, the potential was approximately that of the quinone–hydroquinone couple (+0.43 V), and the increase in the light raised it to values (0.70–0.75 V) close to that of electrochemically produced chl$^{.+}$ a. The photopotential was increased at $-70°C$, but decayed more slowly; at this temperature its appearance was related to that of the form of chl$^{.+}$ a having a band at 470 nm. The photopotential was greatest around pH 4, and fell to negative values above pH 6, where BQ$^{.-}$ is stable and chl$^{.+}$ is not. Very similar phenomena were noted with chl b, where at $-70°C$ the photopotential is associated with the 560 nm-form of oxidized pigment

(Evstigneev and Gavrilova, 1967). The positive photopotentials were found with PQ and UQ as oxidants as well as with BQ (Evstigneev *et al.*, 1968, 1969b).

2. Photoconductivity

Gudkov *et al.* (1973) measured flash photoinduced conductivity of ether solutions of chlorophyll at temperatures from 25°C to −60°C. They ascribed the production of ions to reactions between chlorophyll and the solvent, but Imura *et al.* (1975), who observed photoconductivity in acetonitrile solutions, determined by kinetic analysis that the ions were produced by bimolecular reaction of two triplet chlorophylls. This reaction (26) probably occurs in ether also.

Gudkov *et al.* (1975a,b) have employed photoconductivity to estimate kinetic parameters and the effects of dielectric constant and temperature on them.

3. pH Transients

Quinlan and Fujimori (1967) noted a reversible photoinduced pH decrease when chlorophyll *a* or *b* and BQ in methanol were irradiated with red light at pH 7.5–7.8. In weakly acidic alcohols, Evstigneev and Gavrilova (1968) and Quinlan (1970a) found a much larger pH increase in the light, greatest at pH 4.2. The sign of the transient reversed near pH 6, and the pH decrease was greatest at pH 8.3 (Quinlan, 1970a, 1971). pH transients are recorded with a number of porphyrins and BQ, and so are not dependent on the structure of the pigment (Evstigneev and Chudar, 1972); the pH of maximum transient formation does depend on the solvent, however (Evstigneev and Chudar, 1974).

The positive transient in acid solution is most probably due to protonation of the semiquinone anion, reaction (51), as proposed by Evstigneev and Gavrilova (1968). The negative transient in neutral or weakly basic solution may have to do with photoreduction of the pigment by hydroquinone, as maintained by Evstigneev and Chudar (1974), or, perhaps, to reduction of chl$\cdot^+$ by H_2Q or S.

F. Oxidation of Solvent

The ability of chlorophyll in PS II, after its oxidation by a quinone, to accomplish the oxidation of water to oxygen has naturally excited interest in whether oxidized chlorophylls, produced photochemically *in vitro*, can do the same. Since chlorophyll is not soluble in water, the question has more often concerned the oxidation of an organic solvent, such as ethanol. The absence of a recognizable chl$\cdot^+$ ESR signal in the presence of a photoinduced semiquinone signal strongly suggested removal of the chl$\cdot^+$ by some reaction, and reduction by the solvent is one of the more obvious possibilities.

Chibisov *et al.* (1967) noted a first-order component in the decay of chlorophyll transients in flash photolysis experiments, which they attributed to reduction of chl$\cdot^+$ by the solvent (ethanol) with a rate constant $k_{46} = 860$ M^{-1} sec^{-1}. It was subsequently found that this rate constant was proportional to the BQ concentration, and therefore that chl$\cdot^+$ was decaying not only by (46) with the much smaller rate constants listed in Table I, but also by a reaction (55), for which $k_{55} = 2.0 \times 10^5$ and 1.8×10^5 M^{-1} sec^{-1} for chl a and chl b in methanol, and 4.7×10^5

$$\text{chl}\cdot^+ \xrightarrow{k_{55}[\text{BQ}]} \text{chl} \tag{55}$$

M^{-1} sec^{-1} for chlorophyllin in water (Kuz'min *et al.*, 1971). This reaction is presumably much slower at low temperatures.

Since quinones may be contaminated with perhaps 1% hydroquinone, it is tempting to ascribe the first-order decay of chl$\cdot^+$ to reaction (45) with H$_2$Q in excess. Indeed, Kelly and Porter (1970), having found the much slower *first*-order decay constant listed in Table I for the reaction with DQ in acidic ethanol, ascribe it to reaction (45) with a durohydroquinone impurity. With their value for k_{45}, determined by addition of H$_2$DQ to the system, a 1% impurity would have sufficed.

On the other hand, Hales and Bolton (1972) calculated that reaction with H$_2$BQ was inadequate to account for the rate of disappearance of chl$\cdot^+$ in their system. We note, however, that at their BQ concentration of 10^{-2} M, reaction (55) would have sufficed to account for the limitation on the lifetime of chl$\cdot^+$.

Harbour and Tollin (1974b) considered oxidation of solvent to be responsible for the semiquinone signals seen at room temperature and below, but retained the mechanism of Raman and Tollin (1971) in which chlorophyll or pheophytin somehow sensitizes the direct transfer of an electron from a solvent molecule to a quinone in a ternary complex, reaction (56). The chl$\cdot^+$ ESR signal seen at low temperatures was be-

$$(\text{S} - - - \text{chl}^T - - - \text{Q}) \rightarrow (\text{S}\cdot^+ - - - \text{chl} - - - \text{Q}\cdot^-) \tag{56}$$

lieved to arise from an independent process, perhaps direct photoejection of an electron (Harbour and Tollin, 1974a). White and Tollin (1971b) observed that the recovery of chlorophyll after flash photolysis with BQ in a mixed alcohol solvent increased with BQ concentration, but at a rate measured in seconds, rather than milliseconds as in the work of

Kuz'min *et al.* (1971). Their suggested mechanism presumed direct photochemical oxidation of the alcohol.

Voznyak *et al.* (1974) and Kim *et al.* (1974) follow Kelly and Porter (1970) in assigning the reduction of chl$\cdot^+$ to reaction with hydroquinone (H_2DQ or H_2BQ).

To be sure, hydroquinones will reduce chl$\cdot^+$ if they are present in sufficient quantities, but there is real doubt that the amounts reasonably present can explain the chl$\cdot^+$ transient decay rates as well as the ESR spectra. The reaction between chl$\cdot^+$ and BQ observed by Kuz'min *et al.* (1971) could go far toward explaining events, especially at high BQ concentrations where, for example, the lifetime of chl$\cdot^+$ in ethanol would be less than 10^{-3} sec at $[BQ] = 10^{-2} M$. It is difficult to believe that BQ, but not DQ, is capable of *reducing* chl$\cdot^+$ to chl, which is probably why reaction (55) has not received more attention. However, it is not so difficult to believe that BQ, and not the weaker oxidant DQ, could *oxidize* a radical formed by addition of an alcohol to chl$\cdot^+$ and loss of H^+. The now doubly oxidized chl species could revert to chlorophyll by elimination of aldehyde. This sequence is represented in Eqs. (57)–(59) with ethanol.

$$\text{chl}\cdot^+ + \text{HOC}_2\text{H}_5 \rightarrow (\text{chl-O-C}_2\text{H}_5)\cdot + \text{H}^+ \tag{57}$$

$$(\text{chl-O-C}_2\text{H}_5)\cdot + \text{Q} \rightarrow (\text{chl-O-C}_2\text{H}_5)^+ + \text{Q}\cdot^- \tag{58}$$

$$(\text{chl-O-C}_2\text{H}_5)^+ \rightarrow \text{chl} + \text{CH}_3\text{CHO} + \text{H}^+ \tag{59}$$

Diehn and Seely (1968) have proposed the stabilization of chl$\cdot^+$ by alcohols in another context. The spectroscopy work discussed in Section II,D suggests reaction of chl$\cdot^+$ with the solvent at low temperature, without immediate reduction to chl. If a reaction sequence such as the above exists, the evidence for the direct oxidation of alcohols by reaction (46) is reduced to the limiting rates at $[BQ] = 0$ of Kuz'min *et al.* (1971).

So far as the oxidation of water by chl$\cdot^+$ is concerned, analysis of rate constants for decay of chlorophyllin and chlorin e_6 cation radicals imposes such low upper limits on them that it cannot be of any practical consequence (Chibisov *et al.*, 1969b; Kuz'min *et al.*, 1971). Harbour and Tollin (1974c) inferred the oxidation of water from the appearance of semiquinone ESR signals in acetone only when 5–15% water was added. However, this observation may also be explained by the need for a minimum value of the dielectric constant, or by the need for water as described by Kostikov *et al.* (1974), to effect separation of ion pairs.

The evidence for oxidation of solvents by the pheophytin cation radical, phe$\cdot^+$, is considerably better. It has been repeatedly observed that in the presence of quinones the pheophytin transient is suppressed, but the semiquinone transient and ESR signal appear strongly (Krasnov-skii and Drozdova, 1964; Chibisov *et al.*, 1967; Quinlan, 1970b; Kuz'min *et al.*, 1971; White and Tollin, 1971a,b; Voznyak *et al.*, 1974). Harbour and Tollin (1972, 1974b) detected the ESR signal of $CH_3\dot{C}HOH$ on irradiation of pheophytin with BQ (10^{-2} M) in ethanol at $-140°C$. Tollin (1976) has reported transients of phe^T and $phe^{·+}$ with lifetimes of about 0.4 μsec and 4 μsec, respectively, on laser flash irradiation in alcohol solution with 10^{-3} M BQ.

The first polarographic oxidation potential of pheophytin is cited as ranging from +0.86 to +1.03 V (SCE) depending on the solvent, and is from 0.34 to 0.41 V higher than that of chlorophyll in the same solvent (Stanienda, 1965; Barboi and Dilung, 1969; Kutyurin *et al.*, 1966). This amply accounts for the difference in rates of reaction with solvents, but leaves unsettled the question of to what extent $chl^{·+}$ is capable of this reaction.

Model systems to date have failed signally to reproduce the activity of chlorophyll in PS II to oxidize water, or indeed any particularly difficult substrate. The oxidation potential of $chl^{·+}$, as determined polarographically, is unequal to the task. In retrospect, it appears that prokaryotes, before they could use water successfully as a reductant in photosynthesis, had to come upon a method of raising the oxidation potential of chlorophyll by some tenths of a volt. Perhaps the explanation resides in the photoreactivity of polarized membranes; future model system work might profitably extend in that direction.

G. REACTIVITY OF THE SINGLET EXCITED STATE

A necessary condition for involvement of the lowest singlet excited state in a reaction is the suppression of its fluorescence, and by all accounts quinones are among the best quenchers of fluorescence of chlorophyll and related compounds (Livingston and Ke, 1950; Evstigneev *et al.*, 1950; Drozdova and Krasnovskii, 1967; Kelly and Porter, 1970). But quenching of fluorescence does not necessarily mean electron transfer, however likely that process may seem. Chibisov (1972) noticed that the yield of $chl^{·+}$ fell as the BQ concentration increased into the range where fluorescence was quenched, and ascribed this result to the rapidity of reaction (38), made spin-allowed by the multiplicity of the ion pair. Similarly, Evstigneev *et al.* (1966) had found that the photopotential attained a maximum value at ca. 10^{-4} M BQ, and fell as the quinone concentration was increased, and Quinlan (1972) observed an inhibition of the pH transient at high BQ concentration.

Lamola *et al.* (1975), in their CIDNP study of semiquinone formation in the chl *b*–DMQ system, found the polarized radical ion signal to decrease as the quinone concentration increased above 10^{-2} M. They concluded that reaction of chl* with BQ made no substantial contribution to the production of free semiquinone ions.

Huppert *et al.* (1976) examined the quenching of chl* *a* by 2,6-DMQ by picosecond laser spectroscopy. Although quenching of the singlet state was followed at three wavelengths, there was no indication of formation of ions, or of an ion pair such as ^{1}(chl$\cdot^+$ Q$\cdot^-$), with a lifetime greater than 10 psec.

These results all indicate either that quinones do not quench chl* by electron transfer or that the ion pair formed by electron transfer has a very short lifetime, not much longer than 1 psec. Electron transfer is likely because it is known to be efficient from chl^T and because it is hard to see why an exciplex (chl*Q) would not decay by that route if it were energetically accessible. The lack of detectable charge separation from chl* in the model system is difficult to reconcile with the preponderance of evidence that favors the singlet, but not the triplet, excited state of reaction center chlorophyll as the photochemically active state in nature. The validity of a model systems approach to photosynthesis requires a convincing resolution of this dilemma.

An attempt to provide this is made with the help of Fig. 6, in which energies of the states of the chlorophyll–quinone system are graphed

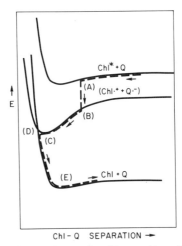

Chl– Q SEPARATION ➔

Fig. 6. Schematic of possible pathway of rapid quenching of ion pair formed from the chlorophyll singlet excited state. Evolution of the molecular separation is indicated by the dashed line. When the molecules diffuse to within the tunneling distance (A), electron transfer occurs to produce an ion pair stabilized only by interaction with the optical component of the dielectric constant of the medium. The pair rapidly collapses to the energy minimum (C) and approaches the point of intersection with the ground state energy curve (D). If (D) is inaccessible, the ions may eventually separate along a curve (not shown) of energy of the ion pair stabilized by the medium of dielectric constant ϵ.

schematically against the chlorophyll–quinone separation. Curves for the ground-state and excited-state molecule pairs are drawn with shallow minima, to indicate weak complexes that are known to exist in similar systems (Hill *et al.*, 1968). At shorter distances the energies of these systems rise rapidly owing to electron cloud repulsion. The energy curve for the ion pair has a deeper minimum, owing to the energy of separation of the oppositely charged ions, and, most important, its minimum is displaced to a shorter molecular separation than that of the neutral molecules [compare Eq. (32)]. The shape of this curve will be little affected by solvent because, in terms of the discussion in Section I,E, it corresponds more nearly to a curve for $U^{(n)}$ than for $U^{(e)}$. Because the attraction of opposite charges counteracts the core repulsion, the curve for the ion pair intersects that of the ground state of the molecules at some separation shorter than the minimum in the latter curve. The stability of the ion pair will depend on the energy difference between the minimum of its curve and the point of intersection with the ground state curve. If this difference is small, as drawn in Fig. 6, the energy of the system will follow a trajectory like that shown as a dashed line in the figure. Since the evolution of $^1[\text{chl}^{\cdot+} \, Q^{\cdot-}]$ to chl + Q is accomplished within what is essentially a vibrational interval, no stable ion pair exists and "lifetimes" of the order of 1 psec or less are to be expected.

With reference to Fig. 6, it should be noted that the greater the oxidation potential of the quinone, the greater the energy difference between chl* + Q and chl$^{\cdot+}$ + Q$^{\cdot-}$, the greater the separation at intersection (D), and the greater the probability of rapid deactivation. If the quinone is a weak oxidant, so that the curve for the ion pair is high in energy, the point of intersection may be at such distance that the energy difference between (C) and (D) is large enough to provide time for the ion pair to come to equilibrium with the medium, and perhaps dissociate into free ion radicals. We are therefore more likely to see photoreaction products when as much energy as possible is conserved in the charge-transfer step, as long as the energy of the triplet state is not exceeded (Section I,D,2). It is interesting to report that Hino *et al.* (1976) have observed exactly this effect in the quenching of pyrene by a series of nitriles, but were unable to account for it in terms of their detailed mechanism.

Two factors may then make photosynthesis possible from the singlet excited state in the plant. First, a fair amount of energy is conserved in the photoreaction, which paradoxically might tend to favor separation of charged products. Second, the reaction-center chlorophylls and the primary acceptors are presumed to be held in a semirigid matrix, which prevents collapse of the ion pair until the charges are borne away by secondary donors and acceptors.

As Lamola *et al*. (1975) pointed out, production of ion pairs by singlet state reactions is not unknown. Our construction in Fig. 6 implies that the production of stable ion pairs is dependent on the energy difference between points (C) and (D), and should be greater for more energetic pairs of ions. There is indeed some evidence for production of ions from chl* and weak oxidants. Seely (1969b) examined the pyrochlorophyll-photosensitized reduction of nitro compounds over concentration ranges of the latter in which pyrochlorophyll fluorescence was quenched. The yield of reduction was diminished as concentration increased, but not as much as the fluorescence. On application of a mechanistic scheme similar to that presented for quinones, it was concluded that radical ions did form, with probabilities of 0.074, 0.030, and 0.015 for *p*-nitrobenzonitrile, *m*-nitroanisole, and *p*-nitrotoluene, per encounter of singlet excited sensitizer with quencher. Since these nitro compounds are much weaker oxidants than most quinones, the implications of the present scheme seem to be supported. It also suggests that among quinones there might be better luck finding radical ion products from the reaction of chl* with anthraquinone derivatives, which are usually rather feeble oxidants.

REFERENCES

Adams, G. E., and Michael, B. D. (1967). *Trans. Faraday Soc.* **63**, 1171–1180.
Bahe, L. W. (1972). *J. Phys. Chem.* **76**, 1062–1071.
Barboi, N. I., and Dilung, I. I. (1969). *Biofizika* **14**, 980–985.
Beddard, G. S., and Porter, G. (1976). *Nature (London)* **260**, 366–367.
Beddard, G. S., Porter, G., and Weese, G. M. (1975). *Proc. R. Soc. London, Ser. A* **342**, 317–325.
Bobrovskii, A. P., and Kholmogorov, V. E. (1973). *Dokl. Akad. Nauk SSSR* **208**, 1472–1475.
Borg, D. C., Fajer, J., Felton, R. H., and Dolphin, D. (1970). *Proc. Natl. Acad. Sci. U.S.A.* **67**, 813–820.
Bowers, P. G., and Porter, G. (1967). *Proc. R. Soc. London, Ser. A* **296**, 435–441.
Bridge, N. K., and Porter, G. (1958). *Proc. R. Soc. London, Ser. A* **244**, 276–288.
Broyde, S. B., and Brody, S. S. (1967). *J. Chem. Phys.* **46**, 3334–3340.
Chessin, M., Livingston, R., and Truscott, T. G. (1966). *Trans. Faraday Soc.* **62**, 1519–1524.
Chibisov, A. K. (1969). *Photochem. Photobiol.* **10**, 331–347.
Chibisov, A. K. (1972). *Dokl. Akad. Nauk SSSR* **205**, 142–145.
Chibisov, A. K., and Barashkov, B. I. (1972). *Dokl. Akad. Nauk SSSR* **204**, 1182–1185.
Chibisov, A. K., Karyakin, A. V., Drozdova, N. N., and Krasnovskii, A. A. (1967). *Dokl. Akad. Nauk SSSR* **175**, 737–740.
Chibisov, A. K., Karyakin, A. V., and Zubrilina, M. E. (1969a). *Biofizika* **14**, 925–927.
Chibisov, A. K., Kuz'min, V. A., and Vinogradov, A. P. (1969b). *Dokl. Akad. Nauk SSSR* **187**, 142–145.
Claes, H. (1961). *Z. Naturforsch., Teil B* **16**, 445–454.
Clarke, R. H., Connors, R. E., Schaafsma, T. J., Kleibeuker, J. F., and Platenkamp, R. J. (1976). *J. Am. Chem. Soc.* **98**, 3674–3677.

Costa, S. M. de B., and Porter, G. (1974). *Proc. R. Soc. London, Ser. A* **341**, 167–176.
Costa, S. M. de B., Froines, J. R., Harris, J. M., Leblanc, R. M., Orger, B. H., and Porter, G. (1972). *Proc. R. Soc. London, Ser. A* **326**, 503–519.
Diehn, B., and Seely, G. R. (1968). *Biochim. Biophys. Acta* **153**, 862–867.
Dreeskamp, H., Koch, E., and Zander, M. (1974). *Ber. Bunsenges. Phys. Chem.* **78**, 1328–1334.
Drozdova, N. N., and Krasnovskii, A. A. (1967). *Mol. Biol.* **1**, 395–409.
Duysens, L. N. M., and Sweers, H. E. (1963). *Plant Cell Physiol.* **4**, 353–372.
Dzhagarov, B. M. (1970). *Opt. Spectrosc. (USSR)* **28**, 33–34.
Emeis, C. A., and Fehder, P. L. (1970). *J. Am. Chem. Soc.* **92**, 2246–2252.
Evstigneev, V. B., and Chudar, V. S. (1972). *Mol. Biol.* **6**, 536–541.
Evstigneev, V. B., and Chudar, V. S. (1974). *Biofizika* **19**, 425–429.
Evstigneev, V. B., and Gavrilova, V. A. (1966). *Biofizika* **11**, 593–600.
Evstigneev, V. B., and Gavrilova, V. A. (1967). *Dokl. Akad. Nauk SSSR* **174**, 476–479.
Evstigneev, V. B., and Gavrilova, V. A. (1968). *Mol. Biol.* **2**, 869–877.
Evstigneev, V. B., and Krasnovskii, A. A. (1948). *Dokl. Akad. Nauk SSSR* **60**, 623–626.
Evstigneev, V. B., and Gavrilova, V. A., and Krasnovskii, A. A. (1950). *Dokl. Akad. Nauk SSSR* **74**, 315–318.
Evstigneev, V. B., Gavrilova, V. A., and Sadovnikova, N. A. (1966). *Biokhimiya* **31**, 1229–1236.
Evstigneev, V. B., Gavrilova, V. A., and Olovyanishnikova, G. D. (1967). *Mol. Biol.* **1**, 59–66.
Evstigneev, V. B., Sadovnikova, N. A., and Olovyanishnikova, G. D. (1968). *Mol. Biol.* **2**, 21–28.
Evstigneev, V. B., Sadovnikova, N. A., and Olovyanishnikova, G. D. (1969a). *Dokl. Akad. Nauk SSSR* **187**, 1184–1187.
Evstigneev, V. B., Olovyanishnikova, G. D., and Sadovnikova, N. A. (1969b). *Mol. Biol.* **3**, 41–48.
Evstigneev, V. B., Sadovnikova, N. A., Kostikov, A. P., Gribova, Z. P., and Kayushin, L. P. (1971). *Biofizika* **16**, 431–435.
Fetisova, Z. G., and Borisov, A. Yu. (1973–1974). *J. Photochem.* **2**, 151–159.
Frenkel, J. (1955). "Kinetic Theory of Liquids," p. 435. Dover, New York.
Friedman, H. L., and Ramanathan, P. S. (1970). *J. Phys. Chem.* **74**, 3756–3765.
Fujimori, E., and Livingston, R. (1957). *Nature (London)* **180**, 1036–1038.
Gordeyev, V. I., Gorshkov, V. K., Evstigneev, V. B., and D'yachenko, A. P. (1973). *Biofizika* **18**, 631–636.
Gouterman, M., and Holten, D. (1977). *Photochem. Photobiol.* **25**, 85–92.
Gudkov, N. D., Stolovitskii, Yu. M., and Evstigneev, V. B. (1973). *Biofizika* **18**, 807–812.
Gudkov, N. D., Stolovitskii, Yu. M., and Evstigneev, V. B. (1975a). *Biofizika* **20**, 214–218.
Gudkov, N. D., Stolovitskii, Yu. M., and Evstigneev, V. B. (1975b). *Biofizika* **20**, 807–811.
Hales, B. J., and Bolton, J. R. (1972). *J. Am. Chem. Soc.* **94**, 3314–3320.
Harbour, J. R., and Tollin, G. (1972). *Proc. Natl. Acad. Sci. U.S.A.* **69**, 2066–2068.
Harbour, J. R., and Tollin, G. (1974a). *Photochem. Photobiol.* **19**, 69–74.
Harbour, J. R., and Tollin, G. (1974b). *Photochem. Photobiol.* **19**, 147–161.
Harbour, J. R., and Tollin, G. (1974c). *Photochem. Photobiol.* **20**, 271–277.
Hill, H. A. O., Macfarlane, A. J., Mann, B. E., and Williams, R. J. P. (1968). *J. Chem. Soc., Chem. Commun.* pp. 123–124.
Hino, T., Akazawa, H., Masuhara, H., and Mataga, N. (1976). *J. Phys. Chem.* **80**, 33–37.
Holten, D., Gouterman, M., Parson, W. W., Windsor, M. W., and Rockley, M. G. (1976). *Photochem. Photobiol.* **23**, 415–423.
Huppert, D., Rentzepis, P. M., and Tollin, G. (1976). *Biochim. Biophys. Acta* **440**, 356–364.

Imura, T., Furutsuka, T., and Kawabe, K. (1975). *Photochem. Photobiol.* **22,** 129–134.

Ke, B., Vernon, L. P., and Shaw, E. R. (1965). *Biochemistry* **4,** 137–144.

Kelly, A. R., and Patterson, L. K. (1971). *Proc. R. Soc. London, Ser. A* **324,** 117–126.

Kelly, J. M., and Porter, G. (1970). *Proc. R. Soc. London, Ser. A* **319,** 319–329.

Kim, V. A., Voznyak, V. M., and Evstigneev, V. B. (1974). *Biofizika* **19,** 992–996.

Kiselev, B. A., Kozlov, Yu. N., and Evstigneev, V. B. (1974). *Biofizika* **19,** 430–434.

Knaff, D. B., and Arnon, D. I. (1969). *Proc. Natl. Acad. Sci. U.S.A.* **63,** 963–969.

Knox, R. S. (1975). *In* "Bioenergetics of Photosynthesis" (Govindjee, ed.), Chapter 4. Academic Press, New York.

Kostikov, A. P., Sadovnikova, N. A., Evstigneev, V. B., and Kayushin, L. P. (1974). *Biofizika* **19,** 244–248.

Krasnovskii, A. A., and Drozdova, N. N. (1961). *Biokhimiya* **26,** 859–871.

Krasnovskii, A. A., and Drozdova, N. N. (1964). *Dokl. Akad. Nauk SSSR* **158,** 730–733.

Krasnovskii, A. A., and Sapozhnikova, I. M. (1966). *Dokl. Akad. Nauk SSSR* **169,** 695–698.

Krasnovskii, A. A., Jr., Lebedev, N. N., and Litvin, F. F. (1974). *Dokl. Akad. Nauk SSSR* **216,** 1406–1409.

Kutyurin, V. M., Solov'ev, V. P., and Grigorovich, V. I. (1966). *Dokl. Akad. Nauk SSSR* **169,** 479–482.

Kutyurin, V. M., Slavnova, T. D., and Chibisov, A. K. (1973a). *Biofizika* **18,** 1004–1007.

Kutyurin, V. M., Artamkina, I. Yu., Korsun, A. D., Anisimova, I. N., and Matveeva, I. V. (1973b). *Dokl. Akad. Nauk SSSR* **212,** 243–245.

Kuz'min, V. A., Chibisov, A. K., and Vinogradov, A. P. (1971). *Dokl. Akad. Nauk SSSR* **197,** 129–132.

Lamola, A. A., Manion, M. L., Roth, H. D., and Tollin, G. (1975). *Proc. Natl. Acad. Sci. U.S.A.* **72,** 3265–3269.

Linschitz, H., and Rennert, J. (1952). *Nature (London)* **169,** 193–194.

Linschitz, H., and Sarkanen, K. (1958). *J. Am. Chem. Soc.* **80,** 4826–4832.

Livingston, R. (1955). *J. Am. Chem. Soc.* **77,** 2179–2182.

Livingston, R. (1960). *Q. Rev., Chem. Soc.* **14,** 174–199.

Livingston, R., and Ke, C.-L. (1950). *J. Am. Chem. Soc.* **72,** 909–915.

Livingston, R., and McCartin, P. J. (1963). *J. Phys. Chem.* **67,** 2511–2513.

Livingston, R., and Owens, K. E. (1956). *J. Am. Chem. Soc.* **78,** 3301–3305.

McGlynn, S. P., Daigre, J., and Smith, F. J. (1963). *J. Chem. Phys.* **39,** 675–679.

Meisel, D., and Fessenden, R. W. (1976). *J. Am. Chem. Soc.* **98,** 7505–7510.

Mukherjee, D. C., Cho, D. H., and Tollin, G. (1969). *Photochem. Photobiol.* **9,** 273–289.

Nakato, Y., Chiyoda, T., and Tsubomura, H. (1974). *Bull. Chem. Soc. Jpn.* **47,** 3001–3005.

Norris, J. R., Uphaus, R. A., and Katz, J. J. (1975). *Chem. Phys. Lett.* **31,** 157–161.

Okayama, S., and Chiba, Y. (1965). *Nature (London)* **205,** 172–174.

Onsager, L. (1936). *J. Am. Chem. Soc.* **58,** 1486–1493.

Parker, C. A., and Joyce, T. A. (1967). *Photochem. Photobiol.* **6,** 395–406.

Quinlan, K. P. (1970a). *J. Phys. Chem.* **74,** 3303–3305.

Quinlan, K. P. (1970b). *Biochim. Biophys. Acta* **216,** 441–444.

Quinlan, K. P. (1971). *Photochem. Photobiol.* **13,** 113–121.

Quinlan, K. P. (1972). *Biochim. Biophys. Acta* **267,** 493–497.

Quinlan, K. P., and Fujimori, E. (1967). *J. Phys. Chem.* **71,** 4154–4155.

Raman, R., and Tollin, G. (1971). *Photochem. Photobiol.* **13,** 135–145.

Rao, P. S., and Hayon, E. (1973). *J. Phys. Chem.* **77,** 2274–2276.

Rikhireva, G. T., Sibel'dina, L. A., Gribova, Z. P., Marinov, B. S., Kayushin, L. P., and Krasnovskii, A. A. (1968). *Dokl. Akad. Nauk SSSR* **181,** 1485–1488.

Seely, G. R. (1966). *In* "The Chlorophylls" (L. P. Vernon and G. R. Seely, eds.), Chapter 17. Academic Press, New York.

Seely, G. R. (1967). *J. Phys. Chem.* **71**, 2091–2102.
Seely, G. R. (1969a). *J. Phys. Chem.* **73**, 117–124.
Seely, G. R. (1969b). *J. Phys. Chem.* **73**, 125–129.
Seely, G. R. (1976). *J. Phys. Chem.* **80**, 441–446.
Seely, G. R., and Jensen, R. G. (1965). *Spectrochim. Acta* **21**, 1835–1845.
Seifert, K., and Witt, H. T. (1969). *Prog. Photosynth. Res., Proc. Int. Congr. [1st], 1968,* Vol. II, pp. 750–756.
Seki, H., Arai, S., Shida, T., and Imamura, M. (1973). *J. Am. Chem. Soc.* **95**, 3404–3405.
Singhal, G. S., and Hevesi, J. (1971). *Photochem. Photobiol.* **14**, 509–514.
Singhal, G. S., Williams, W. P., and Rabinowitch, E. (1968). *J. Phys. Chem.* **72**, 3941–3951.
Smyth, C. P. (1955). "Dielectric Behavior and Structure," p. 105ff. McGraw-Hill, New York.
Stanienda, A. (1965). *Z. Phys. Chem.* **229**, 257–272.
Stanienda, A. (1968). *Z. Naturforsch., Teil B* **23**, 147–152.
Stiehl, H. H., and Witt, H. T. (1969). *Z. Naturforsch., Teil B* **24**, 1588–1598.
Strickler, S. J., and Berg, R. A. (1962). *J. Chem. Phys.* **37**, 814–822.
Sullivan, A. B., and Reynolds, G. F. (1976). *J. Phys. Chem.* **80**, 2671–2674.
Tollin, G. (1976). *J. Phys. Chem.* **80**, 2274–2277.
Tollin, G., and Green, G. (1962). *Biochim. Biophys. Acta* **60**, 524–538.
Tollin, G., Chatterjee, K. K., and Green, G. (1965). *Photochem. Photobiol.* **4**, 593–601.
van Gorkom, H. J. (1974). *Biochim. Biophys. Acta* **347**, 439–442.
Voznyak, V. M., Proskuryakov, I. I., and Evstigneev, V. B. (1974). *Biofizika* **19**, 815–819.
Watson, W. F., and Livingston, R. (1950). *J. Chem. Phys.* **18**, 802–809.
Weikard, J., Müller, A., and Witt, H. T. (1963). *Z. Naturforsch., Teil B* **18**, 139–141.
Weiss, C. (1972). *J. Mol. Spectrosc.* **44**, 37–80.
White, R. A., and Tollin, G. (1971a). *Photochem. Photobiol.* **14**, 15–42.
White, R. A., and Tollin, G. (1971b). *Photochem. Photobiol.* **14**, 43–63.
Willson, R. L. (1971). *Trans. Faraday Soc.* **67**, 3020–3029.
Yamazaki, I., and Piette, L. H. (1965). *J. Am. Chem. Soc.* **87**, 986–990.
Zamaraev, K. I., and Khairutdinov, R. F. (1974). *Chem. Phys.* **4**, 181–195.
Zen'kevich, E. I., Losev, A. P., and Gurinovich, G. P. (1972). *Mol. Biol.* **6**, 824–833.
Zen'kevich, E. I., Losev, A. P., and Gurinovich, G. P. (1975). *Mol. Biol.* **9**, 516–523.

Picosecond Events and Their Measurement

MICHAEL SEIBERT[1]

GTE Laboratories, Inc.
Waltham, Massachusetts

I. Introduction

During the past 5 years, our understanding of the primary processes in photosynthesis has benefited greatly from direct kinetic observations on ultrafast time scales, which 10 years ago were technically impossible to attain. Consequently, this chapter will include a description of the advances in pulsed laser spectroscopy that have enabled the direct observation of fluorescent emission kinetics and absorbance changes in the picosecond (10^{-12} second) time domain. Furthermore, it will discuss

---

[1] Present address: Solar Energy Research Institute, 1536 Cole Blvd., Golden, Colorado 80401.

the recent picosecond studies probing both the absorbing antenna pigments that transfer energy to specialized trapping or reaction center complexes and the reaction center itself, which uses the energy to initiate the electron transfer and energy conversion processes covered in other chapters.

II. Recent Developments in Picosecond Laser Technology

To understand the processes occurring within the very rapid time scales discussed in this chapter, it is necessary to look briefly at the laser technology which has opened up this area of research to the biological community. This section is not meant to be a comprehensive review on lasers, but rather a brief synopsis of the basic principles necessary to understand the operation of solid-state lasers. It will also describe the hardware and techniques that investigators are presently using in their efforts to understand the primary photosynthetic processes, which occur in less than a nanosecond (nsec).

A. OPERATION OF SOLID-STATE LASERS

1. Generation of Laser Pulses

Laser pulses are generated in an optical cavity, which is illustrated in Fig. 1. The cavity consists simply of a rod of lasing material and two mirrors. When the laser rod is exposed to high intensity, broadband light from a flash lamp, it will absorb some of the energy. If sufficient energy is available in this "pumping" action, more active ions (Cr^{3+} in ruby lasers and Nd^{3+} in neodymium lasers) in the laser rod will reside in the upper of the two laser transition energy levels shown in Fig. 2. Such a situation is termed a population inversion, and under this condition spontaneous emission (just normal fluorescence) from one excited ion can stimulate additional emission as the photon passes by other excited ions in the laser rod.

A laser takes repeated advantage of this process by providing "optical feedback," i.e., taking part of the emitted radiation and returning it to the assembly of excited ions to stimulate further emission. The mirrors

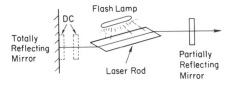

FIG. 1. Components of a laser optical cavity. The dye cell, DC, will be discussed in Sections II,A,2 and II,B,1.

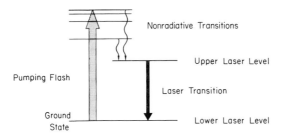

FIG. 2. Energy level diagram for a three-level ruby laser system. Neodymium lasers are an example of a four-level system in which the lower laser level lies above the ground state.

in Fig. 1 provide this feedback function. As the stimulated emission grows in an avalanche manner, a threshold point is reached when the inversion is swept away, and a powerful pulse of radiation is emitted through the partially reflecting mirror. Above threshold the emission of a laser is completely dominated by the stimulated process. Thus, all the well known properties of lasers, such as coherence of the light, its monochromaticity and directionality, are a direct consequence of the dynamics of stimulated emission.

There is one further property of the laser that strongly influences the output, and that is its *mode* structure. Like a taut string the laser cavity can resonate only at discrete frequencies, which form standing waves between the cavity boundaries (the mirrors). Thus, the emission of a "free-running" laser (one in which no control is exercised over which modes can oscillate) consists of a random assembly of discrete frequencies corresponding to the various possible modes. This random character gives rise to fast (microsecond) spikes of emission within the main pulse envelope. The pulse can last hundreds of microseconds or approximately the duration of the pumping flash. Since the length of the pulse is rather long and the peak power eratic, this type of laser has found limited application in biological research. Additional information and a quantitative treatment of the above description is available (Siegman, 1971).

2. Q-Switched Lasers

Soon after the development of the first laser (Schawlow and Townes, 1958; Maiman, 1960), new techniques were devised leading to the availability of pulse widths in the order of 10 nsec (Hellwarth, 1961; McClung and Hellwarth, 1962). The method, termed Q-switching, consists of placing a shutter inside the laser cavity of Fig. 1. The function of the shutter is to prevent laser action during the initial phase of pumping by quenching any stimulated emission. When the radiative lifetime of the upper laser level is long compared with the pumping time, as is the case with ruby or neodymium laser rods, the excited state population can build up to a much higher level than is attainable without the shutter. The excited state population of the laser rod is not depleted because the cavity is not complete. If the shutter is suddenly opened (completing the cavity) at the time of peak excited state buildup, the energy stored in the laser rod while the shutter was closed is rapidly emitted in a single, short, giant pulse.

The shutter function can be accomplished mechanically by rapidly rotating the totally reflecting back mirror in Fig. 1, electrooptically by placing a Kerr cell in the cavity, or optically by placing a dye cell (DC) containing a photobleachable dye in the cavity. Pulse output ($\sim 10^8$ W) in the first case occurs when the mirror rotates to a position perpendicular to the optical axis of the cavity, in the second case when the electrooptical shutter is opened, and in the third case when the light in the cavity is of sufficient power to rapidly bleach the dye. Typical pulse widths are in the order of 10–30 nsec, and as a consequence Q-switched lasers were used to study photosynthetic processes (DeVault, 1964; Chance and DeVault, 1964) soon after their development. Picosecond applications, however, had to await further developments in laser technology. More detailed discussions of Q-switching have been treated by Hellwarth (1966) and DeMaria (1971).

B. Mode-Locked Lasers

During the years from 1966 to 1968, physicists decreased the duration of laser pulses by five orders of magnitude (from 10^{-8} to 10^{-13} second). This progress commenced with the first successful demonstration of mode-locked operation in a ruby laser by Mocker and Collins (1965) and in a Nd:glass laser by DeMaria et al. (1966), although mode-locking was achieved with a gas laser somewhat earlier (Hargrove et al., 1964). The results, of course, led to the feasibility of picosecond spectroscopy.

1. Generation of a Mode-Locked Pulse Train

In normal laser operation, the allowable Fabry–Perot resonance modes (see Fig. 1 in DeMaria et al., 1967), as determined by the geometry of the laser cavity and the spectral bandwidth of the laser transition, are uncoupled and thus have no fixed phase relationship. Interference effects cause the laser to jump from one mode (as many as 6×10^4 modes for a 1.5 m Nd:glass laser) to another resulting in a random output. In a Q-switched laser the number of modes is usually quite small and often can be restricted to one. Mode-locked operation is fundamentally different, though. It takes advantage of all available longitudinal modes which are coupled or locked into phase. Figure 3

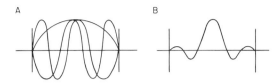

FIG. 3. A schematic illustration of the generation of a short pulse (B) by the addition of 3 modes (A) with fixed-phase relationships. The ordinate scale of (A) is three times that of (B). The coupling of additional modes would result in a narrower pulse width.

shows how the addition of these phase-locked modes can produce a narrow pulse. The coupling of an increasingly large number of modes results in an increasingly narrower pulse.

Figure 1 shows the essential components of a passively mode-locked laser. They consist of an optical cavity containing two mirrors or feedback reflectors, the laser rod, and a dye cell (DC) containing a nonlinear saturable absorber (a photobleachable dye with a much shorter recovery time than the Q-switching dye described in Section II,A,2). In general, mode-locking can be described theoretically in terms of the frequency domain (DeMaria *et al.*, 1967) or the time domain (Letokhov, 1968; Kuznetsova, 1969; Fleck, 1970; Mathieu and Weber, 1971). The latter model has been reviewed by DeMaria (1971), Laubereau and Kaiser (1974), and Eckardt *et al.* (1974). It proceeds as follows.

Spontaneous emission at the laser transition is amplified by the optically pumped laser rod, and many longitudinal modes are excited. During this initial stage of pulse development, the population inversion of the laser transition and the transmission of the bleachable dye are not affected by the radiation in the cavity. As a result, linear amplification of the axial light can occur over many cavity round trips (the saturable absorber at this point transmits a large fraction of the light). Interference effects of laser modes with random-phase relationships lead to fluctuations of the light intensity within the cavity. These fluctuations are periodic with respect to the cavity round-trip transit time $T = 2L/c$, where L is the pathlength of the cavity and c is the speed of light. In time a small number of intensity maxima will exceed the average intensity to a significant extent.

When the intensity fluctuations in the cavity approach the saturation intensity of the saturable absorber, nonlinear transmission occurs in the dye cell. This property in conjunction with the rapid recovery time, τ_D, of the dye ($\tau_D \ll T$) locks the modes into phase by ensuring that the most intense fluctuation that arose during the initial stage is preferentially amplified relative to the less intense fluctuations over many cavity round trips. This occurs because the smaller fluctuations encounter larger absorption in the dye cell than does the most intense one and are

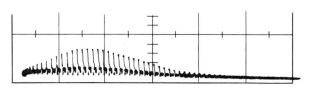

FIG. 4. Emission output pulse train from a mode-locked Nd:glass laser. The time scale is 50 nsec/cm.

effectively suppressed. The fast recovery time of the dye also shortens the pulse duration by absorbing the wings of the pulse more strongly than the peak.

The resultant emission output pulse train for a Nd:glass laser is a series of 20–100 periodic pulses 5–10 picoseconds (psec) wide occurring every 5.5 nsec in the case of Fig. 4. The pulses are significantly broader than the lower limit of ~0.1 psec predicted from the 200 cm^{-1} bandwidth of the Nd:glass laser transition (Penzkofer, 1974). This broadening as reviewed by Penzkofer (1974) is due to spectral narrowing in the linear amplification phase of pulse development, the finite τ_D of the saturable absorber dye (9.3 psec for Kodak No. 9860 dye used in Nd:glass lasers), the effects of self-phase modulation, and dispersion.

2. Isolation of a Single Mode-Locked Pulse

The output of a mode-locked laser is a train of picosecond pulses separated by the cavity round-trip transit time as discussed above. Such a train can be used as an excitation source in picosecond spectroscopy if the relaxation time of the process being measured is short compared with the time between pulses and the state of the system is not significantly altered by the train of pulses. If the process to be studied does not meet these criteria, one must employ a single pulse as an excitation source. There are also additional advantages in using a selected pulse in examining picosecond time scale phenomena, since the results are not influenced by differences in the intensity, width, and frequency of pulses from different positions in the pulse train (DeMaria, 1971; Netzel et al., 1973a; Eckardt et al., 1974; Laubereau and Kaiser, 1974).

An apparatus (see the upper left portion of Fig. 7 for one configuration) which can select a single pulse with well defined pulse properties from a mode-locked train consists of a mode-locked laser, an electrooptical shutter, and an amplifier system. Selection of a single pulse from the train is accomplished by placing a Pockels' cell, energized by a spark gap charged to high voltage, between two crossed polarizers. The two polarizers block the initial pulses. However, part of the train (reflected out of the optical path to the second polarizer) is focused onto the spark gap, which is adjusted so that its breakdown threshold corresponds to the intensity of one of the pulses in the middle of the train. When a pulse of sufficient intensity reaches the spark gap, bieakdown occurs and a high-voltage pulse lasting less than two times the round-trip cavity transit time is applied to the Pockels' cell. At this point the polarization of the next pulse is rotated so that only that pulse can pass through the second polarizer. The remainder of the pulse train is rejected. Since the energy of the single pulse is rather small ($\sim 10^{-4}$ J), it requires amplification. Two passes through an additional 30-cm Nd:glass laser rod provides a factor of 100 amplification, and typical pulse widths are in the order of 6 psec (Laubereau and Kaiser, 1974).

Although picosecond light pulses from mode-locked lasers have thus far been used exclusively in photosynthetic research, it should be noted that other means of producing picosecond light pulses, including the use of pulsed radiolysis methods, are available (Bronskill et al., 1970).

C. Methods Used in Picosecond Fluorescence Measurements

1. Optical Gate Technique

The first reported picosecond spectrophotometric measurements on a photosynthetic system (Seibert et al., 1973) were obtained using the optical Kerr cell or gate method of Duguay and Hansen (1969a,b) and Shimizu and Stoicheff (1969). The apparatus is described schematically in Fig. 5. Light output at 1060 nm from a Nd:glass mode-locked laser passes through a 2-cm potassium–dihydrogen phosphate (KDP) second harmonic generator crystal, which converts as much as 10% of the 1060 nm light to the harmonic at 530 nm. The process is called frequency doubling. Pulse widths of 6–8 psec at 1060 nm and 4–5 psec at 530 nm can be measured by the two-photon fluorescence technique (Giordmaine et al., 1967). The function of the dielectric mirror (M_1) is to separate the

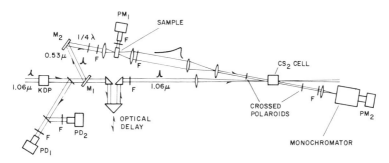

FIG. 5. A schematic diagram of a laser–optical Kerr gate apparatus (Fig. 1 of Seibert and Alfano, 1974) used to measure fluorescence emission kinetics in the picosecond time range. Components: PD, photodiode; M, mirror; F, optical filter; ¼λ, quarter-wave plate; and PM, photomultiplier tube.

pulse train into its component wavelengths and direct the two different wavelength trains along different delay paths.

The sample is excited by the 530 nm train and the resultant fluorescence is collimated through the optical gate. The gate consists of a system of two crossed polaroids with a 1-cm path length cell of CS_2, which is placed in between and serves as an ultrafast light shutter. The 1060-nm train follows a path that can be adjusted by a variable optical delay and is reduced to 1 mm in diameter upon passage through the CS_2 cell. Its function is to induce a short-lived birefringence in the CS_2, thus allowing fluorescent light from the sample cuvette to pass through the crossed polaroid configuration only during the induced birefringence. By adjusting the optical delay (moving the prism) in the 1060 nm path, the fluorescence output can be sampled at various times with respect to the actinic pulse. The zero point of the optical gate is determined by the coincidence in time and space of both the 1060 nm and 530 nm pulses, with buffer solution substituted for the experimental sample in the cuvette. After going through the gate, the fluorescence is passed through a monochromator and detected with a photomultiplier tube. To correct the data for differences in laser output from shot to shot, the intensity of the 1060-nm and 530-nm beams and the total fluorescence intensity were also detected and displayed simultaneously with the corresponding gated fluorescence (at the particular delay time and wavelength) on a dual-beam oscilloscope using appropriate delay cables. Recent improvements in this technique, including the elimination of the quarter-wave plate and modifications in the geometry of the gate (Yu *et al.*, 1975), have increased the sensitivity of the instrument.

The extremely rapid response ($\sim$10 psec) of this instrument is not due to the photomultiplier or its associated electronics (which in fact is relatively slow), but to the extremely short sampling period provided by the optical gate light shutter. It must be emphasized that if a pulse train is used the results obtained by the optical gate method are an average of those for each of the individual pulses in the train. Thus, as stated previously, the technique is suitable for measuring fluorescence kinetics as long as (a) the fluorescent component being observed recovers within the time period between pulses and (b) there are no long-term effects that significantly alter the lifetime of that component.

2. Streak-Camera Techniques

Several groups throughout the world (Shapiro *et al.*, 1975; Paschenko *et al.*, 1975; Beddard *et al.*, 1975) have recently made use of electron-optical image converters or streak cameras in conjunction with mode-locked lasers to investigate rapid fluorescence phenomena in various photosynthetic organisms. Streak cameras have the advantage of allowing the continuous recording of fluorescence as a function of time up to a few nanoseconds and do not require data to be obtained on a point by point basis as in the optical gate method discussed in the last section.

Figure 6 is a diagram of an apparatus employing a streak-camera detecting system. The ruby (1) and Nd:glass lasers (2) and their respective second harmonic generating crystals (3) provide actinic pulse trains at 1060, 694, 530, and 347 nm. Filters (4) and mirrors (8) are used to select the desired wavelength train and focus (20) it onto the sample cuvette (9). Fluorescence can be detected with a photocell (14) connected to a traveling-wave oscilloscope (7), but the time resolution of this combination is about 500 psec. Faster fluorescence changes are monitored with the streak camera (15). The time resolution of this device is about 10 psec, and it works as follows:

Light from the cuvette emitted perpendicular to the actinic pulse is focused (13) onto a photocathode in the camera. Photoelectrons, produced in proportion to the light intensity, are accelerated through an anode and then deflected by a voltage ramp. The voltage ramp is generated when part of the actinic pulse is deflected to trigger a spark gap (12) charged to high voltage. The deflected photoelectrons streak across a phosphorescent screen such that electrons released from the

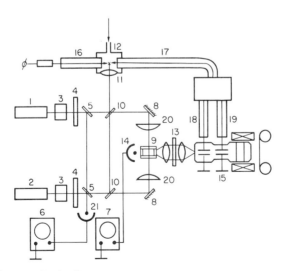

Fig. 6. A picosecond pulse fluorometer employing a streak camera (Fig. 1 of Paschenko et al., 1975). Components: 1 and 2, ruby and Nd:glass lasers; 3, KDP crystals; 4, glass filters; 5 and 10, beam diverting glass plates; 6 and 7, oscilloscopes; 8, mirrors; 9, cuvette containing the solution under investigation; 11, focusing system; 12, spark-gap; 13, system of image transfer; 14 and 21, photocells; 15, streak camera; 16–19, cable lines; 20, cylindrical lenses.

cathode at different times hit the screen at different locations. The resultant phosphorescence can be photographed directly or the sensitivity improved by including additional image-intensifier stages. Densitometer traces of the photographs record the kinetics of the picosecond event under observation. Although fluorescence is generated by each pulse in the train, the fluorescence elicited by only one of the pulses is detected in this system.

Recent modifications in this technique have included excitation of the sample with a single pulse and substitution of a silicon vidicon optical multichannel analyzer (OMA) for the film as a means of detection (Campillo *et al.*, 1976b). The system functions accurately as long as the silicon target is not scanned faster than limits set by lag considerations (Princeton Applied Research Corp., 1975). Picosecond dye lasers are also being used now as excitation sources (Breton and Geacintov, 1976).

D. Methods Used in Picosecond Absorbance-Change Measurements

Two methods have thus far been used to observe the fast absorbance changes occurring on a picosecond time scale in photosynthetic systems. Both employ a single 530-nm pulse as an excitation source. The first method (Rentzepis *et al.*, 1973; Leigh *et al.*, 1974; Kaufmann *et al.*, 1975) uses a stepped-delay echelon to measure the absorbance in a sample at several discrete times using the same initial laser pulse. The second is a point-by-point method (Magde and Windsor, 1974; Rockley *et al.*, 1975) that uses a variable optical delay line to measure absorbance changes at different times after an actinic pulse.

1. Echelon Method

Figure 7 is a diagram of the apparatus used in the echelon method. A single, 1060-nm, 10-psec pulse is selected (3–6) from the output train of a mode-locked Nd:glass laser (1,2), then amplified (7) and frequency doubled (8). The second harmonic 530-nm pulse, which passes through the partial reflector (PR), serves as the actinic light at the sample cuvette (13), and the residual 1060-nm pulse reflected by the PR ultimately provides the measuring or interrogating light. A lens focuses the 1060-nm pulse onto a cell containing *n*-octanol (9) and produces a broadband continuum (Alfano and Shapiro, 1970) of light with a pulse width approximately that of the original 1060-nm pulse. The phenomenon is generally termed self-phase modulation and may involve several nonlinear optical processes (Magde and Windsor, 1974).

The continuum (still a single pulse at this point) is then passed through the echelon (a stack of microscope slides), which delays light hitting different parts of the echelon (11) by preselected discrete amounts. The

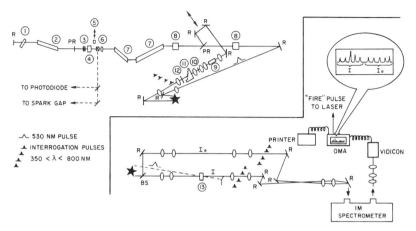

FIG. 7. A double-beam picosecond spectrometer (Fig. 1 of Leigh *et al.*, 1974). Components: 1, mode-locking dye cell; 2, laser oscillator rod; 3, calcite polarizer; 4, Pockels' cell;
5, translatable 90° polarization rotator for 1060-nm radiation; 6, fixed position 90°
polarization rotator; 7, laser amplifier rod; 8, second harmonic (530 nm) generating KDP
crystal; 9, 20-cm octanol cell for generating the interrogation wavelengths; 10, ground-glass
diffuser; 11, index matched glass echelon for producing picosecond optical delays between
the stacked interrogation pulses; 12, vertical polarizer; 13, sample cell; R, reflector; PR,
partial reflector; BS, beam splitter; OMA, optical multichannel analyzer.

resulting interrogation pulse train is split (BS) into a measuring beam, I,
and a reference beam, I_0. The measuring beam passes through the
sample cell with the single 530-nm actinic pulse timed to hit the cell near
the beginning of the I train. Thus, the I pulses monitor the absorbance of
the sample at different times before and after the actinic pulse. I_0 pulses
are directed around the sample and are used to correct for power
variation between individual laser shots. Both I and I_0 are recombined
behind the sample cuvette and imaged beside each other on the entrance
slit of the spectrometer. I and I_0 are then detected on a silicon vidicon
surface and the image stored in an optical multichannel analyzer (OMA).

The circled inset in Fig. 7 shows a schematic display of the data. The
measuring train, I, is on the left and the reference train, I_0, is on the
right. The actinic pulse in this example would occur somewhere between
the second and third pulse in the I train. Changes in absorbance for each
time segment are calculated using Eq. (1) (Leigh *et al.*, 1974)

$$\Delta A = -\log \frac{I^e I_0^{\;u}}{I_0^{\;e} I^u} \qquad (1)$$

where the superscripts, e and u, denote experimental runs with and
without the actinic pulse, respectively.

2. Delay Line Method

The second method is described in Fig. 8A. A single 8-psec pulse from a frequency-doubled, mode-locked Nd:glass laser is split into its component 530-nm and 1060-nm wavelengths by a dichroic beam splitter (DM). The 530-nm pulse (P) serves as an actinic source and travels to the sample cuvette (C) along a variable-length optical path. The residual 1060-nm pulse is focused on a CCl_4 self-phase modulation cell (SPM), and the resultant white continuum measuring pulse (S) is collimated, filtered, passed along a second optical delay line, and focused onto the sample cuvette in the same region as the actinic pulse. Figure 8B emphasizes the geometry of the region. The important point to note is that both the actinic and measuring pulses overlap physically.

Reference volumes both above and below the excited region are sampled simultaneously along with the excited volume. The measuring pulse passing through the sample cuvette enters a spectrograph equipped with a silicon vidicon detector coupled to an optical multichannel analyzer. The multichannel analyzer output is a vertical intensity profile along the S beam in Fig. 8B. In this manner absorbance differences caused by the actinic pulse can be measured on the sample at different times as controlled by the two delay lines.

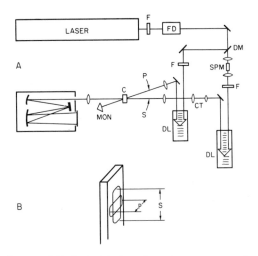

FIG. 8. (A) A picosecond flash photolysis and spectroscopy apparatus (Fig. 1 of Magde and Windsor, 1974). Components: FD, frequency-doubler KDP crystal; DM, dichroic beam splitter; SPM, self-phase modulation cell; DL, optical delay line; CT, cylindrical telescope; P, pump pulse path; S, probe continuum pulse path; C, sample cell; MON, excitation intensity monitor; F, colored glass filters. Mirrors and lenses are not labeled. The spectrograph is a ¾ m grating instrument. (B) Enlargement of the beam geometry at the sample cell to illustrate how S probes both the excited volume and adjacent unexcited regions.

This method has two advantages (Magde and Windsor, 1974) over the echelon technique of Section II,D,1. It is not sensitive to problems in spatial uniformity of the continuum at longer time delays and is not limited to discrete jumps in time delay. Thus, absorbance changes can be monitored anywhere in the 3-psec to 10-nsec range with no equipment changes.

E. DIFFICULTIES ENCOUNTERED IN PICOSECOND SPECTROSCOPY

Needless to say, any experimental technique that involves the sophistication described in the previous sections has its share of problems. Examples include background radiation, lack of reproducibility from one shot to the next, nonlinear effects in the laser rod such as self-focusing, and the presence of extraneous satellite pulses (Garmire and Yariv, 1967; Fleck, 1970). Concave mirrors facilitate optical alignment and improve reproducibility by reducing the stringent requirements of cavity stability. Unintentional longitudinal mode selection is avoided with the help of Brewster's-angle cuts at the laser rod and wedged plates at the dye cell and laser mirrors (Laubereau and Kaiser, 1974). Satellite pulses, resulting from the presence of two light pulses propagating inside the laser cavity in opposite directions, can be suppressed by using a thin dye cell in contact with a mirror (Fig. 1) or by inserting more than one dye cell in the cavity (Eckardt et al., 1974). In addition to the problems associated with the apparatus, care must be taken that the sample material is not exposed to actinic pulse conditions, which cause nonlinear effects (see Section III,B).

A final note of caution should be mentioned before proceeding to a discussion of the recent picosecond studies in photosynthesis. Mode-locked laser pulses present ocular hazards which have received little attention. However, Ham et al. (1974) have exposed rhesus monkey eyes to single 25–35-psec pulses of 1064 nm light from a mode-locked Nd:YAG laser. They determined that threshold injury resulted at a mean pulse energy of 13 ± 3 μJ (the beam diameter was 3.8 mm at the eye), which is up to 3 orders of magnitude lower than the pulse energies that have been used in picosecond spectroscopy. Damage was reported to be highly localized in the photoreceptor and pigmented epithelial cells at the outer retina.

III. Picosecond Studies in Green Plant Photosynthesis

Much interest during the past 20 years has been directed toward the rapid energy transfer processes that proceed between the absorbing, accessory antenna pigments and the specialized chlorophyll (Chl) trapping complexes of Photosystem I (PSI) and Photosystem II (PSII). Since the fluorescence lifetime of the antenna pigments is a direct measure of the time required for excitation energy to reach the trap (Sauer, 1975), many investigators have attempted to observe the kinetics of *in vivo* fluorescence emission and to measure lifetimes. The techniques employed have included direct methods using nanosecond light pulses from flash lamps (Brody and Rabinowitch, 1957; Tomita and Rabinowitch, 1962; Murty and Rabinowitch, 1965; Nicholson and Fortoul, 1967; Singhal and Rabinowitch, 1969) or indirect methods using phase fluorometry (Dmitrievsky *et al.*, 1957; Tumerman *et al.*, 1961; Rubin and Osnitskaya, 1963; Butler and Norris, 1963; Müller *et al.*, 1969; Merkelo *et al.*, 1969; Briantais *et al.*, 1972; Mar *et al.*, 1972; Borisov and Il'ina, 1973). The results were often conflicting owing to the lack of sufficiently short instrumental response time and to the experimental conditions used. However, it was established that the lifetime of fluorescence was dependent on the intensity of the exciting light (Dmitrievsky *et al.*, 1957; Müller *et al.*, 1969; Borisov and Il'ina, 1973) and on photosynthetic activity (Tumerman *et al.*, 1961; Briantais *et al.*, 1972). With the advent of picosecond technology, the direct measurement of the fluorescence kinetics of photosynthetic systems in the time range over which the fluorescence occurred became a reality (Seibert *et al.*, 1973).

A. FLUORESCENCE MEASUREMENTS OBTAINED USING A PULSE TRAIN

All early picosecond fluorescence kinetic studies were performed using the entire pulse train output from a mode-locked laser as an actinic source.[2] Unfortunately, the results were not always consistent, and in many cases the lifetimes obtained were shorter than those determined by older methods. Since the outcome of these studies was potentially subject to long-term (up to several hundred nanoseconds) effects (Seibert and Alfano, 1974), such as the buildup of triplet states (Duysens *et al.*, 1975; Campillo *et al.*, 1976a) or coupling changes between antenna and trap molecules (Mauzerall, 1976), these conditions might affect the observed lifetimes. Recent single-pulse experiments of Mauzerall (1976) and Campillo *et al.* (1976a,b) have pointed out the additional complication that high pulse flux densities can decrease quantum yields and lifetimes in photosynthetic systems. These points will be discussed in Section III,B. However, this section will be devoted to the train experiments because the fluorescence kinetics obtained using pulse trains can be the same as those using single pulses if the train flux density is low enough (Harris *et al.*, 1976) and because the train

---

[2] The excitation wavelength of all picosecond studies reported in this chapter is 530 nm unless otherwise stated.

technique appears to be a promising way to obtain information on the topology of the photosynthetic units (Mauzerall, 1976).

1. Subchloroplast Fractions

Fluorescence emission kinetics have been observed in PSI-enriched fractions of spinach, isolated using both the digitonin (Boardman *et al.*, 1966) and French pressure cell (Sane *et al.*, 1970) methods. Rise times were reported in the order of 5 psec, but the lifetimes ranged between 60 ± 10 psec at 683 nm using the optical-gate method (Yu *et al.*, 1975)[3] and around 130 psec in the red spectral region using a streak camera (Beddard *et al.*, 1975). The risetime of fluorescence in PSII enriched fractions was also around 5 psec, but the lifetime was 200 ± 20 psec (Yu *et al.*, 1975), substantially longer than those observed in PSI fractions.

The quantum yields (ϕ) for fluorescence in PSI and PSII fractions of spinach determined from the picosecond lifetimes can be compared with those measured by conventional means. From the well known relationship $\phi = \tau/\tau_0$, where τ is the measured lifetime and the intrinsic lifetime, τ_0, is 15.2 nsec (Brody and Rabinowitch, 1957), one obtains quantum yields of 0.0039 ± 0.0007 and 0.013 ± 0.001 for the two respective photosystems using the data of Yu *et al.* (1975). These values are very close to the values 0.003 and 0.016 reported by Boardman *et al.* (1966) for PSI and PSII enriched fractions of spinach measured under low light conditions. In addition, Borisov and Il'ina (1973) using phase fluorometric techniques have estimated that the lifetime of PSI fluorescence in pea is in the order of 30 psec or less. Thus, it appeared that, under the experimental conditions employed, the picosecond studies were consistent with previous results. The difference in the lifetimes reported for PSI fractions by Yu *et al.* (1975) and Beddard *et al.* (1975) may be a consequence of the different pulse train energies used. However, they might also be explained if the sample material in the latter case were subjected to conditions that closed some of the traps at the time of observation (see Section III,A,2).

2. Chloroplasts

Several groups have reported the observation of multiple fluorescent components in chloroplasts using the optical-gate (Seibert *et al.*, 1973; Seibert and Alfano, 1974) and streak-camera (Paschenko *et al.*, 1975; Beddard *et al.*, 1975; Breton and Roux, 1975) methods. Seibert and co-

---

[3] The value 6×10^{14} photons/cm^2 for the average actinic flux density per 530-nm pulse stated previously was three times higher than the actual value owing to an underestimate of the size of the focused laser beam spot at the sample cell (W. Yu, personal communication).

workers (1973; Seibert and Alfano, 1974) reported two fluorescing components at 685 nm. The longer-lived component decayed with a lifetime of 220 ± 25 psec in spinach and 320 psec in escarole. It was interpreted as arising from PSII, since it was the source of most of the fluorescent emission and since the quantum yield calculated from the lifetime was close to the measured value for PSII-enriched fractions of spinach (Boardman et al., 1966). The short-lived component, on the other hand, was interpreted as rising from PSI and had a reported lifetime in the order 10 psec. More recent studies with spinach chloroplasts (Yu et al., 1977) confirm that two fluorescing components are apparent at 685 nm. However, fluorescence emission peaks within 10 psec of excitation and decays nonexponentially (see also Yu et al., 1975). The kinetics can be fitted by two exponentially decaying components with lifetimes of 56 and 220 psec. The previously reported lifetime of 10 psec was probably due to an insufficient number of data points and the large statistical fluctuation in laser shots characteristic of the apparatus at that time. Nonlinear effects (at least on PSII fluorescence) due to high intensities of the individual actinic pulses in the train were not noted at the experimental flux densities ($\sim 2 \times 10^{14}$ photons/cm^2 per pulse) since a plot of the total fluorescence elicited by the train as a function of the total train flux density resulted in a straight line passing through the origin (Seibert and Alfano, 1974).

Paschenko et al. (1975), whose results were generally consistent with the above work, identified three fluorescent components in pea chloroplasts with lifetimes of 80 psec (PSI), 300 psec (PSII), and 4500 psec (solubilized monomeric Chl not involved in photosynthesis). The 80-psec component was observed at wavelengths longer than 730 nm, and the slower two components were detected at wavelengths longer than 650 nm. A delay time of 200 psec attributed to energy transfer from carotenoids to Chl was observed for all three components when 530-nm excitation pulses were used. However, no large delay time was reported in the 80-psec component with 694 nm excitation pulses. Paschenko et al. (1975) pointed out that delay times were not consistent with the spinach results (Seibert and Alfano, 1974; Yu et al., 1975) and attributed the conflicting results to differences in the geometry of detection. In view of the otherwise close agreement in the lifetimes as determined by the two methods, it would appear that additional explanation is necessary. For example, differences in carotenoid content of the chloroplast preparations might cause the apparent inconsistency.

In any case, the 60- to 80-psec lifetimes which the above groups observed for PSI fluorescence are consistent with previous observations (Boardman et al., 1966; Borisov and Il'ina, 1973). However, the 200- to 300-psec lifetimes obtained for PSII fluorescence are somewhat lower than previous values obtained in leaves and chloroplasts (see Sections III,B and III,C for possible explanations). For example, Müller *et al.* (1969) reported lifetimes (predominantly PSII fluorescence) as low as 380 psec using a phase fluorometer, and Singhal and Rabinowitch (1969) reported values of 700 ± 200 psec using nanosecond actinic flashes.

Beddard *et al.* (1975) observing over a wide band in the red spectral region also reported two fluorescent species in spinach chloroplasts. The major component decayed with a lifetime of 134 ± 10 psec, but the minor one had a much longer lifetime. Breton and Roux (1975) reported a single 55 ± 10-psec component. These results are probably influenced by high pulse-train intensities (see Section III,B).

Conditions that "close" the traps (prevent electron donation to the electron transport chain) have a significant influence on the lifetime of chloroplast fluorescence, as has been seen using the older techniques mentioned in the beginning of Section III. Paschenko *et al.* (1975) demonstrated that saturating background illumination increases the 80-psec lifetime of PSI fluorescence to 200 psec and that of PSII fluorescence from 300 psec to 600 psec. In addition, 3-(3,4-dichlorophenyl)-1,1-dimethylurea (DCMU), which blocks PSII activity, increased the lifetime of PSII fluorescence to 600 psec but had no effect on PSI fluorescence. These observations agree with previous studies which show that the quantum yield of PSII fractions increases with background light, but they disagree with those that show that the quantum yield of PSI fractions remains constant (Vredenberg and Slooten, 1967). Beddard *et al.* (1975), on the other hand, showed that dithionite, which prereduces the acceptor molecules, increased both the intensity and lifetime of their minor (long-lifetime) component but had no effect on the main component. Assuming that electron transfer occurs from an excited reaction center Chl singlet state, the relatively small increases in the lifetimes reported by Paschenko *et al.* (1975) are not consistent with the high yields of primary photochemistry ≥0.85 in plants (A. Yu. Borisov and M. D. Il'ina, personal communication) and ~1 in bacteria (Wraight and Clayton, 1974). However, this evidence could suggest the possibility of an intermediate state in green plant photosynthesis (Paschenko *et al.*, 1975; see the article by Ke, in this volume, for additional discussion of this point) occurring between the excited trap complex state and the reduction of the classical primary electron acceptor as has been demonstrated in bacterial photosynthesis (see Section IV,A).

3. Algae

The first picosecond studies in algae indicated that the fluorescence lifetimes were very short. Kollman *et al.* (1975) and Shapiro *et al.* (1975), observing at wavelengths beyond 640 nm with a streak camera, reported nonexponential decay of the fluorescence. The measured lifetimes were 74 ± 5 psec in *Anacystis nidulans* and 41 ± 5 psec in *Chlorella pyrenoidosa*. These low values were attributed to concentration quenching on the basis of similar lifetimes found in concentrated solutions of Chl *a* and Chl *b*. However, Beddard *et al.* (1975), who observed lifetimes of 108 ± 10 psec in *Chlorella* and 92 ± 10 psec in *Porphyridium cruentum*, pointed out that although concentration quenching does account for the fluorescence lifetimes of Chl in solution, it is inconsistent with the operation of efficient photosynthesis. In addition to the difference in the lifetime of fluorescence in *Chlorella* recorded by the two groups, a further complication was apparent since the 41- to 108-psec lifetimes observed were much shorter than those obtained by previous methods. For example, the lifetime of both *Anacystis* and *Porphyridium* in weak light was reported as 500 ± 200 psec (Singhal and Rabinowitch, 1969) using nanosecond pulse techniques. That for *Chlorella* was between 350 psec (Müller *et al.*, 1969) and 600 to 700 ± 200 psec (Singhal and Rabinowitch, 1969; Nicholson and Fortoul, 1969) using phase fluorometric and nanosecond pulse techniques, respectively.

B. Exciton Annihilation Processes

The recent studies of Mauzerall (1976) and Campillo *et al.* (1976a) reported a decrease in the fluorescence quantum yield of *Chlorella* as a function of the pulse flux density of a single 7-nsec or a single 20-psec pulse, respectively. Campillo *et al.* (1976b) subsequently showed that the lifetime of 700 nm fluorescence in *Chlorella* was also dependent on the intensity of a single pulse. The phenomenon was explained on the basis of a multitrap model for the photosynthetic unit (Mauzerall, 1976) and singlet exciton–exciton annihilation (Mauzerall, 1976; Campillo *et al.*, 1976a; Swenberg *et al.*, 1976) resulting from the presence of more than one excitation per photosynthetic unit.

Quenching of fluorescence also occurs in spinach chloroplasts (Breton and Geacintov, 1976; Geacintov and Breton, 1976). At low temperature (100°K) the fluorescence yield at both 735 nm (PSI) and 685 nm (PSII) decreases with exposure to an increasing number of 10-psec pulses (3.7 × 10^{16} photons/cm^2 at 610 nm) spaced 5 nsec apart. In addition, the ratio of the fluorescence yields at 735 and 685 nm decreases when chloroplasts are exposed to more than four such pulses. In whole train experiments (~300 pulses at 610 nm) quenching of 735 nm fluorescence appears as the flux density of the *entire* train increases above about 10^{15} photons/cm^2 while that at 685 nm occurs at between 10^{16} and 10^{17} photons/cm^2 (Geacintov and Breton, 1976). At room temperature, quenching also occurs at 685 nm when chloroplasts are exposed to an increasing number of 10-psec pulses, but the effect is less pronounced than at low temperature (Breton and Geacintov, 1976). These results were interpreted as defining two classes of exciton annihilation processes (Geacintov and Breton, 1976). The first class is observed when single, high intensity picosecond pulses are used (as in the algal studies) and involves the quenching of singlet states in the following possible reactions:

$$S_1 + S_1 \rightarrow S_0 + S_n \tag{2}$$

$$S_1 + S_1 \rightarrow N(+) + N(-) \tag{3}$$

S_0 denotes the ground state, S_1, the first excited singlet state, and S_n, a higher excited singlet state. $N(\pm)$ denotes either a positive or negative Chl ion. The second class of annihilation processes is observed when multiple pulses of sufficient flux density are used. They involve long-lived quenching species since singlet states do not last from one picosecond pulse to the next ($\sim$5 nsec). The following are possible quenching reactions:

$$S_1 + T_1 \rightarrow S_0 + T_n \tag{4}$$

$$S_1 + T_1 \rightarrow N(+) + N(-) \tag{5}$$

$$S_1 + N(\pm) \rightarrow S_0 + N(\pm) \tag{6}$$

T_1 and T_n represent first excited and higher excited triplet states, respectively. Triplet states could be formed by intersystem crossing (Mathis, 1969) or by recombination of the paired ions formed in Eq. (3) (Geacintov and Breton, 1976).

Needless to say, all these annihilation phenomena can decrease the observed fluorescence lifetimes and could explain in part some of the results reviewed in Section III,A. For example, the 41- to 108-psec lifetimes reported in the early picosecond studies on algae using pulse trains were anomalously low owing to the use of high pulse-train flux densities. Campillo *et al.* (1976b) have recently estimated that the "true" fluorescence lifetime for *Chlorella* is about 650 ± 150 psec when a single, low-intensity ($<10^{14}$ photons/cm^2) actinic pulse is used. This was accomplished by averaging the lifetimes obtained by an indirect means (535 psec) and by a calculation using a Hartree approximation (800 psec). The value is equal to that obtained in weak light by the nanosecond pulse methods (Singhal and Rabinowitch, 1969; Nicholson and Fortoul, 1969) but higher than that obtained by the phase method

(Müller *et al.*, 1969). In actuality, the fluorescence decay kinetics in *Chlorella* are nonexponential (Campillo *et al.*, 1976b), as has been reported in spinach chloroplasts (Yu *et al.*, 1975).

Recently, Harris *et al.* (1976), working with either single 10^{14} photon/ cm^2 pulses or pulse trains in which only the fluorescence resulting from one 10^{14} photon/cm^2 pulse was monitored, also observed nonexponential decay in *Chlorella*. Since the results of the two excitation conditions were the same, studies using low enough train energies are equivalent to those employing a single pulse because large numbers of long-lived quenchers do not build up. At higher pulse-train energies in which the fluorescence from one of the pulses (10^{15} photons/cm^2) was monitored, they observed an additional 32 psec, exponentially decaying component. The lifetime of the nonexponential component decreased by only 20%. The rapid decaying component was thought to be associated with PSI. However, the possibility that it arose as an artifact of the higher pulse intensities could not be excluded since the sensitivity of the equipment was not sufficient to observe the component when 10^{14} photon/cm^2 pulses were used.

The 60- to 80-psec (PSI) and 200- to 300-psec (PSII) lifetime components observed in chloroplasts and subchloroplast fractions are also influenced by these annihilation processes. PSI fluorescence, in particular, is more sensitive to the buildup of long-lived states than PSII fluorescence, at least at low temperature (Geacintov and Breton, 1976). It should be emphasized, though, that the two components arise from Chl associated with the two respective photosystems because only the shorter component is seen in isolated PSI fractions and only the longer component is seen in isolated PSII fractions. Both components are observed in chloroplasts.

C. Meaning of the Measured Lifetimes

In addition to the possible effects of exciton annihilation processes on fluorescence lifetimes (which are real and must be eliminated if accurate lifetimes are desired) reported in the above sections, one must consider the meaning of the lifetime being measured. For example, the $1/e$ point lifetime measured by the picosecond techniques may not correlate directly with the deconvoluted lifetimes obtained by the older methods, especially if multiple fluorescing components and nonexponential decay are involved. Harris *et al.* (1976) have suggested that the phase method measures a lifetime, τ_M, for nonexponential decay which can be five times longer than the $1/e$ point lifetime observed in *Chlorella*. Thus, the comparison of lifetimes obtained using different methods must consider exactly what lifetime is being measured.

D. GENERAL COMMENTS REGARDING PICOSECOND FLUORESCENCE MEASUREMENTS

The investigations examined in the preceding sections point out the difficulties in performing picosecond fluorescence studies in photosynthetic systems. Besides the physical problems characteristic of the apparatus (see Section II,E), problems associated with the environmental or measuring conditions can lead to difficulties in interpretation. For example, fluorescence yields and emission kinetics can be influenced by (1) the lack of uniform irradiation at the sample surface, (2) the optical density of the sample, (3) conditions that close the traps or inactivate their primary function (DCMU, background light, low redox potential, or high redox potential), (4) the flux density of a single picosecond excitation pulse, (5) the number of picosecond pulses, and (6) the flux density of an entire pulse train. In addition, the comparison of the lifetimes obtained by picosecond techniques and those observed using the older methods may be difficult.

All these points must be considered when using picosecond light pulses to observe fluorescence emission. Excitation with a single pulse can simplify the interpretation of the observed fluorescence measurements. However, the comparison of results using different numbers of picosecond pulses and pulse flux densities opens up a new means of exploring the photosynthetic unit. Future studies will most certainly have impact on the exact nature of the energy transfer processes that occur between the antenna pigments and the trapping centers. Both weak and strong exciton coupling interactions may be involved (Borisov and Godik, 1973; Sauer, 1975; Knox, 1975).

IV. Picosecond Studies in Bacterial Photosynthesis

The primary photochemical reaction in bacterial photosynthesis involves the transfer of an electron from a bacteriochlorophyll pigment complex to an acceptor molecule (Clayton, 1973; Parson, 1974). Progress in defining the details of this reaction has benefited greatly from the availability of detergent-isolated "photochemical reaction center" preparations obtained from the R-26 strain of *Rhodopseudomonas sphaeroides* (Reed and Clayton, 1968; Clayton and Wang, 1971; Dutton *et al.*, 1973b). The reaction center protein complex (RC) has been characterized (Reed, 1969; Clayton and Wang, 1971; Feher, 1971; Reed and Peters, 1972; Okamura *et al.*, 1974) and contains four bacteriochlorophylls (BChl), two bacteriopheophytins (BPh), a ubiquinone, and a nonheme iron moiety. Since the complex is devoid of the light-gathering, antenna BChl and carotenoids which funnel energy into the RC as well as the redox pigments involved in the slower electron transfer reactions, it has served as a valuable simplified model system in which to study primary photochemistry. See the article by Ke, elsewhere in this volume, for a more extended discussion of the primary photochemical reactions of bacterial photosynthesis.

A. ELECTRON TRANSFER TO THE ACCEPTOR —INVOLVEMENT OF AN INTERMEDIATE STATE

Absorbance changes in the 435 nm, 600 nm, 800–890 nm, and 1250 nm regions occur when the reaction center BChl loses an electron (Duysens *et al.*, 1956; Arnold and Clayton, 1960; Parson, 1968; Reed, 1969; Seibert and DeVault, 1971; Clayton, 1973). In addition, the lifetime of the lowest excited singlet state of reaction center BChl has been estimated at 5–10 psec based on calculations from measurements of the fluorescent yield (Zankel *et al.*, 1968; Slooten, 1972) in RC preparations at moderate redox potentials. If the primary photochemical event involves transfer of an electron directly from an excited BChl molecule to the acceptor, as has been long assumed (Clayton, 1971), one would expect that the absorbance changes associated with the oxidation of reaction center BChl would also occur in the 5- to 10-psec range (Borisov and Godik, 1973) since the kinetics of fluorescence is a direct measure of the population of the fluorescing state. The first picosecond studies in bacterial reaction center preparations (Netzel *et al.*, 1973b) were consistent with this model since a 10-psec actinic pulse at 530 nm (one of the BPh bands) led to the bleaching of the 865-nm BChl band in 7 ± 2 psec.

Although this interpretation explained the picosecond data, evidence indicating that the transfer of an electron to the acceptor involved more than the singlet state of a BChl molecule was appearing in many laboratories. For example, Clayton *et al.* (1972) observed that the quantum yield of RC fluorescence increased by a factor of three when the reaction centers were inactivated at low redox potential. This rather small increase corresponds to a calculated photochemical quantum efficiency of about 0.7, much lower than the measured value of 1.02 ± 0.04 (Wraight and Clayton, 1974). Assuming that the inhibition of electron transfer to the acceptor influences only the rate constant of that particular reaction, this inconsistency could be explained if an additional intermediate state (Kamen, 1963; Leigh *et al.*, 1974; Parson *et al.*, 1975) before the primary acceptor, X, were necessary for electron transfer to occur. In addition, low-temperature electron spin resonance (ESR) studies at low redox potentials have identified a triplet state associated with reaction-center BChl (Dutton *et al.*, 1972, 1973a). Wraight *et al.* (1974) showed that the quantum yield of the triplet state was the same as that for photooxidation of the reaction center BChl, at least at <20°K where the ESR measurements could be made. This work stimulated renewed interest in older suggestions that Chl triplet states may be involved in photosynthesis (Franck and Rosenberg, 1964; Clayton, 1965; Robinson, 1966). A further complication appeared with the realization that more than one reaction center BChl was probably involved in the electron transfer process. On the basis of appreciably lower zero-field splitting parameters observed in RC triplets than in isolated Chl (Leigh and Dutton, 1974) or BChl (Uphaus *et al.*, 1974), the RC triplet was found to be shared by two BChl molecules. This was consistent with earlier ESR (Norris *et al.*, 1971; McElroy *et al.*, 1972) and electron nuclear double resonance (ENDOR) (Norris *et al.*, 1973; Feher *et al.*, 1973) studies showing that the BChl radical cation was also delocalized over two BChl molecules. Finally, Parson *et al.* (1975) and Cogdell *et al.* (1975) observed a reaction center transient state in the 420-nm and 540-nm regions which they termed P[F]. This transient state was observed at low redox potential where electron transfer to the acceptor was inhibited and appeared to have a quantum yield of about one. Since its decay half-time was 10 nsec, Parson and co-workers concluded that P[F] could not be the lowest excited singlet state of the reaction center BChl, but suggested that it could be an intermediate in the electron transfer to X.

This brief picture of the state of knowledge in early 1975, set the stage for the important picosecond studies of Kaufmann *et al.* (1975) and Rockley *et al.* (1975). Both groups reported the discovery of a transient absorbance change in *R. sphaeroides* R-26 reaction center preparations under conditions in which X was oxidized prior to excitation with a single 8- to 10-psec pulse of 530-nm light. This transient state appeared within 13–20 psec and decayed exponentially with a decay time, τ, of about 150 psec at 540 and 640 nm (Kaufmann *et al.*, 1975) or 246 ± 10 psec at 680 nm (Rockley *et al.*, 1975).[4] The spectral changes (13–20 psec after excitation) accompanying the formation of the transient were characterized by absorbance increases near 500 and 680 nm, by absorbance decreases near 540, 600, 760, and 870 nm, and by a blue shift of a band in the 800-nm region. However, spectra obtained 240–250 psec after excitation did not reveal the pronounced absorbance changes at 540, 680, and 760 nm. Instead they showed all the features characteristic of oxidized reaction center BChl familiar from measurements on slower time scales (Reed, 1969; Clayton, 1973). At low redox potentials where electron transfer to X does not occur since X is already reduced, the

---

[4] The discrepency is probably due to the fact that the former lifetime was obtained assuming that the absorbance changes at 540 and 640 nm decayed to zero after a few hundred picoseconds. However, the BChl radical cation does contribute to the absorbance changes at these wavelengths and could have led to a slight underestimate of the lifetime.

absorbance changes at 13 psec were the same as those observed at higher potentials (Kaufmann *et al.*, 1975). In this case, though, the spectrum did not change after 250 psec (Kaufmann *et al.*, 1975) and appeared to be identical to that of P^F (Rockley *et al.*, 1975; Parson *et al.*, 1975; Cogdell *et al.*, 1975).

All this evidence led to the identification of the transient state as an intermediate in the electron transfer process from reaction center BChl to X. The picosecond fluorescence studies of Kononenko *et al.* (1976) are consistent with the rapid formation and subsequent decay of the transient intermediate from a reaction center BChl singlet state. They observed two RC fluorescent components with 694 nm actinic light, one with a lifetime of 15 ± 8 psec (850–950 nm) and the other with a lifetime of 250 psec (710–950 nm). These lifetimes correspond to the appearance and decay of the transient intermediate (Paschenko *et al.*, 1977).

What about the nature of the transient intermediate? As indicated above, fluorescence yield and kinetic studies eliminate its identification as a reaction center BChl singlet state. In addition, it is not a triplet state since the triplet state (P^R in Parson's terminology) probably forms from the transient intermediate (Cogdell *et al.*, 1975; Rockley *et al.*, 1975; Dutton *et al.*, 1975; Thurnauer *et al.*, 1975). The rapid bleaching of the 540- and 760-nm bands within 20 psec (Kaufmann *et al.*, 1975; Rockley *et al.*, 1975), on the other hand, implicated the involvement of BPh. However, the concomitant appearance of absorbance decreases at 870 nm (Netzel *et al.*, 1973b; Kaufmann *et al.*, 1975; Rockley *et al.*, 1975) led to the conclusion that the transient intermediate was a state of the RC complex involving both BPh and BChl chromophores (Kaufmann *et al.*, 1975). Moreover, the ESR and ENDOR studies mentioned earlier suggested the possible participation of a BChl dimer. Finally, the fact that the blue shift of the 800-nm band occurred more rapidly than the reduction of X indicated that the formation of the transient intermediate state involved electron transfer (Rockley *et al.*, 1975). Dutton *et al.* (1975) clarified the situation when they reported that the 1250-nm band (Reed, 1969; Clayton, 1973), which is unique to the oxidized reaction center BChl dimer (Fajer *et al.*, 1974, 1975; Dutton *et al.*, 1975), forms within 20 psec of the excitation flash. This is also true in the case of *Rhodopseudomonas viridis* and *Chromatium vinosum* (Netzel *et al.*, 1977). Chemically reducing the acceptor did not prevent formation of the band, in contrast to chemically oxidizing the reaction center BChl (Dutton *et al.*, 1975).

The conclusion then was that the transient intermediate consisted of an oxidized BChl dimer radical [BChl—BChl]$^{\overset{+}{\cdot}}$ and a reduced component called "I" (a chemical species which may take part in the initial photochemistry and acts as an electron transfer species between the dimer and X). I has been identified as BPh on the basis of the 540 and 760 nm spectral changes, its low midpoint potential (Fajer *et al.*, 1975; Shuvalov and Klimov, 1976) and the lack of picosecond absorbance changes at 1000 nm (Dutton *et al.*, 1975). This last result is consistent with the formation of BPh$^{-\cdot}$ but not BChl$^{-\cdot}$ (Fajer *et al.*, 1973, 1975). In addition, Clayton and Yamamoto (1976) and Kaufmann *et al.* (1976) determined that the longer wavelength BPh species absorbing at 542 nm underwent reduction in the formation of the proposed [BChl$^{+\cdot}$—BChl BPh^{-}]X state.

In spite of this evidence, there are inconsistencies in the bleaching kinetics in the 600-nm and 800-nm regions (Rockley *et al.*, 1975) that have not been totally explained. This may mean that I cannot be identified as a single species at this point (Shuvalov and Klimov, 1976; Dutton *et al.*, 1976), and caution must be emphasized in identifying it as BPh alone, as has been done for the sake of illustration in Fig. 9.

Figure 9 summarizes current ideas regarding the room-temperature picosecond time scale reactions leading to the reduction of X in reaction center preparations. Light is absorbed by BChl in step (a) leading to the very rapid formation of an excited singlet state. Excitation transfer from the antenna pigments to drive this step may be rate limiting in *R. sphaeroides* chromatophores, though, since the fluorescent lifetime of the antenna pigments is 200 psec in the 1760-I strain (Paschenko *et al.*, 1977) and 300 psec in the R-26 strain (Campillo, *et al.*, 1977). The radical cation BChl dimer is formed in step (b) when an electron from the dimer reduces the long-wavelength BPh molecule. In step (c) X is reduced in

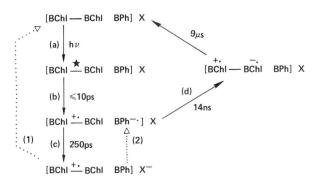

FIG. 9. A working scheme outlining the room-temperature picosecond time-scale reactions leading to the reduction of X in reaction center preparations. The rate at which each step occurs is expressed in terms of its $1/e$ time. Step (d) occurs when X is reduced prior to a flash, since step (c) cannot take place. Steps (1) and (2) are possible back reaction pathways. The identification of BPh as I in this figure is tentative since I may involve an additional species. See the text for more details. This figure was modified from Fig. 4 in Dutton *et al.* (1975) and incorporates data taken from Cogdell *et al.* (1975).

an electron transfer reaction from the BPh radical anion. Further electron transfer involving cytochromes and secondary acceptors would occur after this step in the intact organism. The dotted lines (1) and (2) represent possible reverse reactions (Dutton *et al.*, 1975; Prince *et al.*, 1976) since electron transfer beyond step (c) is not possible in RC preparations. Present evidence strongly favors reaction (1), especially at low temperature (Dutton *et al.*, 1976). Step (d) can occur at room temperature but is kinetically much less favorable than (c). However, under conditions when X is reduced prior to a flash, step (d) can be observed spectrally (Parson *et al.*, 1975) although with a low quantum efficiency. This pathway presumably involves the BChl dimer as a biradical or spin-polarized triplet state [BChl$^+$—BChl$^-$] and is the species responsible for the ESR triplet signal observable at low temperature (Dutton *et al.*, 1972, 1973a).

The identity of X, the classical primary electron acceptor, is discussed by Ke elsewhere in this volume but picosecond studies (Kaufmann *et al.*, 1976) indicate that removal of ubiquinone by the method of Okamura *et al.* (1975) has no detectable effect on the rise time of the transient intermediate. On the other hand, removal does considerably lengthen the decay time. This is consistent with the 10-nsec decay time of the transient intermediate state when the acceptor is in the reduced form prior to excitation (Cogdell *et al.*, 1975).

B. MODEL STUDIES IN SOLUTION

Recent picosecond model system studies with porphyrin (Magde *et al.*, 1974), BPh (Holten *et al.*, 1976), and Chl (Huppert *et al.*, 1976) solutions have provided insight into the charge transfer processes involved in photosynthesis. For example, the transient intermediate [BChl$^+$—BChl BPh$^-$]X singlet charge transfer state is necessary in bacterial photosynthesis because its rather high electronic energy gap (Fajer *et al.*, 1975) results in a slow rate of internal conversion (Siebrand, 1966). Thus electron transfer to the acceptor can compete favorably with internal conversion. In contrast, singlet charge transfer states of BPh (Holton *et al.*, 1976) or Chl (Huppert *et al.*, 1976) and quinones in solution, if they occur at all, do not lead to separated ions presumably due to their small energy gaps.

C. CAROTENOID TRIPLET STATES

Rapid bandshifts attributable to carotenoid excitation in chloroplast membranes have been known to occur in less than 25 nsec (Jackson and Crofts, 1969; Wolff *et al.*, 1969; Witt and Wolff, 1970). When chromatophores isolated from the carotenoid-containing Ga strain of *R. sphae-*

roides were exposed to a 10-psec pulse of 530-nm light, an absorbance increase appeared at 515 nm within 10 psec (Leigh *et al.*, 1974). The absorbance change was attributed to carotenoid excitation since its spectrum was similar to that observed for carotenoid responses measured in the microsecond and millisecond time range and since the picosecond spectral changes were not observed in chromatophores isolated from *R. sphaeroides* R-26 (which is carotenoid-free). Nevertheless, the picosecond absorbance changes were about 20 times larger than those reported on slower time scales. Leigh *et al.* (1974) suggested that the response was an electrochromic effect caused by the generation of large electric fields during the initial charge separation act in the excited RC complex because chemically reducing X had no effect on the response, while chemically oxidizing the primary electron donor abolished the response.

Kung and DeVault (1976) and Monger *et al.* (1976) have subsequently shown on slower time scales that the carotenoid responses are present when the primary donor is oxidized (this has been verified in the picosecond time range) and when high flash intensities are used. Since the decay time of the carotenoid response decreased in the presence of oxygen, both groups concluded that the absorbance changes were due to antenna carotenoid triplet states arising from antenna BChl triplets. The large absorbance increase at 515 nm was also consistent with the large molar absorptivity reported for carotenoid triplets (Land *et al.*, 1970; Mathis, 1970). The much faster rise time reported for the carotenoid triplet in the picosecond studies compared with the nanosecond work is not as yet understood, although both Kung and DeVault (1976) and Monger *et al.* (1976) have intimated that the picosecond studies may involve the direct excitation of carotenoids. In any case, these results are consistent with a protective role for carotenoids that prevent photodegradation of BChl by oxygen (Griffiths *et al.*, 1955) when the photosynthetic apparatus is oversaturated with light.

Nanosecond flash-induced transients observed by a number of investigators (Parson, 1967; Seibert *et al.*, 1971; Seibert and DeVault, 1971; Cogdell *et al.*, 1975) at relatively low flash intensities and low redox potentials have also been identified as carotenoid triplet states. These, however, are coupled to the decay of the transient intermediate or P^F state (Cogdell *et al.*, 1975).

V. Summary and Future Developments

Recent advances in pulsed laser spectroscopy have made the observation of fluorescent emission and absorbance change kinetics possible on a picosecond time scale. Fluorescent studies in various photosynthetic systems have probed the antenna pigments that funnel energy to the trapping or reaction center complexes. Excitation with single, low-intensity pulses or low-intensity pulse trains can result in the measurement of fluorescent lifetimes that are consistent with those obtained under low light conditions using the older phase and nanosecond-pulse methods. Excitation with high-intensity pulses or pulse trains have identified a number of potential exciton annihilation processes that occur when more than one excitation is present per photosynthetic unit or when long-lived triplet or photoionization states build up. Absorbance change measurements have contributed greatly to the identification and characterization of the rapid reaction center events that occur in bacteria prior to the reduction of the X acceptor. Application of these same techniques in green plant systems is likely to yield similarly enlightening results.

Future developments will include the use of picosecond spectroscopy as a tool to probe many of the recently isolated fractions containing the photosynthetic (both green plant and bacterial) apparatus in various states of complexity. Mutant strains will be of particular interest and the control of redox potential will be important. The availability of picosecond pulses from frequency-tunable dye lasers (Bradley, 1974) will greatly increase the choice of excitation wavelengths. Finally, the time resolution of laser spectroscopy will improve by at least a factor of magnitude with the introduction of subpicosecond pulses (Penzkofer, 1974; Shank et al., 1976) and fast-streak cameras (Bradley, 1974).

Addendum

Since this article was written a number of groups have reported additional progress in understanding the picosecond events of photosynthesis.

New room-temperature studies with choloroplasts (Yu et al., 1977) and photosystem-enriched fractions (Searle et al., 1977) of spinach have confirmed the earlier work (Seibert and Alfano, 1974; Yu et al., 1975; Paschenko et al., 1975) showing that the fluorescence lifetime of the pigments associated with PSI is shorter than that for the pigments associated with PSII. In addition, Searle et al. (1977) demonstrated that the lifetime in PSI fractions is less than 100 psec and is independent of the intensity of a single 6-psec pulse. In contrast, the lifetime of the pigments in PSII fractions depends on the flux density of a single pulse and reportedly approaches 500 psec at the low range of 5×10^{13} photons/cm^2. The quantitative difference in the lifetime of PSII fluorescence obtained with single pulses and with pulse trains has been explained on the basis of quenching due to the buildup of triplet states in the latter case. The fact that the fluorescence lifetimes and yields did not appear to change over the flux density ranges used in the train experiments (Seibert and Alfano, 1974; Yu *et al.*, 1975; Beddard *et al.*, 1975; Yu *et al.*, 1977) was explained by virtue of the fact that singlet–triplet rather than singlet–singlet annihilation would predominate over a wide range of excitation intensities (Porter *et al.*, 1977) or because the effects of the annihilation processes which shorten the lifetime were balanced by the effects of closing the traps which would lengthen the lifetime (Yu *et al.*, 1977).

Lowering the temperature in spinach systems to under 100°K was reported to increase the fluorescence lifetime of both photosystems in single pulse experiments (Searle *et al.*, 1977) but only the long-wavelength 730-nm component (PSI) in train experiments (Yu *et al.*, 1977). Comparison of these results is complicated by the fact that the former investigators monitored the fluorescence over a broad range in the red spectral region. The single-pulse results of Geacintov *et al.* (1977), on the other hand, seem consistent with a relatively minor change in the lifetime of 685–690 nm fluorescence (light-harvesting and PSII pigments) at low temperatures. As was the case at room temperature, kinetic studies at 77°K showed that the lifetime of PSI fluorescence was independent of the intensity of a single pulse, while that of the 690-nm component was subject to singlet–singlet quenching (Geacintov *et al.*, 1977). The low-temperature absolute lifetimes measured by the different groups vary widely [the reported values for PSI and PSII fluorescence, respectively, are: 600 and 200 psec (using a pulse train), Yu *et al.* (1977); 1100 and 700 psec (using a single pulse), Geacintov *et al.* (1977); 1900–2850 and 900–2100 psec (using a single pulse), Searle *et al.* (1977)] and can probably be rationalized on the basis of previous arguments. Nevertheless, the results of Geacintov *et al.* (1977) are most consistent with those of Hervo *et al.* (1975), who used photon-counting techniques.

Additional low-temperature quantum yield studies of Geacintov *et al.* (1977) have shown that 735-nm fluorescence (the authors warn that this fluorescent component may not be associated with pigments directly connected with P-700 reaction centers of PSI) preferentially accumulates triplet states. They also proposed that singlet–singlet annihilation occurs at the level of the light-harvesting antenna–chlorophyll complex which then passes excitons on the 735-nm fluorescence band. The fact that singlet–singlet annihilation does not occur in the PSI pigments would explain why the lifetime of 735-nm fluorescence does not depend on the intensity of a single pulse.

In bacterial systems, Netzel *et al.* (1977) addressed the possibility of an alternate intermediate acceptor prior to I. Spectral observations of the BChl dimer in the 1200–1300 nm region as well as EPR assays of the biradical/triplet were made as a function of the reduction state of the I/I^-· couple. I was obtained in a stable reduced form by both chemical and photochemical means in reaction-center cytochrome complexes of *Chromatium vinosum* and *Rhodopseudomonas viridis*. Since the reduction state of I alone determined the ability of the BChl dimer to undergo photo-oxidation using both assay methods, the authors concluded that I was the immediate and sole electron acceptor for the dimer.

The nature of I is still somewhat unclear, but two recent lines of evidence have been reported which argue against the participation of the 800-nm absorbing form of BChl. The 800-nm band is known to undergo bleaching in the I^-· minus I spectrum of several BChl *a* containing species (Dutton *et al.*, 1976; Shuvalov and Klimov, 1976). However, in *R. viridis* the analogous BChl *b* band at 830 nm does not show any evidence of bleaching (Netzel *et al.*, 1977). In addition, Paschenko *et al.* (1977) have obtained fluorescence spectra of their 250-psec decaying component (associated with the disappearance of the transient intermediate state) in reaction centers of *Rhodopseudomonas spheroides* 1760-1. They observed bands at 890 nm (dimer) and at 772 nm (BPh) but saw no fluorescence band attributable to the 800-nm BChl.

The threshold flux density for single-pulse fluorescence quenching has been studied in four strains of *R. spheroides* chromatophores (Campillo *et al.*, 1977) and was found to vary with the strain. In addition, the bacterial quenching thresholds (10^{14}–10^{15} photons/cm^2) were higher than those noted in *Chlorella*. The interpretation was that both the absorption cross section at the actinic wavelength and the singlet–singlet annihilation rate constant varies in different organisms. The involvement of charged states in the reaction centers was ruled out as the main mechanism for quenching because the PM-8 strain of *R. spheroides* (which lacks reaction centers) displayed a low quenching threshold compared with the other strains.

In every strain tested (Campillo *et al.*, 1977), the measured fluorescence lifetime, obtained with low-intensity pulses, agreed with the lifetime calculated from the quantum yield obtained using classical techniques. This correlation has also been observed in spinach systems.

Finally, Hindman *et al.* (1977) have demonstrated that high flux density flashes (in the range used in picosecond photosynthetic studies) can cause lasing of Chl, BChl, and BPh (in pyridine) in cavities with short photon decay times such as are used in the picosecond investigations. As a result of this work, the authors suggest the possibility of alternative photophysical processes which can lead to the quenching of fluorescence at high single-pulse intensities. However, it should be noted that stimulated emission of *in vivo* chlorophylls has not as yet been demonstrated, possibly because the spectral resolution of published picosecond emission spectra (Seibert and Alfano, 1974; Yu *et al.*,

1977) are not sufficiently narrow to distinguish conclusively between fluorescence and stimulated emission. Also, Gordon Gould is now recognized by the U.S. Patent Office for his contribution to the invention of the laser (Wade, 1977).

ACKNOWLEDGMENTS

The kindness of Drs. A. Yu. Borisov, P. L. Dutton, N. E. Geacintov, M. D. Il'ina, W. W. Parson, A. B. Rubin, S. L. Shapiro, and W. Yu, who provided unpublished data and manuscripts, is greatfully appreciated. Discussions with Drs. A. Lempicki and R. R. Alfano regarding Section II were very helpful.

REFERENCES

Alfano, R. R., and Shapiro, S. L. (1970). *Phys. Rev. Lett.* **24**, 592–594.
Arnold, W., and Clayton, R. K. (1960) *Proc. Natl. Acad. Sci. U.S.A.* **46**, 769–776.
Beddard, G. S., Porter, G., Tredwell, C. J., and Barber, J. (1975). *Nature (London)* **258**, 166–168.
Boardman, N. K., Thorne, S. W., and Anderson, J. M. (1966). *Proc. Natl. Acad. Sci. U.S.A.* **56**, 586–593.
Borisov, A. Yu., and Godik, V. I. (1973). *Biochim. Biophys. Acta* **301**, 227–248.
Borisov, A. Yu., and Il'ina, M. D. (1973). *Biochim. Biophys. Acta* **305**, 364–371.
Bradley, D. J. (1974). *Opto-electronics* **6**, 25–42.
Breton, J., and Geacintov, N. E. (1976). *FEBS Lett.* **69**, 86–89.
Breton, J., and Roux, E. (1975). *In* "Lasers in Physical Chemistry and Biophysics" (J. Joussot-Dubien, ed.), pp. 379–388. Elsevier, Amsterdam.
Briantais, J. M., Merkelo, H., and Govindjee. (1972). *Photosynthetica* **6**, 133–141.
Brody, S. S., and Rabinowitch, E. (1957). *Science* **125**, 555.
Bronskill, M. J., Taylor, W. B., Wolff, R. K., and Hunt, J. W. (1970). *Rev. Sci. Instrum.* **41**, 333–340.
Butler, W. L., and Norris, K. H. (1963). *Biochim. Biophys. Acta* **66**, 72–77.
Campillo, A. J., Shapiro, S. L., Kollman, V. H., Winn, K. R., and Hyer, R. C. (1976a). *Biophys. J.* **16**, 93–97.
Campillo, A. J., Kollman, V. H., and Shapiro, S. L. (1976b). *Science* **193**, 227–229.
Campillo, A. J., Hyer, R. C., Monger, T. G., Parson, W. W., and Shapiro, S. L. (1977). *Proc. Natl. Acad. Sci. U.S.A.* **74**, 1997–2001.
Chance, B., and DeVault, D. C. (1964). *Ber. Bunsenges. Phys. Chem.* **68**, 722–726.
Clayton, R. K. (1965). "Molecular Physics in Photosynthesis." Ginn (Blaisdell), Boston, Massachusetts.
Clayton, R. K. (1971). "Light and Living Matter," Vol. I. McGraw-Hill, New York.
Clayton, R. K. (1973). *Annu. Rev. Biophys. Bioeng.* **2**, 131–156.
Clayton, R. K., and Wang, R. T. (1971). *In* "Methods in Enzymology" (A. San Pietro, ed.), Vol. 23, pp. 696–704. Academic Press, New York.
Clayton, R. K., and Yamamoto, T. (1976). *Photochem. Photobiol.* **24**, 67–70.
Clayton, R. K., Fleming, H., and Szuts, E. Z. (1972). *Biophys. J.* **12**, 46–63.
Cogdell, R. J., Monger, T. G., and Parson, W. W. (1975). *Biochim. Biophys. Acta* **408**, 189–199.
DeMaria, A. J. (1971). *In* "Progress in Optics" (E. Wolf, ed.), pp. 31–71. North-Holland Publ., Amsterdam.
DeMaria, A. J., Stetser, D. A., and Heynau, H. (1966). *Appl. Phys. Lett.* **8**, 174–176.
DeMaria, A. J., Stetser, D. A., and Glenn, W. H., Jr. (1967). *Science* **156**, 1557–1568.

DeVault, D. C. (1964). *In* "Rapid Mixing and Sampling Techniques in Biochemistry" (B. Chance *et al.*, eds.), pp. 165–174. Academic Press, New York.

Dmitrievsky, O., Ermolaev, V., and Terenin, A. (1957). *Dokl. Akad. Nauk SSSR (Engl. Transl.)* 114, 468.

Duguay, M. A., and Hansen, J. W. (1969a). *Appl. Phys. Lett.* 15, 192–194.

Duguay, M. A., and Hansen, J. W. (1969b). *Opt. Commun.* 1, 254–256.

Dutton, P. L., Leigh, J. S., and Seibert, M. (1972). *Biochem. Biophys. Res. Commun.* 46, 406–413.

Dutton, P. L., Leigh, J. S., and Reed, D. W. (1973a). *Biochim. Biophys. Acta* 292, 654–664.

Dutton, P. L., Leigh, J. S., and Wraight, C. A. (1973b). *FEBS Lett.* 36, 169–173.

Dutton, P. L., Kaufmann, K. J., Chance, B., and Rentzepis, P. M. (1975). *FEBS Lett.* 60, 275–280.

Dutton, P. L., Prince, R. C., Tiede, D. M., Petty, K. M., Kaufmann, K. J., Netzel, T. L., and Rentzepis, P. M. (1976). *Brookhaven Symp. Biol.* 28, 213–237.

Duysens, L. N. M., Huiskamp, W. J., Vos, J. J., and van der Hart, J. M. (1956). *Biochim. Biophys. Acta* 19, 188–190.

Duysens, L. N. M., den Haan, G. A., and Van Best, J. A. (1975). *Congr. Photosyn., 1974* Vol. 1, pp. 1–12.

Eckardt, R. C., Lee, C. H., and Bradford, J. N. (1974). *Opto-electronics* 6, 67–85.

Fajer, J., Borg, D. C., Forman, A., Dolphin, D., and Felton, R. H. (1973). *J. Am. Chem. Soc.* 95, 2739–2741.

Fajer, J., Borg, D. C., Forman, A., Felton, R. H., Dolphin, D., and Vegh, L. (1974). *Proc. Natl. Acad. Sci. U.S.A.* 71, 994–998.

Fajer, J., Brune, D. C., Davis, M. S., Forman, A., and Spaulding, L. D. (1975). *Proc. Natl. Acad. Sci. U.S.A.* 72, 4956–4960.

Feher, G. (1971). *Photochem. Photobiol.* 14, 373–387.

Feher, G., Hoff, A. J., Isaacson, R. A., and McElroy, J. D. (1973). *Biophys. J.* 13, 61a.

Fleck, J. A., Jr. (1970). *Phys. Rev. B* 1, 84–100.

Franck, J., and Rosenberg, J. L. (1964). *J. Theor. Biol.* 7, 276–301.

Garmire, E., and Yariv, A. (1967). *IEEE J. Quantum Electron.* qe-3, 222–226.

Geacintov, N. E., and Breton, J. (1977). *Biophys. J.* 17, 1–15.

Geacintov, N. E., Breton, J., Swenberg, C., Campillo, A. J., Hyer, R. C., and Shapiro, S. L. (1977). *Biochim. Biophys. Acta* 461, 306–312.

Giordmaine, J. A., Rentzepis, P. M., Shapiro, S. L., and Wecht, K. W. (1967). *Appl. Phys. Lett.* 11, 216–218.

Griffiths, M., Sistrom, W. R., Cohen-Bazire, G., and Stanier, R. Y. (1955). *Nature (London)* 176, 1211–1214.

Ham, W. T., Jr., Mueller, H. A., Goldman, A. I., Newman, B. E., Holland, L. M., and Kuwabara, T. (1974). *Science* 185, 362–363.

Hargrove, L. E., Fork, R. L., and Pollack, M. A. (1964). *Appl. Phys. Lett.* 5, 4–5.

Harris, L., Porter, G., Synowiec, J. A., Tredwell, C. J., and Barber, J. (1976). *Biochim. Biophys. Acta* 449, 329–339.

Hellwarth, R. W. (1961). *In* "Advances in Quantum Electronics" (J. R. Singer, ed.), pp. 334–341. Columbia Univ. Press, New York.

Hellwarth, R. W. (1966). *In* "Lasers" (A. K. Levine, ed.), pp. 253–294. Dekker, New York.

Hervo, G., Paillotin, G., Thiery, J., and Breuze, G. (1975) *J. Chim. Phys.* 72, 761–766.

Hindman, J. C., Kugel, R., Svirmickas, A., and Katz, J. J. (1977). *Proc. Natl. Acad. Sci. U.S.A.* 74, 5–9.

Holten, D., Gouterman, M., Parson, W. W., Windsor, M. W., and Rockley, M. G. (1976). *Photochem. Photobiol.* 23, 415–423.

Huppert, D., Rentzepis, P. M., and Tollin, G. (1976). *Biochim. Biophys. Acta* **440**, 356–364.

Jackson, J. B., and Crofts, A. R. (1969). *FEBS Lett.* **4**, 185–189.

Kamen, M. (1963). "Primary Processes in Photosynthesis." Academic Press, New York.

Kaufmann, K. J., Dutton, P. L., Netzel, T. L., Leigh, J. S., and Rentzepis, P. M. (1975). *Science* **188**, 1301–1304.

Kaufmann, K. J., Petty, K. M., Dutton, P. L., and Rentzepis, P. M. (1976). *Biochem. Biophys. Res. Commun.* **70**, 839–845.

Knox, R. S. (1975). *In* "Bioenergetics of Photosynthesis" (Govindjee, ed.), pp. 183–221. Academic Press, New York.

Kollman, V. H., Shapiro, S. L., and Campillo, A. J. (1975). *Biochem. Biophys. Res. Commun.* **63**, 917–923.

Kononenko, A. A., Knox, P. P., Adamova, N. P., Paschenko, V. Z., Timofeev, K. N., Rubin, A. B., and Morita, S. (1976). *Stud. Biophys.* **55**, 183–198.

Kung, M. C., and DeVault, D. (1976). *Photochem. Photobiol.* **24**, 87–91.

Kuznetsova, T. I. (1969). *Sov. Phys.—JETP* **28**, 1303–1305.

Land, E. J., Sykes, A., and Truscott, T. G. (1970). *Chem. Commun.* p. 332.

Laubereau, A., and Kaiser, W. (1974). *Opto-electronics* **6**, 1–24.

Leigh, J. S., and Dutton, P. L. (1974). *Biochim. Biophys. Acta* **357**, 67–77.

Leigh, J. S., Netzel, T. L., Dutton, P. L., and Rentzepis, P. M. (1974). *FEBS Lett.* **48**, 136–140.

Letokhov, V. S. (1968). *Sov. Phys.—JETP* **27**, 746–751.

Magde, D., and Windsor, M. W. (1974). *Chem. Phys. Lett.* **27**, 31–36.

Magde, D., Windsor, M. W., Holton, D., and Gouterman, M. (1974). *Chem. Phys. Lett.* **29**, 183–188.

Maiman, T. H. (1960). *Nature (London)* **187**, 493–494.

Mar, T., Govindjee, Singhal, G. S., and Merkelo, H. (1972). *Biophys. J.* **12**, 797–808.

Mathieu, E., and Weber, H. (1971). *Z. Angew. Math. Phys.* **22**, 458–460.

Mathis, P. (1969). *Prog. Photosynth. Res., Proc. Int. Congr. [1st], 1968* Vol. I, pp. 818–822.

Mathis, P. (1970). Thèse, Paris–Orsay, France.

Mauzerall, D. (1976). *Biophys. J.* **16**, 87–91.

McClung, F. J., and Hellwarth, R. W. (1962). *J. Appl. Phys.* **33**, 828–829.

McElroy, J. D., Feher, G., and Mauzerall, D. C. (1972). *Biochim. Biophys. Acta* **267**, 363–374.

Merkelo, H., Hartman, S. R., Mar, T., Singhal, G. S., and Govindjee. (1969). *Science* **164**, 301–302.

Mocker, H. W., and Collins, R. J. (1965). *Appl. Phys. Lett.* **7**, 270–273.

Monger, T. G., Cogdell, R. J., and Parson, W. W. (1976). *Biochim. Biophys. Acta* **449**, 136–153.

Müller, A., Lumry, R., and Walker, M. S. (1969). *Photochem. Photobiol.* **9**, 113–126.

Murty, N. R., and Rabinowitch, E. (1965). *Biophys. J.* **5**, 655–661.

Netzel, T. L., Struve, W. S., and Rentzepis, P. M. (1973a). *Annu. Rev. Phys. Chem.* **24**, 473–492.

Netzel, T. L., Rentzepis, P. M., and Leigh, J. S. (1973b). *Science* **182**, 238–241.

Netzel, T. L., Rentzepis, P. M., Tiede, D. M., Prince, R. C., and Dutton, P. L. (1977). *Biochim. Biophys. Acta* **460**, 467–479.

Nicholson, W. J., and Fortoul, J. I. (1969). *Biochim. Biophys. Acta* **143**, 577–582.

Norris, J. R., Uphaus, R. A., Crespi, H. L., and Katz, J. J. (1971). *Proc. Natl. Acad. Sci. U.S.A.* **68**, 625–628.

Norris, J. R., Druyan, M. E., and Katz, J. J. (1973). *J. Am. Chem. Soc.* **95**, 1680–1682.

Okamura, M. Y., Steiner, L. A., and Feher, G. (1974). *Biochemistry* **13**, 1394–1403.

Okamura, M. Y., Isaacson, R. A., and Feher, G. (1975). *Proc. Natl. Acad. Sci. U.S.A.* **72**, 3491–3495.

Parson, W. W. (1967). *Biochim. Biophys. Acta* **131**, 154–172.

Parson, W. W. (1968). *Biochim. Biophys. Acta* **153**, 248–259.

Parson, W. W. (1974). *Annu. Rev. Microbiol.* **28**, 41–59.

Parson, W. W., Clayton, R. K., and Cogdell, R. J. (1975). *Biochim. Biophys. Acta* **387**, 265–278.

Paschenko, V. Z., Protasov, S. P., Rubin, A. B., Timofeev, K. N., Zamazova, L. M., and Rubin, L. B. (1975). *Biochim. Biophys. Acta* **408**, 143–153.

Paschenko, V. Z., Kononenko, A. A., Protasov, S. P., Rubin, A. B., Rubin, L. B., and Uspenskaya, N. Ya. (1977). *Biochim. Biophys. Acta* **461**, 403–412.

Penzkofer, A. (1974). *Opto-electronics* **6**, 87–98.

Porter, G., Synowiec, J. A., and Tredwell, C. J. (1977). *Biochim. Biophys. Acta* **459**, 329–336.

Prince, R. C., Leigh, J. S., and Dutton, P. L. (1976). *Biochim. Biophys. Acta* **440**, 622–636.

Princeton Applied Research Corp. (1975). "OMA Catalog," T336-20M-7/75-PB. Princeton Appl. Res. Corp., Princeton, New Jersey.

Reed, D. W. (1969). *J. Biol. Chem.* **244**, 4936–4941.

Reed, D. W., and Clayton, R. K. (1968). *Biochem. Biophys. Res. Commun.* **30**, 471–475.

Reed, D. W., and Peters, G. A. (1972). *J. Biol. Chem.* **247**, 7148–7152.

Rentzepis, P. M., Jones, R. P., and Jortner, J. (1973). *J. Chem. Phys.* **59**, 766–773.

Robinson, G. W. (1966). *Brookhaven Symp. Biol.* **19**, 16–44.

Rockley, M. G., Windsor, M. W., Cogdell, R. J., and Parson, W. W. (1975). *Proc. Natl. Acad. Sci. U.S.A.* **72**, 2251–2255.

Rubin, A. B., and Osnitskaya, L. K. (1963). *Mikrobiologiya* **32**, 200–203.

Sane, P. V., Goodchild, D. J., and Park, R. B. (1970). *Biochim. Biophys. Acta* **216**, 162–178.

Sauer, K. (1975). *In* "Bioenergetics of Photosynthesis" (Govindjee, ed.), pp. 115–181. Academic Press, New York.

Schawlow, A. L., and Townes, C. H. (1958). *Phys. Rev.* [2] **112**, 1940–1949.

Searle, G. F. W., Barber, J., Harris, L., Porter, G., and Tredwell, C. J. (1977). *Biochim. Biophys. Acta* **459**, 390–401.

Seibert, M., and Alfano, R. R. (1974). *Biophys. J.* **14**, 269–283.

Seibert, M., and DeVault, D. (1971). *Biochim. Biophys. Acta* **153**, 396–411.

Seibert, M., Dutton, P. L., and DeVault, D. (1971). *Biochim. Biophys. Acta* **226**, 189–192.

Seibert, M., Alfano, R. R., and Shapiro, S. L. (1973). *Biochim. Biophys. Acta* **292**, 493–495.

Shank, C. V., Ippen, E. P., and Bersohn, R. (1976). *Science* **193**, 50–51.

Shapiro, S. L., Kollman, V. H., and Campillo, A. J. (1975). *FEBS Lett.* **54**, 358–362.

Shimizu, F., and Stoicheff, B. P. (1969). *IEEE J. Quantum Electron.* **-5**, 544–546.

Shuvalov, V. A., and Klimov, V. V. (1976). *Biochim. Biophys. Acta* **440**, 587–599.

Siebrand, W. (1966). *J. Chem. Phys.* **44**, 4055–4057.

Siegman, A. E. (1971). "An Introduction to Lasers and Masers." McGraw-Hill, New York.

Singhal, G. S., and Rabinowitch, E. (1969). *Biophys. J.* **9**, 586–591.

Slooten, L. (1972). *Biochim. Biophys. Acta* **256**, 452–466.

Swenberg, C. E., Geacintov, N. E., and Pope, M. (1976). *Biophys. J.* **16**, 1447–1452.

Thurnauer, M. C., Katz, J. J., and Norris, J. R. (1975). *Proc. Natl. Acad. Sci. U.S.A.* **72**, 3270–3274.

Tomita, G., and Rabinowitch, E. (1962). *Biophys. J.* **2**, 483–499.

Tumerman, L. A., Borisova, O. F., and Rubin, A. B. (1961). *Biophysics (Engl. Transl.)* **6**, 723–728.

Uphaus, R. A., Norris, J. R., and Katz, J. J. (1974). *Biochem. Biophys. Res. Commun.* **61**, 1057–1063.

Vredenberg, W. J., and Slooten, L. (1967). *Biochim. Biophys. Acta* **143**, 583–594.

Wade, N. (1977). *Science* **198**, 379–381.

Witt, K., and Wolff, C. (1970). *Z. Naturforsch., Teil B* **25**, 387–388.

Wolff, C., Buchwald, H. E., Ruppel, H., Witt, K., and Witt, H. T. (1969). *Z. Naturforsch., Teil B* **24**, 1038.

Wraight, C. A., and Clayton, R. K. (1974). *Biochim. Biophys. Acta* **333**, 246–260.

Wraight, C. A., Leigh, J. S., Dutton, P. L., and Clayton, R. K. (1974). *Biochim. Biophys. Acta* **333**, 401–403.

Yu, W., Ho, P. P., Alfano, R. R., and Seibert, M. (1975). *Biochim. Biophys. Acta* **387**, 159–164.

Yu, W., Pellegrino, F., and Alfano, R. R. (1977). *Biochim. Biophys. Acta* **460**, 171–181.

Zankel, K. L., Reed, D. W., and Clayton, R. K. (1968). *Proc. Natl. Acad. Sci. U.S.A.* **61**, 1243–1249.

The Primary Electron Acceptors in Green-Plant Photosystem I and Photosynthetic Bacteria[1]

BACON KE

*Charles F. Kettering Research Laboratory
Yellow Springs, Ohio*

The primary photochemical reaction in green plants and photosynthetic bacteria follows the absorption of light energy by the antenna chlorophyll molecules and transfer of the electronic excitation to the reaction-center chlorophyll (a special pair of chlorophyll molecules). In the excited state, an electron is ejected from the reaction-center chlorophyll pair (or the primary donor) to a nearby primary acceptor. The net

[1] Contribution No. 581 from the Charles F. Kettering Research Laboratory.

result of the primary photochemical reaction is a conversion of light energy into chemical energy in the form of two oppositely charged species, the primary oxidant and the primary reductant.

Significant advances in our understanding of these primary reactants have been made during the past decade, especially those in green-plant photosystem I and photosynthetic bacteria. While it is generally considered that the primary electron acceptor in each photosystem is an initially reduced stable species, more recent work has revealed that the electron ejected by the primary donor may reside very briefly in an intermediary transient electron acceptor prior to its transfer to the more stable primary acceptor. Items covered by the present review will be divided among the primary reactants as major topics, and confined to recent developments on the reaction-center components and primary photochemistry, with emphasis on advances made since 1975. Recent reviews on related subjects have been prepared by Sauer (1975), Parson and Cogdell (1975), Ke (1973), Bearden and Malkin (1975), and others.

I. The Primary Electron Donor of Photosystem I

P700, the primary electron donor in the photochemical act of Photosystem I (PSI), was discovered by Kok in 1956. The photooxidation of P700 is characterized by a reversible bleaching of its absorption spectrum at 430 and 700 nm (see Fig. 4) as well as the appearance of a free-radical electron paramagnetic resonance (EPR) signal. The unique band position of P700 at 700 nm, which is red-shifted from that of monomer chlorophyll near ~670 nm, had variously been attributed to some sort of interaction between the pigment and lipids or proteins, or to some kind of perturbations by the environment. A better understanding of the physical state of P700 has become possible only recently through various spectroscopic studies.

A. The Dimer Configuration

Extensive studies of chlorophyll model systems have been made by Katz and his co-workers (see Katz and Norris, 1973). Results from optical and magnetic-resonance spectroscopic techniques led them to assign a special pair of chlorophyll a molecules to the PS I reaction-center chlorophyll, P700. The assignment to a dimer is consistent with the fact that monomeric chlorophyll always absorbs light at wavelengths shorter than 700 nm, and that free radicals of a chlorophyll–ligand combination produced in the laboratory have a much broader line than the light-induced EPR free-radical sign from PS I reaction centers.

Katz's studies established that chlorophyll–water interaction causes both a red shift and EPR-line narrowing. One of the most interesting species prepared in the laboratory is a chlorophyll–water adduct (Chl $a \cdot H_2O$), which has a visible absorption spectrum that is highly red-shifted (740 nm), a remarkably narrow EPR signal linewidth (ΔH_{pp} = 1 G); it is the only synthetic chlorophyll model system that is photoactive (Katz et al., 1968). The unusually narrow EPR line was rationalized by a process of spin delocalization over many chlorophyll molecules. Delocalization of an unpaired spin over N chlorophyll molecules should narrow the signal linewidth of monomeric chlorophyll Chl^+ by $1/\sqrt{N}$. Thus a (Chl $a \cdot H_2O$)$_n$ adduct with an EPR linewidth near 1 G should contain at least 100 chlorophyll molecules. A similar consideration applied to the in vivo P700 signal showed that the linewidth can very well be accounted for by assuming that N = 2. As (Chl $a \cdot H_2O$) and (BChl·H_2O) adducts are the only photoactive species prepared in the laboratory so far, it is reasonable to assign a (Chl·H_2O·Chl) configuration to the in vivo P700. In this model, two chlorophyll molecules are linked via a water molecule that simultaneously coordinates to the Mg of one chlorophyll molecule and forms hydrogen bonds to both ring-V keto C=O and the carbomethoxy C=O of the other chlorophyll molecule.

The special dimer model was further tested by comparing the hyperfine splitting in chlorophyll free radicals measured by the technique of electron-nuclear double-resonance (ENDOR) spectroscopy. ENDOR spectra of in vitro and in vivo chlorophyll species confirmed the predictions that in the special dimer model the EPR line shape would be narrowed by a factor of $1/\sqrt{2}$ relative to the monomer Chl^+, and the electron-nuclear hyperfine splitting constant correspondingly smaller by a factor of ½ (Norris et al., 1974; Feher et al., 1975).

It was also pointed out by Katz that application of the exciton formula (Hochstrasser and Kasha, 1964) suggests that the entity (Chl·H_2O·Chl) would be expected to absorb near 700 nm, half way between the 664 nm maximum of monomeric Chl·H_2O and the 743 nm maximum of the infinite array. Although water was used in the special dimer model, bifunctional ligands other than water or even a protein structural matrix conceivably could be implicated as the required orienting mechanism. Most recently a new structure for the special-pair chlorophyll has been proposed (Shipman et al., 1976). In this new model, two chlorophyll a molecules are held together by (a) two strong ring-V keto C=O··H—O (or keto C=O··H—N, or keto C=O··H—S) hydrogen bonds and by (b) π–π van der Waals stacking interactions between the two chlorophyll a macrocycles. Macrocycle stacking provides the intermolecular π–π overlap necessary to promote the delocalization of the unpaired electron of the in vivo radical over two chlorophyll π systems. The new model provides an explicit role for the participation of protein in the formation of the chlorophyll special pair.

B. EXCITON INTERACTION IN P700

The conclusion derived from EPR and ENDOR studies that P700 consists of strongly coupled chlorophyll molecules was complemented by absorption and circular dichroism (CD) results of Phillipson et al. (1972). The bleaching near 700 and 680 nm in the light-minus-dark difference spectrum of P700 (cf. Fig. 4) had previously led other workers (Döring et al., 1968; Vernon et al., 1969; Murata and Takamiya, 1969) to suggest that two chlorophyll molecules are involved in the PS I reaction-center chlorophyll P700. Phillipson et al. (1972) further proposed that these molecules are apparently coupled by an exciton interaction. As a result of this interaction, a splitting of energy levels occurs, with the number of new absorption bands equaling the number of molecules involved. Upon illumination, the oxidation of one of the chlorophyll a molecules within the reaction center results in the loss of resonant coupling. The light-minus-dark difference spectrum can be accounted for by the light-induced disappearance of two observable exciton components and the associated appearance of a new absorption band due to the remaining unoxidized chlorophyll in the P700 dimer.

The CD of PS I subchloroplasts is contributed mainly by the bulk chlorophyll molecules and provides little information about the reaction-center chlorophyll. The light-minus-dark CD difference spectrum contributed by P700 exhibits a positive band at 696.5 nm and a negative band at 688 nm; both bands are approximately equal in area and are reversible in the dark. The appreciable magnitude of the light-minus-dark CD difference also implies that at least two chlorophyll a molecules are involved in the PS I center, and that the chlorophyll molecules are apparently coupled by exciton interaction. The exciton contribution to the rotational strength is conservative, i.e., its contribution to the rotational strength should sum to zero over the exciton band. Thus, the exciton interaction for the different components within an electronic band is characterized by the appearance of both positive and negative bands. The conservative light-minus-dark difference spectrum thus reflects the disappearance of exciton interaction as a result of light oxidation.

C. ORIENTATION OF P700 IN THE PHOTOSYNTHETIC MEMBRANE

Information about the orientation of pigments in a solid matrix can be measured with polarized light in a variety of ways. The prerequisite for such measurements is that the solid matrix must be oriented. Orientation of the photosynthetic pigments has been achieved by orienting the pigment-containing membrane either by mechanical means [e.g., by air drying or spreading (Breton and Roux, 1971); by sedimentation (Thomas *et al.*, 1967); in a flow gradient (Sauer, 1965)] or by an electric field (Sauer and Calvin, 1962). Because of a low concentration of the reaction-center chlorophylls in chloroplasts, linear dichroism per se does not have adequate resolution to provide information on the orientation of these pigments. For this reason, Junge and Eckhof (1973) reasoned that the orientation of the reaction-center chlorophyll, P700, can best be measured from the linear dichroism of a property that is exclusively linked to it—for instance, the light-induced absorption change associated with P700 photooxidation. Instead of using oriented chloroplasts that require an external orienting force, Junge and Eckhof (1973) used the novel technique of "photoselection," which in essence selectively observes only those pigment ensembles that are more or less parallel to each other. As illustrated in Fig. 1, the excitation and measuring lights

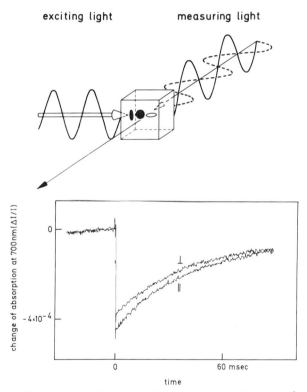

FIG. 1. *Top:* The geometry of photoselection experiments. The orientations of the E-vector of the exciting and measuring beams are indicated by sinusoidal waves. Filled dots inside the cube represent membrane orientation which leads to excitation; the open circle represents membrane orientations that do not lead to excitation. *Bottom:* Flash-induced absorption changes due to P700 photooxidation at 705 nm. Orientation of measuring beam as indicated; excitation intensity was 20% of saturation. From Junge (1974).

are both linearly polarized, and the two lights impinge at right angles to each other across the sample. The E-vector of the excitation light is vertical, and that of the measuring light could be either vertical or horizontal. Illumination of a sample containing randomly oriented chromophores with linearly polarized excitation light causes a preferential excitation of those chromophores whose transition moments are parallel with the E-vector of the excitation flash (filled circles). When the excited chromophore undergoes a photochemical change, an absorption change is produced, and the extent of this change depends on whether the E-vector of the measuring light is parallel or perpendicular to that of the excitation light.

The most important factor to be considered in the photoselection experiments with chloroplasts is that a photon absorbed by one of the antenna chlorophyll molecules undergoes about 100 resonance transfers (Borisov and Il'ina, 1973) among the antenna chlorophyll molecules before it is trapped by the reaction-center chlorophyll, P700. If the antenna chlorophyll molecules were randomly oriented in the thylakoid membrane, information on the original polarization of the excitation light would have been lost, and thus no dichroism would be observed. However, previous linear-dichroism studies of oriented chloroplasts (e.g., Breton and Roux, 1971, and others) have revealed that some long-wavelength absorbing antenna chlorophyll molecules are highly oriented in the plane of the thylakoid membrane. Thus, as shown in Fig. 1, bottom, excitation with linearly polarized light at a wavelength longer than 690 nm produced a greater absorbance change at 705 nm (as well as 430 nm) when the E-vector of the measuring light was parallel than that which was perpendicular to the excitation light.

The second factor involves possible Brownian rotation of chlorophyll molecules in the membrane or tumbling of whole membranes, both of which would effectively decrease the apparent linear dichroism. However, both are considered improbable. Note that the flash energy used in the photoselection experiments was only about 20% of the saturation intensity, as no dichroism was observed at saturating intensity. Other factors affecting the apparent dichroic ratio are discussed by Junge and Eckhof (1974) and Junge (1974).

The fact that a dichroic ratio ($\Delta A_\parallel : \Delta A_\perp$) of 1.16 was measured for absorption changes at both 430 and 705 nm indicates that all transition moments which characterize the transformation of P700 to P700$^+$ are inclined at less than 30° in the plane of the membrane. The orientation of the respective transition moments of the individual chlorophyll molecules at 430 and 705 nm in the coordinate system of the dimer is unknown. However, in monomeric chlorophyll molecules the Soret and Q-band transition moments are expected to lie roughly perpendicular to each other in the plane of the porphyrin ring. If the absorption changes at 430 and 705 nm represent the bleaching of a single chlorophyll molecule, then the rather appreciable and equal dichroic ratio of the two transition moments implies that the whole porphyrin ring is very parallel to the membrane plane.

Similar results were obtained by Breton *et al.* (1975), who oriented the chloroplasts in a magnetic field and measured the dichroism of light-induced absorption changes due to P700 photooxidation at 700 nm. As shown in Fig. 2, the P700 absorption changes monitored with a measuring beam whose electric vector was polarized parallel to the plane of the photosynthetic membrane was much larger than that monitored with a beam whose electric vector was polarized perpendicular to the membrane plane. The large dichroic ratio (2.3 ± 0.1) indicates that the transition oscillator involved at this wavelength is nearly parallel to the plane of the photosynthetic membrane. The identical kinetics for both polarizations is a good indication that both correspond to a single absorption change. When no magnetic field was applied, the absorption changes were identical for both polarizations.

Vermeglio *et al.* (1976) extended these studies by orienting chloroplasts (suspended in a viscous medium) in a magnetic field at room temperature (cf. Geacintov *et al.*, 1971) followed by slowly freezing the sample in the magnetic field to preserve the orientation. Linear dichroism measured at −50° and −170°C indicates that the membranes oriented by a magnetic field at room temperature remain oriented at lower temperatures. Flash-induced absorption changes due to P700 photooxidation are also consistent with the notion that all transition moments are parallel to the membrane plane.

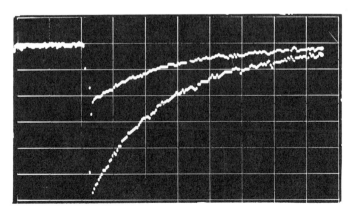

FIG. 2. Flash-induced absorption changes due to P700 photooxidation at 700 nm. Top trace, $\Delta A_\perp$; bottom trace, $\Delta A_\parallel$. Vertical scale: $A = 3.5 \times 10^{-5}$ per division; horizontal scale: 20 msec per division. From Breton *et al.* (1975).

D. The Differential Molar Extinction Coefficient of P700

The molar extinction coefficient of P700 is not known at present, since the absorption spectra of separate reduced and oxidized forms of P700 are unknown. However, the differential molar extinction coefficient between the reduced and oxidized forms of P700 can serve the purpose in the quantitative calculations of P700 concentrations, as P700 is invariably observed as a difference signal produced by light-induced oxidation or reduction. The differential molar extinction coefficient may be estimated if a P700 reaction is utilized and if the extinction coefficient of the species interacting with P700 is also known. On this basis, the differential molar extinction has been measured from a directly coupled reaction between P700$^+$ and reduced cytochrome c (Ke et $al.$, 1971; Hiyama and Ke, 1972b) and also from a reaction where oxidized and reduced forms of TMPD interact with P430$^-$ and P700$^+$, respectively, through a cyclic electron-transfer reaction (Hiyama and Ke, 1972b). The latter reaction yielded a differential molar extinction coefficient of 64 meq^{-1}·cm^{-1} at 700 nm for PS I subchloroplast particles fractionated from spinach by digitonin and 70 meq^{-1}·cm^{-1} for Triton-fractionated PS I particles from $Anabaena$ $variablis$.

Subsequently Haehnel (1973) used the "flash titration" technique to estimate the differential extinction of P700 in intact chloroplasts from oxygen yield. In this procedure, the chloroplasts were preilluminated with high-intensity far-red light so that all electron carriers between the two photosystems are oxidized. After turning off the far-red light, a group of saturating flashes which activate both PS I and PS II are applied. PS II generates electrons with a half-time of 0.6 msec, but the oxidation of plastohydroquinone takes place with a half-time of 20 msec. As a result, all electrons generated in PS II by the closely spaced group of flashes are accumulated in the plastoquinone pool. After the group of flashes, the absorption changes associated with the reoxidation of plastohydroquinone and the simultaneous reduction of P700 are then measured. The number of electrons generated by a flash can be determined by the oxygen yield per flash. In the first such measurement, Haehnel (1973) reported a differential extinction coefficient of 5.4 × 10^4 M^{-1}·cm^{-1} for P700 at 700 nm. It was subsequently found that only 75–85% of the total P700 was coupled to PS II (Haehnel, 1976). Taking this into consideration, and again by measuring the oxygen yield per flash, the differential extinction coefficient of P700 in spinach chloroplasts has been found to be (6.7 ± 0.7) × 10^4 M^{-1}·cm^{-1} at 703 nm. The good agreement of the extinction value determined with intact chloroplasts with that measured with subchloroplast particles also added support to the assumption of a quantitative electron transport from water via plastoquinone to the coupled P700.

II. The Primary Electron Acceptor of Photosystem I

Prior to 1971, much had been speculated regarding the nature of the primary electron acceptor of PS I [see Ke (1973) for a more detailed account]; these range from a chlorophyll dimer (Kamen, 1961, 1963), ferredoxin (Arnon, 1965; Vernon and Ke, 1966; Kassner and Kamen, 1967), a pteridine derivative (Fuller and Nugent, 1969), a flavin-type compound (Wang, 1970), and, finally, the "ferredoxin-reducing substance" (Yocum and San Pietro, 1969; cf. Tsujimoto et al., 1973). In 1971, a spectral species, P430, detected by flash kinetic spectroscopy (Hiyama and Ke, 1971a), and a bound ferredoxin, detected by EPR spectroscopy (Malkin and Bearden, 1971), were separately proposed to be the primary electron acceptor of PS I. This section will summarize developments during the past several years on these latter two components, on a possible identity between the two, and on a possible transient intermediate that may precede the stable primary electron acceptor in the photochemical charge separation in PS I.

A. P430

Since the earlier developments on P430 have been presented in a recent review (Ke, 1973), we will only briefly summarize these earlier results and bring the subject up to date.

1. Kinetic Behavior of P430

The elucidation of the spectral species and the role it plays as the primary electron acceptor in PS I of green plants were primarily based on spectroscopic evidence obtained from the kinetics of light-induced absorption changes involving both the primary donor, P700, and the primary acceptor, P430. The kinetic evidence can most conveniently be discussed by considering the five possible reaction courses taken by the primary photooxidant, $P700^+$, and the primary photoreductant, $P430^-$, after cessation of photoactivation. In each of the five reaction courses, the fate of the primary photooxidant and photoreductant is defined, and the kinetics of absorption changes and those of the interacting secondary donor and acceptor should be consistent with the expected reaction course, and this turned out to be true for all cases examined.

The five reaction courses presented in Fig. 3 are: (1) recombination of the primary photooxidant and photoreductant; (2) cyclic electron flow; (3) noncyclic electron flow; (4) accumulation of the primary photooxidant $P700^+$; and (5) accumulation of the primary photoreductant, $P430^-$. The five cases will be discussed only briefly below; the interested reader is referred to an earlier review (Ke, 1973) for a more detailed discussion as well as illustrations.

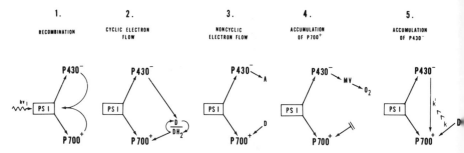

FIG. 3. Five possible reaction routes for the primary photooxidant P700$^+$ and the primary photoreductant P430$^-$ formed in the photochemical reaction in photosystem I. D stands for a secondary donor and A a secondary acceptor. From Ke (1973).

Recombination of P700$^+$ and P430$^-$ occurs when no secondary donor or acceptor is present. In this reaction course, the absorption changes of both P700$^+$ and P430$^-$ are kinetically identical. The kinetics of the recombination reaction can be fitted to a linear plot by the application of a special case of a second-order reaction where the two reactants exist in equal amounts throughout the reaction course. Such a linear plot also indicates the simultaneous production of equimolar quantities of P700$^+$ and P430$^-$ in the primary photochemical reaction in PS I. The $t_{1/2}$ of the recombination reaction is about 45 msec.

Cyclic electron flow occurs when both the oxidized and reduced forms of a suitable electron carrier [e.g., N,N,N',N'-tetramethyl-p-phenylene-diamine (TMPD)] are present. Thus TMPD (reduced) can serve as the secondary donor by donating an electron to P700$^+$, and TMPD$^+$ serves as the secondary acceptor by accepting an electron from P430$^-$. As expected, the kinetics of absorption changes monitored at wavelengths appropriate for the various electron carriers was found to be consistent with this simple cyclic electron flow in which the oxidized and reduced forms of TMPD interact with the primary photoreductant and photooxidant, respectively.

In *noncyclic electron flow*, a different secondary donor (e.g., TMPD) and acceptor (e.g., methylviologen) are interacting with the primary photooxidant and photoreductant of PS I. Experimentally, this was demonstrated by an identical kinetics of P700$^+$ re-reduction and TMPD oxidation, and likewise, that of P430$^-$ reoxidation and methylviologen reduction. These absorption changes showed for the first time a direct interaction of this highly electronegative (-446 mV) dye with the primary photoreductant (P430$^-$) formed in PS I. The noncyclic electron flow was further illustrated by using safranine T as the secondary electron acceptor and by NADP$^+$ reduction mediated by soluble ferredoxin and ferredoxin–NADP reductase (Hiyama and Ke, 1972a). It should also be pointed out that noncyclic electron flow involving an appropriate secondary donor and acceptor often results in quite different rates in the back reaction of $P700^+$ and $P430^-$. And it was such a kinetic differentiation that led to the discovery of the spectral species, P430 (Hiyama and Ke, 1971a,b, 1972a; Ke, 1973).

The three reaction courses discussed thus far are all characterized by the restoration of the primary reactants to their original uncharged state by interacting with appropriate secondary electron carriers. A unique situation of partial restoration of one of the primary reactants, $P700^+$ or $P430^-$, is possible if only a secondary acceptor is present in the former case, and only a secondary donor is present in the latter. The low-potential viologens can serve as secondary electron acceptors for PS I which efficiently traps electrons from $P430^-$, and the reduced viologens are also rapidly turned over by oxygen, thus leaving the primary photooxidant, $P700^+$, accumulated (Ke, 1972a). Such a "one-way" electron discharge by PS I has also been observed by Rumberg (1964) in chloroplasts where the endogenous secondary electron donor coming from PS II was blocked by DCMU. The $P700^+$-accumulation regime provides a simple and convenient method for quantitative estimation of P700 content in chloroplasts or subchloroplast particles.

Conversely, *accumulation of P430$^-$* would be possible if an appropriate secondary electron donor that can compete with the recombination reaction is present. $P430^-$ accumulation was demonstrated by using phenazine methosulfate reduced by dithiothreitol under anaerobic conditions as the secondary donor. Under this condition, $P700^+$ recovers rapidly, but approximately one-fifth of the total absorption decrease at 430 nm, which presumably represents P430 photoreduction, recovered very slowly (Ke, 1973). Addition of a secondary acceptor, say, benzyl-viologen, to this system immediately caused a rapid recovery of the entire absorption change.

2. Difference Spectrum

It was through the noncyclic-electron-flow type reaction in PS I particles, where the dark decay kinetics of the primary photooxidant $P700^+$ and photoreductant $P430^-$ can be readily differentiated, that light-minus-dark difference spectra for the two species in the visible region were first constructed (Hiyama and Ke, 1971a,b). Similar difference spectra were found for PS I subchloroplasts from spinach as well as several blue-green algae (Hiyama and Ke, 1971a). Subsequently, measurements were extended to the near-infrared (Hiyama and Ke, 1972b) and the ultraviolet region (Ke, 1972b). The complete spectra for the two species are presented in Fig. 4.

The P700 spectrum shows major changes at 430 and 700 nm, a broad

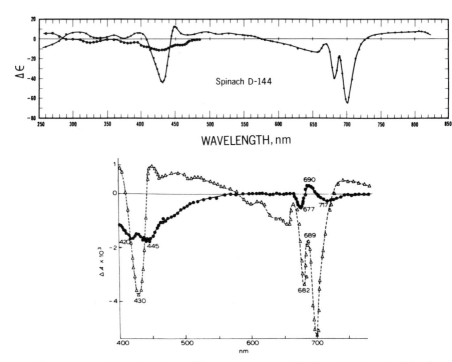

WAVELENGTH, nm

FIG. 4. *Top:* Light-minus-dark difference spectra of P700$^+$ and P430$^-$ in digitonin-fractionated photosystem I (PSI) particles. From Ke (1973). *Bottom:* Same difference spectra of pea PS I subchloroplasts fractionated by treatment with digitonin and Triton. From Shuvalov *et al.* (1976a).

positive band beyond 730 nm, and one below 400 nm that crosses zero at 280 nm. The difference spectrum of P430 shows a broad band in the blue region, with a maximum near 430 nm (hence the initial designation of P430 for this newly found spectral species), but no discernible changes beyond 500 nm. Negative bands were observed at 380 and 325 nm and a positive band at 270 nm, and a zero crossing near 300 nm. The difference spectrum of P430 resembles that of soluble spinach ferredoxin in both the visible and ultraviolet regions. A differential extinction coefficient of P430 at 430 nm calculated from the ratio of absorption changes to P700 is, however, much greater than that of soluble ferredoxin.

Three isobestic points at 408, 442, and 575 nm can be noted in the difference spectrum of P700. These wavelength positions are very useful for kinetic analysis where changes of P430 can be isolated without the interference by that of P700. At 575 nm, such secondary electron carriers as TMPD and 2,6-dichlorophenolindophenol (DICP) may be measured with little interference by either P700 or P430.

More recently, Shuvalov *et al.* (1976a) utilized the P700$^+$-accumulation and P430$^-$-accumulation reactions and measured the difference spectra of the two species in PS I particles. Using a phosphoroscopic photometer (Karapetyan and Klimov, 1970), the authors not only confirmed previous measurements reported for the blue region (Hiyama and Ke, 1971a), but also detected new, interesting changes in the red region. As shown in Fig. 4, bottom, the blue region of the P430 spectrum is now resolved into bands at 430 and 445 nm, and the extinction coefficient appears to be even greater than that reported earlier by Hiyama and Ke (1971b). Few absorption changes occur between 550 and 670 nm. In the red region, the authors found a negative band at 717 nm and a spectral shift from 677 to 690 nm. The spectrum composed of the 420, 445, and 717 nm bands bears a more definitive resemblance to that of soluble spinach ferredoxin (Rawlings et al., 1974). The wavelength shift at 680 nm was interpreted to indicate either a participation of a chlorophyll molecule in the acceptor reaction or a perturbation of the spectrum of a chlorophyll molecule by the local electrical field of P430$^-$.

3. Risetimes, Quantum Requirement, and Thermal Stability

Consistent with the notion that P700 and P430 participate in the primary photochemical reaction, the formation of the charged species should be very rapid. Using 20-nsec ruby-laser flashes as excitation source, the *risetimes* of P700 photooxidation and P430 photoreduction in a composite absorption-change signal at 430 nm were found to be equal to or less than 0.1 μsec, which was limited by instrument time resolution.

The photochemical formation of P700$^+$ and P430$^-$ show parallel response in their yield toward excitation intensity. The absorption changes of P700 to P430 maintained a constant ratio of 4 over the entire excitation-intensity range.

The *quantum requirement* for P700 photooxidation and P430 photo-reduction was measured in a similar manner as that for the light-saturation measurements, except that more monochromatic light sources were used. Using 671 nm excitation light, the quantum requirement was near 2, whereas that for the 703 nm light was unity. These results show that the primary photochemical reaction generating P700$^+$ and P430$^-$ in PS I is an extremely efficient process when excited in the spectral region where P700 absorbs. Some of the 671 nm quanta absorbed by the bulk chlorophyll are presumably not efficiently utilized.

Shuvalov et al. (1976) examined the *thermal stability* of P430 by directly monitoring the light-induced absorption changes at 444 nm and found that the P430 signal disappeared after the sample was heated between 60° and 67°C, whereas P700 signal disappeared until 70°–75°C. The different thermal stability of P700 and P430 and the persistence of

P700 reaction after complete inactivation of P430 suggested the existence of an "alternate" acceptor for P700.

4. Fluorescence-Yield Changes and Delayed Light Emission in Photosystem I

Light-induced fluorescence-yield changes of chlorophyll in photosynthetic organisms is attributable to a temporary loss of ability by the reaction centers to utilize the excitation energy absorbed by chlorophyll as a result of accumulation of charged primary electron carriers. This phenomenon was recognized earlier by Duysens and Sweers (1963) for PS II, where the fluorescence-yield increase was attributed to the reduction of the primary electron acceptor. More recently, fluorescence yield increase as a result of photooxidation of the primary electron donor in PS II has also been reported (see Butler, 1973).

Light-induced fluorescence-yield changes accompanying the changes in redox states of the primary electron carriers in PS I have been reported recently. Karapetyan et al. (1973) and Shuvalov et al. (1976a) demonstrated a light-induced fluorescence yield increase in PS I particles that can be related to either the photooxidation of P700 or the photoreduction of P430. Figure 5 is a summary of results obtained under

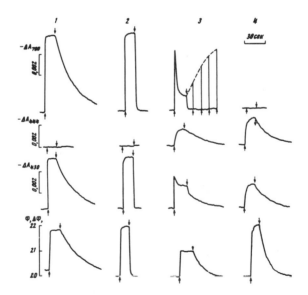

FIG. 5. Light-induced absorption changes (ΔA) at 700, 430, and 444 nm, and fluorescence-yield changes ($\Delta \phi$) in photosystem I subchloroplasts fractionated from pea by treatment with digitonin and Triton. The sample contained 13 μg/ml Chl plus: (1) 1 mM ascorbate; (2) 1 mM ascorbate and 10 μM 2,6-dichlorophenolindophenol; (3) 1 mg/ml dithionite; and (4) 1 mg/ml dithionite and 10 μM neutral red. From Shuvalov et al. (1976a).

a variety of redox conditions: light-induced absorption changes at 700, 444, and 430 nm represent, respectively, changes by P700, P430, and a combination of the two; ΔF is the PS I fluorescence-yield change. In the presence of ascorbate alone or ascorbate plus DCIP, the kinetics of absorption changes at 700 and 430 nm, and the fluorescence-yield change are very similar, indicating that the fluorescence-yield increase is caused by the photooxidation of P700. In the presence of dithionite, the photooxidized P700$^+$ is rapidly reduced even while the actinic light was still on, but became completely reduced as soon as light was turned off. On the other hand, the light-induced fluorescence-yield first increased to a steady level and then relaxed rather slowly after the actinic light was turned off. The kinetics of the fluorescence-yield change resembles that of the absorption change at 444 nm, which reflects the accumulation of P430$^-$. In the presence of dithionite plus neutral red, the reduction of photooxidized P700 was so rapid that it could not be detected at the available time resolution of the instrument. However, the kinetics of the fluorescence-yield change and the absorption change at 444 nm again are similar.

Delayed light emission of PS I and its dependence on the redox state of the primary electron carriers have been reported recently by Shuvalov (1976). When the primary electron donor P700 is maintained in the active state, a flash-induced delayed luminescence signal with a half-decay time of 20 msec was observed (Fig. 6A). When an exogenous acceptor (methylviologen) or an exogenous donor (reduced neutral red) is present, the decay of the delayed luminescence becomes accelerated, and the acceleration parallels that of P430$^-$ reoxidation in the former case and P700$^+$ re-reduction in the latter (Fig. 6, B and C). These results indicate that the luminescence intensity is dependent on the concentrations of P700$^+$ and P430$^-$, and that the luminescence most likely originates from the recombination of the two charged species.

When P430 is either reduced or is inactivated by heat, flashing of PS I generates triplet-state chlorophyll with a half-time of about 500 μsec. The triplet state has been measured by delayed fluorescence at 293 and 77°K or by phosphorescence at 77°K. The delayed fluorescence (710 nm) at 293°K arises from thermal activation of the triplet state up to the singlet level of chlorophyll. At 77°K the luminescence spectrum consists of bands at 960 and 740 nm. The 960 nm band has been attributed to chlorophyll phosphorescence, and the 740 nm band to triplet–triplet annihilation. When P430 is inactivated by heat, and at the same time P700 is oxidized, no triplet formation can be observed. It has been suggested that the triplet–triplet annihilation at 77°K is related to the strong interaction between chlorophyll molecules in the PS I reaction-center complex in which the interaction of triplet exciton can take place.

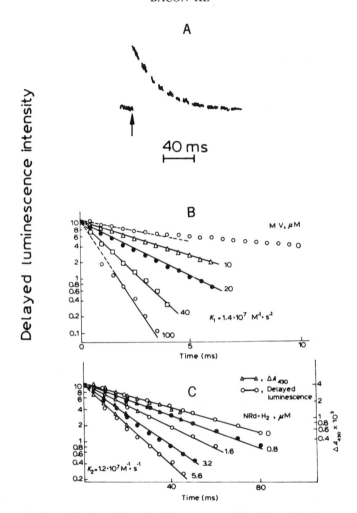

FIG. 6. (A) Decay kinetics of the "slow" component of the delayed luminescence in pea photosystem I particles after a single 2-μsec flash (at arrow). Sample contained 10 μg/ml chlorophyll, 1 mM ascorbate, and 5 μM 2,6-dichlorophenolindophenol. (B) Plot of kinetics of delayed-luminescence decay. Sample composition same as in (A) except it also contained various amounts of methylviologen. (C) Plot of kinetics of delayed-luminescence decay and absorption change at 430 nm (due to P700). Sample contained 10 μg/ml chlorophyll, 1 mg/ml dithionite, and various amounts of neutral red. From Shuvalov (1976).

5. Redox Titration of the Photosystem I Primary Acceptor

The high reducing power of PS I has long been recognized, and much has been reported on this subject in the literature. Previously it was only possible to make an indirect estimate from the reduction of viologen dyes (Kok *et al.*, 1965; Zweig and Avron, 1965; Black, 1966). After the kinetics of P430 absorption changes were established, the interaction of similar viologen dyes with reduced P430 was reexamined and the reaction rate constants were estimated. From these rate constants and the known redox potential of the dyes, the redox potential of P430 was estimated to be about -500 mV (Hiyama and Ke, 1971b).

The first potentiometric titration of P430 was made by an indirect method of measuring the magnitude of light-induced absorption changes of P700 as a function of the redox potential poised onto the reaction medium containing the PS I particles. These experiments were based on the premise that the amplitude of the absorption change representing the primary reaction depends on the fraction of P430 present in the oxidized state. Thus, increasing reduction of P430 would decrease the amplitude of absorption change. The first series of experiments were carried out at room temperature, where interaction of reduced mediators with oxidized $P700^+$ caused premature attenuation of the amplitude of the absorption change and thus led to an underestimated redox-potential value of -470 mV (Ke, 1972b).

Subsequently, redox titrations of the PS I primary acceptor were carried out by monitoring P700 photooxidation at liquid-nitrogen temperature (Ke, 1974; Lozier and Butler, 1974). Although the two studies used different methods of reduction, one using conventional low-potential viologen dyes reduced by dithionite at high pH (Ke, 1974) and the other using H_2/H^+ as the reductant catalyzed by hydrogenase (Lozier and Butler, 1974), both groups found a midpoint potential at about -530 mV and the number of electron change near one.

A chemical titration curve obtained at pH 11.0 by monitoring the light-induced absorption changes of P700 at 86°K is shown in Fig. 7A. Figure 7B shows several individual absorption-change transients which are attenuated in amplitude with decreasing potentials. The titration curves in Fig. 7A and B show an anomaly, i.e., a small and rapidly-decaying absorption change could not be titrated out completely even at potentials more negative than -600 mV. The reversible absorption changes measured in a sample poised at -600 mV at three characteristic wavelengths of P700 (Fig. 7C) indicate that the residual changes belong to P700. A possible explanation for this anomaly will be discussed in the following section.

Electrochemical titration of the PS I primary acceptor at pH 8.0 monitored by low-temperature absorption changes of P700 have also been carried out in a specially constructed "thin-pathlength" electrochemical cell and the results were very similar to those shown in Fig. 7A (Hawkridge and Ke, 1977).

Shuvalov *et al.* (1976a) carried out preliminary redox titration of P430

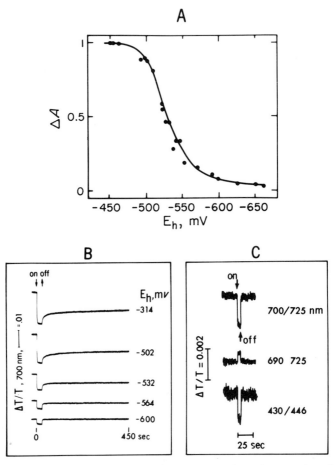

FIG. 7. Redox (chemical) titration of photosystem I subchloroplasts monitored by light-induced absorption changes at 86°K. (A) Maximum extent of the light-induced absorption changes due to P700 photooxidation vs. sample potential. (B) Individual light-induced absorption-change signals at selected potentials. (C) Absorption-change signals measured at approximately −600 mV at three selected wavelengths. Arrows indicate light on and off. In (B): measuring wavelength, 700 nm; reference wavelength, 725 nm. In (C): measuring and reference wavelengths as indicated. From Ke *et al.* (1977).

by directly monitoring the light-induced absorption changes of P430 at 444 nm at room temperature as well as fluorescence yield changes. The midpoint potential of the titration curve was near −550 mV.

More recently, Ke *et al.* (1977) also carried out electrochemical titrations of PS I subchloroplast particles near neutral pH and monitored the light-induced P700$^+$ by EPR spectroscopy at 90°K. As in the case of the optical signal, the amplitude of the $P700^+$ free-radical signal became attenuated with lowering potentials (Fig. 8A). Results from electrochemical titrations at pH 8.0 and 9.5 and monitored by EPR spectroscopy at 90°K are plotted in Fig. 8,B. As seen here, at both pH 8.0 and 9.5, the residual signal remained almost at a constant level of ~20% of the initial level down to a potential of −705 mV.

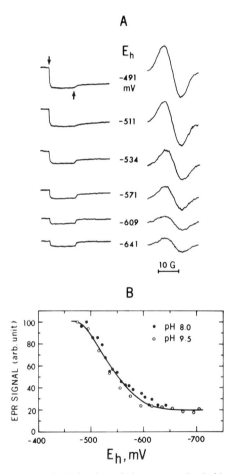

Fig. 8. Redox (electrochemical) titration of photosystem I subchloroplasts monitored by electron paramagnetic resonance spectroscopy at 90°K. (A) *Left:* Light-induced EPR changes monitored at the high-field peak of the first-derivative EPR signal at g = 2.0026 for samples poised at selected potentials. Down- and upward arrows indicate light on and off. *Right:* The corresponding light-minus-dark difference spectra of the $P700^+$ free radicals at g = 2.0026. (B) Titration curve from EPR results for samples titrated at pH 8.0 (●) and 9.5 (○). From Ke *et al.* (1977).

B. BOUND IRON–SULFUR PROTEINS

At about the same time that the spectral species P430 was proposed to be the primary electron acceptor of PS I, a chloroplast-bound ferredoxin detected by EPR spectroscopy was separately proposed by Malkin and Bearden (1971) to serve the same role. This latter proposal was based on the fact that the ferredoxin-like EPR signal was not caused by soluble ferredoxin, and it was produced by photoreduction at cryogenic temperatures. Figure 9 shows the EPR signals of PS I particles kept in the dark under a mild reducing condition (top) and subsequently exposed to actinic light (bottom). It was found later that these bound iron–sulfur proteins are exclusively associated with PS I (Bearden and Malkin, 1972a; Leigh and Dutton, 1972; Evans *et al.*, 1972; Ke *et al.*, 1974a).

Figure 10 shows EPR signals of iron–sulfur proteins in chloroplasts or PS I subchloroplast particles from several green plants and algae. The high-P700 particles prepared from spinach (trace A) or bean (trace B), and chloroplasts derived from the blue-green alga, *Anabaena cylindrica* (trace C), all show typical iron–sulfur protein signals. Chloroplasts prepared from the No. 8 mutant of the green alga, *Scenedesmus obliquus*, which is known to be deficient in P700, showed no EPR signal typical of chloroplast-bound iron–sulfur proteins (trace D). Resonance lines observed at $g = 2.05, 2.02, 1.94$, and 1.92 resemble those observed for the reduced iron–sulfur centers of the NADH dehydrogenase of beef heart and *Candida utilis* (Orme-Johnson *et al.*, 1971; Ohnishi *et al.*, 1972). The PS II reaction-center particles (or TSF-2a) also showed no iron–sulfur-protein spectrum, but one ascribable to manganese (trace E) (Ke *et al.*, 1974a). The PS II particles reduced by dithionite under anaerobic conditions showed a large free-radical signal, the origin of which is unknown.

Our (unpublished) measurements on the P700 concentrations by

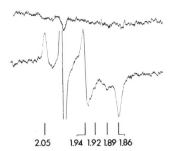

2.05   1.94 1.92 1.89 1.86

FIG. 9. Electron paramagnetic resonance spectra measured at 13°K from a dark-adapted Triton photosystem I subchloroplast (top) and from the same sample after illumination at 13°K with several white-light flashes (B. Ke and H. Beinert, unpublished experiments, 1972).

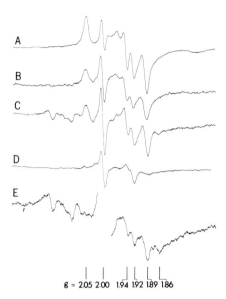

FIG. 10. Electron paramagnetic resonance spectra of several chloroplasts or subchloro-plasts derived from green plants and algae and reduced by dithionite in the presence of trace amount of methylviologen. (A) HP700 particles derived from spinach (cf. item No. 6 in Fig. 11); (B) HP700 particles derived from bean (cf. item No. 7 in Fig. 11); (C) chloroplasts from *Anabaena* (item No. 8 in Fig. 11); (D) chloroplasts derived from mutant No. 8 of *Scenedesmus* (item No. 9 in Fig. 11); (E) a photosystem II reaction-center particle (TSF-IIa). From B. Ke and H. Beinert, unpublished experiments (1973).

optical spectroscopy and iron–sulfur protein concentrations by EPR spectroscopy in chloroplasts and subchloroplast particles from a wide variety of sources showed a linear relationship between the two (Fig. 11). Quantitative analysis by Malkin and Bearden (1971) showed both whole and broken chloroplasts contain about 30 nmoles of nonheme iron and 30 nmoles of acid-labile sulfur per milligram of chlorophyll. Whether the chloroplast-bound iron–sulfur proteins contain the 2Fe–2S or 4Fe–4S centers, the iron–sulfur proteins are present far in excess of P700 in all organisms examined. In a subsequent analysis, Bearden and Malkin (1972a) found that the concentrations of nonheme iron and acid-labile sulfur (and presumably also P700) in the digitonin-fractionated PS I subchloroplast particles (D-144) were enriched about 3-fold. More re-cently, Golbeck *et al.* (1976) reported a ratio of nonheme iron to P700 of 8–10:1 in a spinach high-P700 particle (P700:Chl = 1:25). Bearden and Malkin (1972b) made a quantitative EPR study of the PS I reaction at 25°K using either whole chloroplasts or digitonin-fractionated PS I particles, and found the spin concentrations of photooxidized P700$^+$ and photoreduced iron–sulfur protein (with g values 2.05, 1.94, and 1.86) to

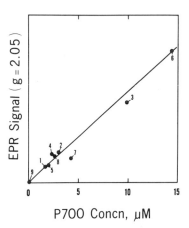

P700 Concn, μM

FIG. 11. Plot of the electron paramagnetic resonance (EPR) signal height at $g = 2.05$ vs. P700 concentration in several chloroplasts or subchloroplasts from green plants and algae as described in table below. From B. Ke and H. Beinert, unpublished experiments (1973).

No.	Sample	Chl $10^{-4} M$	P700/Chl ($\times 10^3$)	P700 $10^{-6} M$
1	Unfractionated chloroplasts	5.05	3.38	1.7
2	Digitonin-fractionated PS I particle (D144)	5.32	5.62	3.0
3	Triton-fractionated PS I particle (TSF-1)	4.45	22.2	9.9
4	Digitonin-fractionated PS I particle from the stroma	3.92	6.25	2.5
5	Digitonin-fractionated PS I particle from the grana	4.13	4.33	1.8
6	HP700 particles (spinach)	6.05	23.75	14.4
7	HP700 particles (bean)	1.80	22.96	4.1
8	Chloroplasts (*Anabaena*)	2.66	9.59	2.6
9	Chloroplasts (mutant No. 8 of *Scenedesmus*)	11.0	0	0
10	PS II reaction-center particle (TSF-2a)	4.44	0	0

be nearly the same. These results support the notion that these two components function as a donor-acceptor pair in the primary photochemical reaction of PS I.

1. Chemical Characterization

Very little is known at present about the chemistry of these chloroplast-bound iron–sulfur proteins. An attempt was made by Malkin *et al.* (1974) to isolate the membrane-bound iron–sulfur proteins by acetone extraction of chloroplasts previously freed of soluble ferredoxin. The isolated iron–sulfur protein was estimated to have a molecular weight of about 8000, and to contain 4 Fe and 4 S per molecule. The protein shows a featureless absorption spectrum typical of those of bacterial-type ferredoxins. This protein could be photoreduced but became rapidly reoxidized in the dark. It could not function in place of soluble ferredoxin in NADP reduction, but could catalyze photoreduction of mammalian cytochrome c by water. The EPR spectrum of the reduced form of the isolated iron–sulfur protein is quite different from the protein bound to the native membrane. Thus, the relationship between the isolated iron–sulfur protein and those present in the native state remains unclear.

A simpler approach for distinguishing the types of iron–sulfur centers in the PS I membrane-bound iron–sulfur proteins was used by Cammack and Evans (1975). It is known (Cammack, 1973) that guanidine-HCl and dimethyl sulfoxide (DMSO) can unfold the protein reversibly, while the iron–sulfur centers remain intact. However, depending on whether the iron–sulfur centers are of the 2Fe–2S or 4Fe–4S type, the iron–sulfur protein undergoes a reversible change in structure in 80% DMSO solution, and the changes have characteristic line shapes and temperature dependence. The 2Fe–2S centers give rise to a rather isotropic signal which is readily detected at temperatures up to 150°K, whereas the 4Fe–4S centers give rise to an axial or near axial signal which is detected only below 35°K, indicating a more rapid electron spin relaxation. DMSO treatment and reduction of PS I subchloroplast particles gave an EPR spectrum very similar to that obtained with a bacterial ferredoxin containing 4Fe–4S centers (from *Clostridium pasteurianum*), and that the spectrum appeared only below 35°K (no signal characteristic of 2Fe–2S centers were detected at 77°K) and was readily saturated with microwave power.

The bound iron–sulfur proteins are apparently susceptible to denaturation or destruction by certain chemical agents. Nelson *et al*. (1975) found that treatment of the digitonin-fractionated subchloroplast particles with increasing concentrations of sodium dodecyl sulfate (SDS) gradually decreased the NADP-reduction activity and diminished the iron–sulfur protein content as detected by EPR spectroscopy. However, photooxidation of P700 in the presence of phenazine methosulfate as detected by absorption changes at 430 nm was not affected by similar SDS treatment. The single polypeptide containing P700 produced by 0.5% SDS treatment completely lost the NADP-reduction activity and was devoid of any bound iron–sulfur proteins.

Similarly Malkin *et al*. (1976) found the "P700-chlorophyll-*a*-protein complexes" isolated from a eukaryotic organism, *Phormidium luridum,* one fractionated by treatment with SDS and one with Triton X-100, to have different properties. Both complexes contain about one P700 per 40 chlorophyll *a* molecules, and both show light-induced photooxidation at room temperature, with different reaction kinetics and quantum requirements. However, the Triton complex was photochemically active at 15°K as evidenced by oxidized $P700^+$ and reduced iron–sulfur protein produced by illumination and detected by EPR spectroscopy. On the other hand, the SDS complex was photochemically inactive at 15°K, as neither illumination at 15°K nor freezing the sample under illumination produced any EPR signals of the P700 free radicals or reduced iron-sulfur proteins. Furthermore, chemical reduction by dithionite (in the presence of methylviologen) also showed only the Triton complex to contain bound iron–sulfur proteins. These results were interpreted to support the assignment of the iron–sulfur protein as the primary electron acceptor of PS I.

Selective biochemical inactivation of membrane-bound iron–sulfur proteins in PS I particles was carried out by urea treatment in the presence of ferricyanide (Golbeck *et al.*, 1976). Such a treatment produces a time-dependent conversion of acid-labile sulfide to the zero-valent sulfur (Petering *et al.*, 1971) in the membrane-bound iron–sulfur proteins. The integrity of P700 in the urea–ferricyanide treated particles appeared to remain intact as measured by a chemical difference (oxidized-minus-reduced) spectrum. However, with increasing incubation in urea–ferricyanide, the particles showed a parallel loss in the P700 photooxidation activity and in the labile-sulfide content. These results suggested that the major fraction of the photochemically active P700 was related to the amount of intact iron–sulfur proteins remaining, and thus supported their roles as the primary electron acceptor.

2. Redox Titrations

Soon after the membrane-bound iron–sulfur proteins were suggested to be the primary electron acceptor of PS I, it became evident that knowledge on the redox potential would be extremely important, as it would complement the proposed role of a primary acceptor and also further the understanding of their precise roles. Potentiometric titration of PS I subchloroplasts, using dithionite as the reductant and appropriate redox mediators including several low-potential viologens, were carried out by Ke *et al.* (1973b) and Evans *et al.* (1974). As the lower limit of the redox potential that can be achieved depends on the pH of the medium, pH 10 and 9 had to be used in our initial titrations. We found, however, that alkaline pH as high as 11 had little influence on the magnitude or kinetics of light-induced P700 photooxidation at room temperature.

The development of EPR spectra of reduced iron–sulfur proteins during the course of a reductive titration is shown in Fig. 12. An EPR

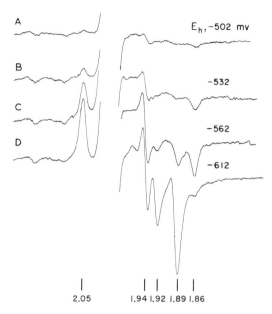

FIG. 12. Electron paramagnetic resonance spectra of Triton photosystem I subchloroplasts at different stages of reductive titration at pH 10. The poised redox potentials are expressed relative to the hydrogen electrode. From Ke *et al*. (1973b).

spectrum similar to that of soluble ferredoxin, with major lines at g of 2.05, 1.94, and 1.86 began to emerge at about -500 mV. The large free-radical signal was mostly due to the dipyridyl radicals formed from the viologen mediators. With lowering potential, all lines seen above increased further in magnitude. At about -560 mV, two new lines at $g = 1.92$ and 1.89 began to appear. Below -560 mV, the lines at $g = 2.05$, 1.92, and 1.89 continued to increase, the line at $g = 1.94$ changed very little, and the peak at $g = 1.86$ began to decrease. As seen from the last spectrum in Fig. 12, at the most negative potential (-612 mV) obtainable, the lines at $g = 2.05$, 1.92, and 1.89 reached the maximum and that at $g = 1.86$ the minimum level.

The changes in the EPR spectra during the titration at pH 10 are plotted for the various peaks vs. the potential in Fig. 13, left. The changes in resonance lines were initially attributed to possibly three iron–sulfur centers: one with lines at $g = 2.05$ and 1.94; one at $g = 2.05$, 1.92, and 1.89; and one for which only a line at $g = 1.86$ has been resolved. The midpoint potentials of the iron–sulfur species fall into two regions: the $g = 1.94$ signal, the first segment of the $g = 2.05$ plot, and the rise phase of the $g = 1.86$ signal had a value of -530 mV; the upper part of the $g = 2.05$ plot, the decrease phase of the $g = 1.86$ signal, and

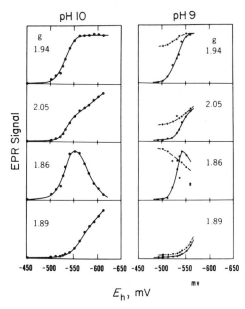

FIG. 13. Plot of electron paramagnetic resonance (EPR)-signal heights of four prominent peaks for redox titration at pH 10 and 9. Each set of points was derived from a separate spectrum of the kind shown in Fig. 12. Each curve was constructed from data in a single extended titration series. Dashed curves represent total signal heights generated by chemical reduction and that formed after subsequent illumination at 77° K. From Ke *et al.* (1973b).

the $g = 1.89$ and 1.92 profile had a midpoint potential estimated to be $\leqq$ -580 mV. The oxidation–reduction reaction of each of the bound iron–sulfur species, as represented by the changes of the EPR spectra, was reversible and apparently involved more than one electron, as determined from the curve slope.

The EPR spectra measured by Evans *et al.* (1974) during the titration of a subchloroplast particle prepared by French press were in close agreement with those shown in Fig. 13. The data were interpreted slightly differently, however. Because of the similarities in potential dependence for the appearance and disappearance of the $g = 1.86$ signal to those of the other signals, Evans *et al.* suggested that the $g = 1.86$ line is a component of the higher-potential iron–sulfur species, which upon further reduction either broadens or undergoes a shift so that it becomes inseparable from the line at $g = 1.89$. Such a shift also implies an interaction between the two iron–sulfur centers. By means of curve fitting, and assuming that the two centers undergo independent one-electron changes, the authors reported the redox potentials as -553 and -594 mV (±20 mV), respectively. The suggested cooperation between the two centers presumably could explain the apparent electron change (the n value in the Nernst equation) being greater than 1, as observed in a straightforward plot shown in Fig. 13.

Titration at pH 9 could only be carried to -560 mV, and essentially only the first half of the titration behavior as found at pH 10 was seen. At any given potential more positive than -560 mV, the part of the iron-sulfur protein that was not reduced chemically could be reduced photochemically, but apparently only to the maximum extent reduced chemically at -560 mV (Fig. 13, right). This observation has been used to support the suggestion that the iron-sulfur protein with a midpoint potential of -530 mV (cf. Sections II,A,5 and II,C; see also Bearden and Malkin, 1972b) is the primary electron acceptor. In view of the importance of this question regarding the identity of the primary electron acceptor, this question should be reexamined in more detail. The recently developed electrochemical method of titration may be of advantage in this regard, as it allows the titration to be carried out to a more extended potential range near neutral pH.

In this connection it may be of interest to note that the more negative iron-sulfur protein has been found to be present in the bound state in a cytochrome complex fractionated from the Triton PS I subchloroplast particles (TSF-I). However, the more negative iron-sulfur protein was not quantitatively removed from the remaining high-P700 particles. A possible involvement of the complex containing cytochrome f, cytochrome b_6, bound iron-sulfur protein, and bound plastocyanin in cyclic photophosphorylation has been proposed (Ke et al., 1975).

Chloroplasts illuminated at room temperature and frozen while still being illuminated developed an EPR signal similar to that produced by chemical reduction at -610 mV. Illumination at or below 77°K of a dark-adapted sample poised under mild reducing condition (e.g., in the presence of ascorbate and TMPD) did not bring about photoreduction beyond that accomplished by chemical reduction at about -560 mV. Dithionite alone in the dark under anaerobic condition brought about a partial reduction to the extent of the first chemical titration step. Dithionite plus illumination at room temperature or dithionite plus methylviologen in the dark produced the maximum signal. EPR spectra due to chemically reduced iron-sulfur proteins showed no detectable decay for at least 3 days when the samples were stored in the dark at 77°K (Ke et al., 1973b).

C. Correlation between P430 and Bound Iron-Sulfur Proteins

The simultaneous but separate findings of P430 by optical spectroscopy and chloroplast-bound iron-sulfur protein by EPR spectroscopy, and the assignments of an identical functional role to the two components as the primary electron acceptor of PS I naturally prompted us to try to correlate the two components and to see if they are one and the same.

The light-minus-dark difference spectrum measured for P430 in the visible and ultraviolet regions (Hiyama and Ke, 1971a,b; Ke, 1972b) (see Fig. 4) already suggested an iron–sulfur protein to be a possible candidate. However, a definitive identification of P430 with an iron–sulfur protein was difficult because the shape of the absorption band was slightly different from that of soluble spinach ferredoxin, and the magnitude of the differential extinction was also greater than that known for soluble spinach ferredoxin. The light-minus-dark difference spectrum measured more recently by Shuvalov et al. (1976a), who utilized the P430-accumulation reaction (see Fig. 4), further strengthened this identification, as the newly measured difference spectrum had better spectral resolution in the Soret region and additional characteristic features of ferredoxin-type compounds were also found. The differential molar absorptivity reported by Shuvalov et al. (1976a) for P430 appeared to be even greater than the value measured by us. The substantial difference between the absorptivities of bound and soluble ferredoxins is probably attributable to an influence due to binding to the chloroplast membrane.

The first experimental attempt in a correlation of the two components by simultaneous optical and EPR measurements was reported by Ke and Beinert in 1973. As described earlier, after illumination of PS I subchloroplast particles in the presence of an autoxidizable secondary electron acceptor, such as methylviologen, subsequent flash illumination caused no transient absorption change due to P700 photooxidation. During preillumination under such a condition, the photoreduced P430⁻ presumably discharged electrons to the secondary acceptor methylviologen, which was in turn autoxidized by air, and the net result was an accumulation of oxidized P700⁺. When a suitable secondary electron donor, such as TMPD, was added to the subchloroplasts containing methylviologen, photochemical activity could be restored and the photochemical reaction following noncyclic electron flow could be demonstrated. EPR spectra consistent with the optical results were measured with PS I subchloroplast particles which had been preilluminated in the presence of methylviologen: a large P700⁺ signal was produced whereas bound iron–sulfur protein was not detected (Fig. 14). Subsequent illumination of the same sample at 77°K did not change the EPR spectrum. However, if the preilluminated subchloroplasts were allowed to recover at room temperature by standing in the dark for 10 minutes or by addition of a chemical reductant, subsequent illumination of the sample at 77°K yielded an EPR spectrum consisting of signals due to both P700⁺ and reduced iron–sulfur protein.

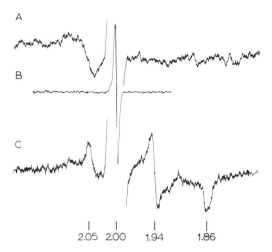

2.05 2.00   1.94   1.86

FIG. 14. (A) Electron paramagnetic resonance (EPR) spectra of photosystem I subchloroplasts preilluminated in the presence of methylviologen and then frozen immediately in liquid nitrogen. (B) Same sample as in (A) but recorded at 10 μW microwave power instead of 3 mW for (A). (C) Same sample as in (A) but stored in the dark for 10 minutes and then frozen in the dark in liquid nitrogen. Before the EPR spectrum was recorded, the sample tube was illuminated with five white flashes at 77°K. From Ke and Beinert (1973).

Illumination of subchloroplast particles in the presence of dithionite yielded maximum EPR signal of the reduced iron–sulfur protein (Ke *et al.*, 1973b). This reaction may be viewed as an EPR confirmation of the optical experiment of P430 accumulation, thus providing another route for correlating P430 and the iron–sulfur protein.

Although the other three reaction courses—recombination, noncyclic and cyclic electron flows—should also provide additional routes for correlating P430 and iron–sulfur proteins, the fact that the reduced form of bound iron–sulfur proteins is detectable by EPR spectroscopy only at very low temperatures ($< -25°$K) would mean that very rapid cycling of sample temperatures between that desired for the reaction and that necessary for observation is required. In principle, these kinetic spectroscopic experiments may be performed if suitable modifications can be made for the rapid mixing/rapid quenching technique, which has been used successfully for studying enzyme kinetics (Bray *et al.*, 1964). Unfortunately, owing to technical difficulties, we have not yet succeeded in several attempts.

Up to two years ago, literature information on PS I reaction at low temperatures was quite contradictory (see Ke *et al.*, 1976b, for a summary review), ranging from practically irreversible to partially reversible. This situation prompted us to make a systematic examination. Earlier preliminary examination of the light-induced P700 absorption changes at low temperatures (Ke, 1974) revealed some complex decay kinetics which were temperature dependent. From these results, it would seem desirable to examine the decay kinetics of the light-induced PS I reactions at low temperatures also by EPR spectroscopy. Figure 15 shows that the kinetics of the EPR free-radical signals representing P700⁺ (left) have characteristics very similar to the absorption-change signals (right). The formation of $P700^+$ free radicals was within the combined time resolution of the instruments (0.25 second), but the decay followed a temperature-dependent multiphasic course. Upon cessation of illumination, not only the decay rate varied with temperature, but more of the reaction became irreversible (as observed within 20 minutes) with lowering temperatures. It was also found that the reversible portion of the absorption change and the EPR signal can undergo reversible bleaching again upon a second illumination.

It was reasoned that if an iron–sulfur protein were indeed the reaction partner of P700 in the primary photochemical act of PS I, the onset time for the formation of the two charged species and their decay kinetics by

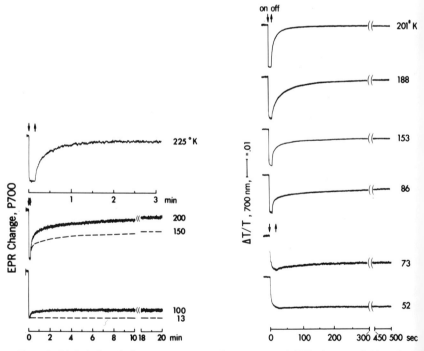

FIG. 15. Light-induced electron paramagnetic resonance (EPR) changes (left) and absorption changes (right) due to P700 photooxidation and re-reduction at various temperatures. From Ke *et al.* (1974b, 1976b).

way of recombination should be identical. As demonstrated in Fig. 16, the onset time for P700 photooxidation and that for the photoreduction of iron–sulfur protein (monitored at $g = 1.86$) at 13°K were both rapid and not resolvable at the time resolution available, and both changes were practically irreversible.

Because the EPR signals of reduced iron–sulfur proteins are detectable only at temperatures below 25°K, continuous monitoring of the decay kinetics at other temperatures could not be made. Considering the fact that the EPR signals of P700$^+$ and reduced iron–sulfur proteins were completely stable at 13°K, one could study the decay of both P700$^+$ and photoreduced iron–sulfur proteins by exposing the EPR samples to other temperatures at which partial decay was allowed (cf. Fig. 15), and then rapidly refreezing the sample to 13°K, and the remaining P700$^+$ and reduced iron–sulfur protein measured. Figure 17, top, shows the EPR spectra of P700$^+$ and reduced iron–sulfur protein produced in a subchloroplast sample after illumination at 13°K for 20 seconds (top row) and after dark decay at 175°K for 6 minutes (bottom row). Both signals recovered 63%, as calculated from these spectra, which are in good agreement with the expected value derived from the continuous decay kinetics of P700$^+$ shown in Fig. 15.

Additional determinations were made for exposures to a number of temperatures ranging from 75 to 225°K for times ranging from 25 seconds to 10 minutes. In Fig. 17, bottom, the percentages of recovery derived from signals of the P700$^+$ and reduced iron–sulfur protein are

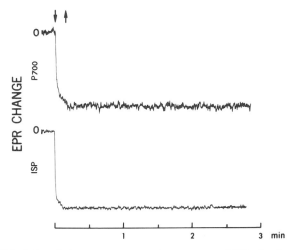

FIG. 16. Light-induced electron paramagnetic resonance (EPR) signal changes due to P700 photooxidation (top) and iron–sulfur protein photoreduction (bottom) at 13°K. From Ke *et al.* (1974b).

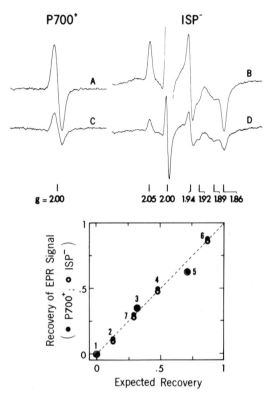

FIG. 17. *Top:* Electron paramagnetic resonance (EPR) spectra of P700⁺ (A) and reduced iron–sulfur protein (B) after the subchloroplast sample was illuminated at 13°K for 20 seconds. (C) and (D) are the corresponding spectra (recorded at 13°K) after the sample was maintained at 175°K for 6 minutes. *Bottom:* Plot of the actual fraction of decay of the EPR signal of P700 (filled circles) and reduced iron–sulfur protein (open circles) vs. the expected extent of decay. Points 1, 2, and 3 are for samples maintained for 10 minutes at 13, 75, and 125°K, respectively; points 4 and 5, 6 minutes at 150 and 175°K, respectively; point 6, 90 seconds at 225°K; and point 7, 25 seconds at 150°K. From Ke *et al.* (1974b).

plotted vs. the expected values derived from the continuously monitored decay kinetics (see Ke *et al.*, 1974b, for more details). Very similar and complementary results were also obtained by Visser *et al.* (1974) and by Bearden and Malkin (1974). Thus, the close match of the onset kinetics of the photoproduction of P700⁺ and reduced iron–sulfur proteins as well as their dark decays strongly support the concept that an iron–sulfur protein is the reaction partner of P700 in the primary photochemical act of PS I.

Additional indirect evidence for correlating P430 with an iron–sulfur protein is provided by the redox titration of the chloroplast-bound iron–sulfur proteins (Fig. 13) on the one hand, and the titration of the amplitude of light-induced P700 absorption changes (Figs. 7A, 8B, and 13; also cf. Lozier and Butler, 1974) on the other. The close agreement of the midpoint potentials of the photometric titration curves and that of the first-stage EPR titration of the bound iron–sulfur proteins of PS I and the fact that illumination of dark-adapted subchloroplast particles at low temperatures only leads to reduction of the more positive iron–sulfur protein, appear to suggest that the less negative of the iron–sulfur proteins is likely the primary electron acceptor of PS I.

D. ELECTROCHEMICAL AND KINETIC EVIDENCE FOR AN INTERMEDIARY ELECTRON ACCEPTOR

In contrast to the evidence presented above, which supports the assignment of an iron–sulfur protein (or P430) as the primary electron acceptor of PS I, other investigators, relying primarily on low-temperature EPR data, have suggested that a transient intermediate might exist between P700 and the stable primary electron acceptor. The essential experimental observation that led to this contention was the finding of a small amount of P700 photooxidation at cryogenic temperature under conditions where the iron–sulfur proteins are reduced and cannot undergo further light-induced changes.

McIntosh *et al.* (1975) reported a small reversibility (1–5%) in the P700 EPR signal but failed to find any reversibility in the signal of the bound iron–sulfur proteins. Instead, a reversible EPR signal with g values of 1.75 and 2.07 was detected using high microwave power, high gain, and large-modulation amplitude. The authors noticed a resemblance in the kinetic behavior between the new EPR signal and that of P700, and claimed that a species of unknown chemical nature, represented by this new EPR signal, is a strong candidate for the primary electron acceptor of PS I. These observations were extended by Evans *et al.* (1975), who showed that after bound ferredoxins in PS I subchloroplast particles were chemically reduced, illumination at low temperature produced a reversible photooxidation of P700 without any change in the EPR signal of the bound ferredoxins. In the same sample, a new EPR signal at $g = 1.76$ and 1.86 was seen at 9°K with 100 mW microwave power, and both the P700 signal and the $g = 1.76$ signal were found to be reversible.

Evidence presented in the previous section on the parallel decay of P700 and bound iron–sulfur protein at temperatures near 77°K and the inverse relationship in the change of the extent of light-induced P700 absorption change and the extent of chemical reduction of bound iron–sulfur proteins both inferred a direct reaction between P700 and an iron–sulfur protein in the photochemical charge separation in PS I. However, an anomaly had been noted in our redox titration experiments (see Figs. 7 and 8), namely, the persistence of a small reversible P700 signal at potentials sufficiently negative that all bound iron–sulfur proteins should have been nearly completely reduced. In view of this observation and the suggested possibility of the existence of another intermediary electron carrier, we have recently performed further redox titrations of PS I subchloroplast particles at liquid-helium temperature with the aim of clarifying this question.

New redox titrations were carried out by the electrochemical method, which allows the reductive titrations to be carried out near neutral pH, and the titrated samples were examined by EPR spectroscopy at 15°K. Interesting and significant differences in several respects were observed. The amplitude of P700$^+$ formation, monitored either by light-minus-dark difference spectra at pH 8 (Fig. 18, left), or by light-induced P700 EPR changes at pH 10 (Fig. 18, right), is relatively constant. On the other hand, as seen in Fig. 18, right, the extent of the signal decay after light was turned off was greater at lower potentials, i.e., the extent of decay was directly proportional to the extent of (electro)chemical reduction of

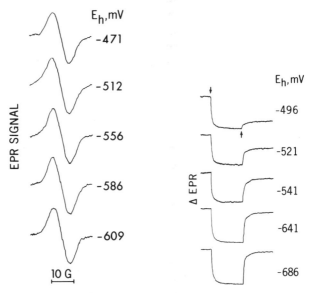

FIG. 18. Electron paramagnetic resonance (EPR) difference spectra (g = 2.0026) of light-induced P700$^+$ formation in PS I subchloroplasts titrated to various potentials at pH 8.0. The difference spectra were recorded in a Fabri-tek model 1062 signal averager by subtracting the dark signal from the light-induced signal. All signals were recorded by a single scan. *Right:* Light-induced EPR changes in PS I subchloroplasts titrated to various potentials at pH 10. The EPR changes were monitored at the high-field peak of the first derivative of P700$^+$ signal at g = 2.0026 at 15° K. Illumination time, 15 seconds; down- and upward arrows indicate light on and off. All signals were recorded with a single scan. From Ke *et al.* (1977).

the iron–sulfur proteins. The dependence of the extent of reversibility was practically the same at pH 8, 9, and 10 (Fig. 19, left). It is worth noting that not only the shapes of the curves are similar, but they occupy the same position relative to the solution potential and are independent of pH. With increasing pH, however, more and more negative potentials could be reached: -640 mV at pH 8, -700 mV at pH 9, and -740 mV at pH 10. However, the reversible fraction at these more negative potentials all remained virtually constant at about 90%.

The initial amplitude of $P700^+$ formation for pH 8, 9, and 10 is also relatively constant over a wide potential range (cf. Fig. 18), and the results are plotted in Fig. 19, right. Unlike the results in Fig. 18 for the fraction of dark decay, when the solution potential becomes more negative than -700 mV, a definite decrease in the initial amplitude of $P700^+$ signal could be noted. More electrochemical studies are needed to extend the potential range in order to ascertain the midpoint potential of the intermediate electron carrier. However, data shown in Fig. 19, right, indicates that the midpoint potential of this transition is near -730 mV.

In retrospect, it should be noted that photoreduction of bound "ferredoxin" at cryogenic temperature is a necessary, but not a sufficient, condition for assigning to bound ferredoxin the role of the true "primary" electron acceptor of PS I. Other spectral kinetic and low-

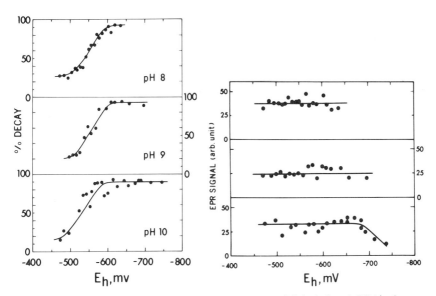

FIG. 19. *Left:* Plot of the extent of the dark decay of light-induced $P700^+$ electron paramagnetic resonance (EPR) changes at 15°K for photosystem I subchloroplasts titrated at pH 8, 9, and 10. *Right:* Plot of the initial magnitude of the light-induced EPR changes at 15°K for PS I subchloroplasts titrated at pH 8, 9, and 10. From Ke *et al.* (1977).

temperature EPR data on iron–sulfur proteins also did not entirely preclude the possible existence of another intermediary electron carrier, although the rapid risetime ($\leq 10^{-7}$ second) of formation of P700$^+$ and P430$^-$ (Ke, 1972b) appeared to make its existence unlikely at that time. These new results on the redox titration of PS I subchloroplasts monitored by EPR spectroscopy at 15°K confirmed and extended the observations of McIntosh *et al.* (1975) and Evans *et al.* (1975). Whereas a reversible P700 EPR signal had previously been observed only under one limiting condition, namely, near complete reduction of all iron–sulfur proteins, the newer results of a titration sequence provided continuity and a more extensive and quantitative view of the redox events in the PS I reaction center.

These newer results can be interpreted to indicate that, subsequent to absorption of a photon by PS I, P700 loses an electron to the intermediary electron carrier, I, which in turn transfers the electron to the iron–sulfur proteins. The EPR spectra of the latter are irreversible at 15°K after light is turned off, indicating that the iron–sulfur proteins are stable electron traps at this low temperature. When all iron–sulfur proteins are (electro)chemically reduced, they are no longer able to serve as electron traps for I, the electron from I$^-$ recombines with P700$^+$, as manifested at 15°K by an increasing extent of dark decay in the light-induced P700$^+$ EPR signal at the more negative potentials (Fig. 18, right). Furthermore, this picture would also predict that when sufficiently negative potentials are imposed on PS I, the intermediary acceptor would eventually also be reduced, and the extent of P700$^+$ formation would become attenuated. Data in Fig. 19, right, for the titration at pH 10 are consistent with this notion; the initial amplitude due to P700$^+$ formation began to decrease beyond -700 mV. From the available data, a midpoint potential near -730 mV has been estimated for the transient intermediary electron acceptor, I. Such a situation had presumably been observed previously under a limiting condition, namely, when PS I is illuminated in the presence of a strong reductant and then frozen while the sample is being continually illuminated (Lozier and Butler, 1974; Ke *et al.*, 1976b).

At the moment, the chemical nature of I is completely unknown. The light-induced EPR signal with *g* values of 1.88 and 1.76 (McIntosh *et al.*, 1975; Evans *et al.*, 1975) has been claimed to represent I. However, further clarification and characterization of this signal are needed. Blankenship *et al.* (1975) recently observed a transient EPR emission signal when chloroplasts were flash-excited at room temperature. The authors suggested that the emission signal may represent the counterradical produced by P700 photooxidation in PS I.

Even though little is known about I, available electrochemical and kinetic evidence indicates that it is probably a transient intermediate, whereas the lifetimes of P430 (or iron–sulfur protein) may extend from 10^{-4} to 10^{-2} second depending on the secondary electron acceptor with which it is interacting. In analogy with the current picture of the primary photochemistry in photosynthetic bacteria presented in the following sections, the designation "primary electron acceptor" to P430 or the iron–sulfur protein is deemed appropriate.

Although available evidence is consistent with the picture that I is an intermediary electron carrier between P700 and the stable primary acceptor, P430, it may be premature to preclude the possibility that it may simply be an alternate electron acceptor not linked to P430 or the iron–sulfur protein. Further studies of this problem by optical spectroscopic techniques are needed.

Other unanswered questions also remain. One is the discrepancy between the data obtained by optical and EPR methods near liquid-nitrogen temperatures (Figs. 7A and 8A) and those by EPR spectroscopy at 15°K (Fig. 19). It is reasonable to attribute this difference to the difference in temperatures employed in the two sets of experiments. It is expected that the rapid decay observed at 15°K would be much faster at liquid-nitrogen temperatures. Under such circumstances, the rapidly decaying portion in the absorption-change signals may not be observed simply because of an inadequate time resolution of the measuring instrument. Taking this view, the attenuation of the total signal amplitude at liquid-nitrogen temperature may then be considered as reflecting the decrease in the irreversible portion observed at 15°K. Rapid-kinetic studies by optical methods may also throw some light on this question.

III. The Primary Electron Donor of Photosynthetic Bacteria

The state of knowledge on the primary photochemical reaction in bacterial photosynthesis up to 1974 has been presented in an excellent review by Parson and Cogdell (1975). In the intervening period, tremendous progress has been made in this field, particularly on the primary electron acceptor and the newly discovered intermediate acceptor, which is thought to act between the primary donor and the stable primary electron acceptor. Recent advances in bacterial photosynthesis have very much been facilitated by the availability of the purified reaction-center preparations and by the application of the rapid (picosecond) kinetic spectroscopy (see the chapter by Seibert in this volume). Because of space limitation and also because excellent reviews on previous developments are readily available, this section will deal only with those developments evolved more recently on the subjects of the primary electron donor and the transient and stable primary electron acceptors in photosynthetic bacterial reaction centers.

A. THE PHOTOSYNTHETIC REACTION CENTER: ISOLATION AND PROPERTIES

The light-induced absorption change associated with the photooxidation of the reaction-center (RC) bacteriochlorophyll was first reported by Duysens in 1954. Clayton (1963) isolated the photochemical RC from pheophytinized chromatophores of *Rhodopseudomonas spheroides* as a spectrophotometric entity. Reed and Clayton (1968) first succeeded in isolating a purified RC preparation from a carotenoidless mutant (strain R-26) of *R. spheroides* by Triton treatment and fractionation. Subsequent refinements of the fractionation procedure (Clayton and Wang, 1971; Feher, 1971; Reed and Peters, 1972) have led to RC preparations of high purity and well-defined composition. At this writing, RC preparations have been isolated from all bacteriochlorophyll-*a*-containing bacteria, including, more recently, *Chromatium* (Lin and Thornber, 1975), and the bacteriochlorophyll *b*-containing bacterium, *Rhodopseudomonas viridis* (Trosper et al., 1977). The RC preparations from the latter two bacteria retain the membrane-bound *c*-type cytochromes. The availability of these RC preparations have already helped greatly in recent advances in our knowledge about the primary photochemistry of bacterial photosynthesis and, undoubtedly, will continue to be of value in future studies.

The RC complex contains four bacteriochlorophyll and two bacteriopheophytin molecules (Reed and Peters, 1972; Straley et al., 1973). Photooxidation of P870 takes place in less than 10^{-11} second (Netzel et al., 1973) with a quantum efficiency near 100% (Wraight and Clayton, 1974). The midpoint potential of the RC bacteriochlorophyll P870/P870$^+$ is +0.45 V in *R. spheroides* (Dutton and Jackson, 1972) and +0.49 V in *Chromatium* (Cusanovich et al., 1968; Dutton, 1971). By coupling mammalian ferrocytochrome *c* to the photooxidized P870$^+$, the differential molar extinction coefficient due to P870 bleaching in the *R. spheroides* reaction center was found to be 112 mM^{-1}·cm^{-1} at 865 nm (Straley et al., 1973; cf. Ke et al., 1970).

The photooxidation of RC bacteriochlorophyll produces a free-radical EPR signal at g = 2.0025. The linewidth of the EPR signal suggests that the bacteriochlorophylls are present as a dimer and that the unpaired electron in the oxidized form is delocalized among both chlorophyll molecules (Norris et al., 1971). The dimer model is also supported by the ENDOR spectrum (Norris et al., 1974; Feher et al., 1975). The CD data (Reed and Ke, 1973; Phillipson and Sauer, 1973) provide evidence that there is a strong exciton interaction between the pigment molecules. Photooxidation of the RC chlorophyll produces a large change in the interaction of the pigments, as evidenced by changes in the CD spectrum.

B. Spectral Properties

Figure 20 shows the absorption spectrum of the reduced and oxidized forms of the RC particle from the mutant *R. spheroides* (top) and their difference spectrum (bottom). The absorption bands near 600, 803, and 865 nm belong to the bacteriochlorophyll; the 532 and 757 nm bands belong to the bacteriopheophytin; and the unresolved band at 365 nm belongs to both. Upon photooxidation, the absorption band at 865 nm is mostly bleached, and the 800 nm and 375 nm bands undergo a blue shift. The 600 nm band is partially bleached, and the remaining absorption band is shifted toward the blue. The bacteriopheophytin bands in the oxidized reaction center are shifted toward the red. Upon oxidation, new absorption bands develop at 430 and 1250 nm.

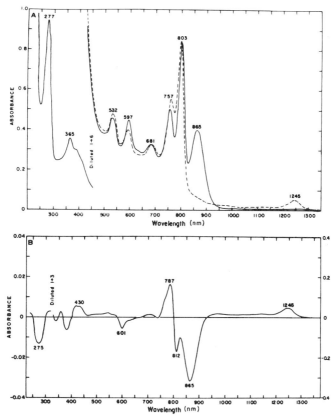

Fig. 20. Purified reaction-center complex from *Rhodopseudomonas spheroides*, R-26 mutant. (A) Absorption spectra taken in the dark (solid) and after illumination (dashed curves). (B) Light-minus-dark difference spectrum between the illuminated and unilluminated sample. From Reed (1969).

More recently, Vermeglio and Clayton (1977) obtained evidence from linear-dichroism studies that are at variance with the current interpretation of the light-induced absorption change near 800 nm as due to a blue shift of a single band. The authors used chromatophores from *R. spheroides* and treated them with K_2IrCl_6 to bleach the absorption bands of the antenna bacteriochlorophyll completely but allow the reaction centers to remain intact. The chromatophores were oriented onto microscope slides so that the photosynthetic membranes lie mainly parallel to the slide but with random angular orientations in the plane. The linear-dichroism spectrum showed strong polarizations in all the long-wavelength absorption bands. The linear dichroism measured with the K_2IrCl_6-treated chromatophores is very similar to the spectrum reported by Penna *et al.* (1974) obtained from oriented RC particles. However, the RC spectrum gave a much smaller dichroic ratio, presumably the larger chromatophores orient much better than the smaller RC particles. The linear dichroism results suggested that the long-wavelength transition moments of the RC bacteriochlorophyll are nearly parallel to the plane of the membrane, whereas the long-wavelength transition moments of the bacteriopheophytin molecules are oriented out of plane.

Figure 21 shows the light-minus-dark difference spectrum measured separately using vertically and horizontally polarized measuring beams. The difference spectrum shows typical changes associated with photoox-

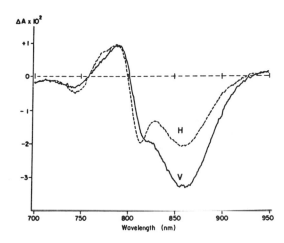

FIG. 21. Light-minus-dark difference spectra of a dried film of chloroiridate-treated chromatophores derived from *Rhodopseudomonas spheroides*. Solid curve: measuring light was vertically polarized; dashed curve: horizontal polarization. The intensity of the excitation light was at 50% saturation. From Vermeglio and Clayton (1977).

idation of the RC bacteriochlorophyll. However, the relative amplitudes of these changes depend on the polarization of the measuring beam. The changes at 860 and 785 nm are greater for the vertically polarized measuring beam, whereas the changes at 812, 770, and 745 nm are greater for the horizontally polarized beam. The zero crossing near 800 nm is slightly different for the two polarizations. Measurements of absorption changes at individual wavelengths showed all changes to have the same kinetics regardless of the polarization. Absorption changes at 600 and 1250 nm also showed strong dichroic effect. Untreated chromatophores yielded similar linear dichroism in the absorption spectrum and light-induced absorption changes, indicating that the bleaching of the antenna chlorophylls by K_2IrCl_6 did not alter the orientation of the reaction centers with respect to the membrane.

The opposite polarizations at 810 and 790 nm appear to be inconsistent with the notion of a spectral shift of a single absorption band at 803 nm, as equal dichroic values would then be expected. Vermeglio and Clayton proposed an alternative interpretation, namely, that the absorption decreases at 870 and 810 nm are due to bleaching of two absorption bands of the bacteriochlorophyll dimer, the primary electron donor. The absorption increase at 790 nm was ascribed to an appearance of a new absorption band due to the singly oxidized dimer $(BChl)_2^+$, a spectral property closer to that of monomeric bacteriochlorophyll.

Vermeglio and Clayton pointed out that the molecular exciton model (Kasha, 1963) predicts that two absorption bands can result in the neutral dimer $(BChl)_2$ as a result of splitting of excited-state energy levels in place of one for the monomer. Dratz *et al.* (1967) reported earlier that dimeric chlorophyll *in vitro* shows a double CD spectrum which reverses sign near the center of the absorption band. This behavior is due to the degenerate exciton interaction of the long-wavelength transition moments of the two monomers in the dimer (Tinoco, 1963). The magnitude and sign of the degenerate CD component is related to the geometry of the dimer. The CD spectra measured with *R. spheroides* RCs (Sauer *et al.*, 1968; Reed and Ke, 1973) are consistent with the notion of two transition moments at 865 and 810 nm for the dimer complex, and that exciton interaction disappears upon photooxidation.

The 810 nm absorption band proposed by Vermeglio and Clayton calls to mind a 810-nm peak reported by Feher (1971) in the second-derivative absorption spectrum measured in the reduced form of the RC from *R. spheroides* at 77°K, which was absent in the oxidized RC. As shown in Fig. 22, Vermeglio and Clayton also demonstrated the presence of an 810-nm band in the reduced form of the *R. spheroides* RC by measuring the first-derivative spectrum at 35°K; it disappears upon photooxidation.

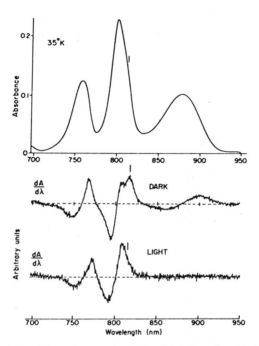

FIG. 22. Upper trace: Absorption spectrum of a dried film of purified reaction centers derived from *Rhodopseudomonas spheroides*. Middle trace: first-derivative spectrum recorded with a weak measuring light. Lower trace: First-derivative spectrum recorded under strong illumination. The vertical bars indicate the 810 nm component. Temperature, 35°K. From Vermeglio and Clayton (1977).

IV. The Primary Electron Acceptor of Photosynthetic Bacteria

The light-induced absorption change in the ultraviolet region had been attributed to the photoreduction of the primary electron acceptor, a ubiquinone (UQ), in photosynthetic bacteria (Clayton, 1963; Bales and Vernon, 1962; Ke *et al.*, 1968). In the early 1970s, a number of studies on the extraction of UQs from bacterial chromatophores cast some doubt on the role of UQ as the primary acceptor, as removal of UQs by extraction did not impair the photooxidation of the primary electron donor, P870 (Takamiya and Takamiya, 1970; Nöel *et al.*, 1972; Ke *et al.*, 1973a; Halsey and Parson, 1974).

A. RECENT DEVELOPMENTS TOWARD AN IRON–UBIQUINONE COMPLEX

In the meantime, studies on the possible involvement of UQ continued; these studies may be grouped into those using the EPR technique and those using optical spectroscopy for chemical identification. Illumination of *R. spheroides* RC particles yielded a difference spectrum with a prominent decrease at 270 nm and increases at 320 and 450 nm (Clayton and Straley, 1970, 1972; Slooten, 1972). These difference spectra bear a close resemblance to that of ubisemiquinone-10 anion formed by radiolysis of the quinone in alkaline methanol (Bensasson and Land, 1973). Additional evidence for ubisemiquinone formation came from the EPR studies with illuminated subchromatophore preparations from iron-deficient *Rhodospirillum rubrum* (Loach and Hall, 1972) or RC particles from *Rhodopseudomonas spheroides* which had been treated with SDS to deplete iron (Feher *et al.*, 1972). Both EPR spectra had *g* values and linewidths appropriate for a semiquinone radical. Bolton and Cost (1973) measured the EPR spectrum of untreated *R. spheroides* RCs at room temperature and attributed it to a ubisemiquinone radical. By extracting the RCs with organic solvent or by blocking the reaction pathway with *o*-phenanthroline, the kinetic behavior of the signal was found to be consistent with the suggestion that UQ was the primary acceptor.

All three groups suggested that a complex of UQ and iron may be the primary acceptor. Subsequently, Feher *et al.* (1974) reported that when iron is partially replaced by manganese in the RCs of *R. spheroides*, their photochemical property was unaffected. They also reported that Mössbauer studies showed that the iron atoms were present in the valence state of 2 regardless of the redox state of the RC. These two findings appear to suggest that iron alone cannot be the primary acceptor.

At the same time, Leigh and Dutton (1972) observed in *R. spheroides* RCs an EPR signal at *g* of 1.68, 1.82, and 2.00, which suggested an iron compound to be the primary acceptor. The kinetic behavior of this EPR signal was also consistent with that of the primary electron acceptor. For instance, when cytochrome oxidation was prevented, the photoinduced Fe-EPR signal decayed with the same kinetics as P870$^+$, as expected for recombination between P870$^+$ and the reduced primary electron acceptor. When cytochrome oxidation occurred, the Fe-EPR signal became irreversible. When the reaction center was reduced with dithionite, whereby the primary electron acceptor was presumably reduced chemically, the light-induced EPR signal of *g* = 1.68 and 1.82 was not observed (Dutton *et al.*, 1973a). Instead a new EPR signal comprised of both absorption and emission bands attributable to the bacteriochlorophyll triplet state was produced upon illumination.

Further clarification of the role of UQ as the primary electron acceptor as well as the secondary acceptor has been reported by Cogdell *et al.* (1974), who selectively extracted the more tightly and less tightly bound UQ molecules. When the *R. spheroides* RC preparation was extracted with isooctane, the decay kinetics of the photoinduced P870$^+$ absorption-change signal became much faster than in the unextracted preparation. Presumably the rapid decay indicates that the secondary-acceptor UQ was removed by extraction, and P870$^+$ recombines with the reduced primary electron acceptor directly. When the RC preparation was further extracted with isooctane containing 0.1–0.3% methanol, the photochemical activity was completely eliminated, presumably the more tightly bound UQ that serves as the primary acceptor was completely removed. The photochemical activity of the extracted particle could be restored by recondensing the UQ extract unto the extracted particle. The reconstituted particles had almost the same activity as the native ones.

Subsequently, Okamura *et al.* (1975) developed a method for removing and re-adding UQ to the RC preparation without the use of organic solvents. These studies have yielded the most definitive identification of UQ as the primary electron acceptor and also achieved an almost perfect reconstitution.

Removal of UQ was carried out by incubating the RC particles in the presence of the detergent laurel dimethyl amine oxide (LDAO) and *o*-phenanthroline. The extent of UQ removal depends on the relative concentrations of the RC protein, the detergent and *o*-phenanthroline, as well as incubation time. It was found that one UQ per RC (the particle contained 2 UQ molecules per RC initially) was readily removed, say, at an RC concentration corresponding to an optical density of 2 at 800 nm, in the presence of 4% LDAO and 10^{-3} M *o*-phenanthroline. The less tightly bound UQ probably represents the secondary electron acceptor. The ease of removal of the first UQ in the presence of *o*-phenanthroline is consistent with its inhibitory effect on electron transport between the primary and secondary electron acceptors (Parson and Case, 1970). One other possible explanation for the effect of *o*-phenanthroline in UQ removal is its ability to chelate Fe^{2+}, thereby displacing the UQ molecule. This suggests that both the primary- and secondary-acceptor UQ molecules are in close proximity of Fe.

Removal of the second UQ molecule requires a more rigorous condition, say, at a RC concentration corresponding to an OD of only 0.2, in the presence of 4% LDAO and 10^{-2} M *o*-phenanthroline. RC particles extracted twice under these conditions retain about 0.05 UQ/RC and 1 Fe/RC (designated as "UQ-depleted" particles).

Reconstitution of UQ-depleted particles were studied by incubating in 1% LADO and ^{14}C-labeled UQ for 10 hours at 4°C. Incorporation of one UQ/RC requires only a stoichiometric amount of added UQ in the incubation medium, whereas incorporation of two molecules requires 5–10 times excess of added UQ.

The photochemical activity of the unextracted, UQ-depleted, and the reconstituted RC particles were examined by measuring the light-induced absorption change associated with the photooxidation of P870 and the formation of EPR signals corresponding to the photooxidized primary donor, P870$^+$ (g = 2.0026) and the photoreduced primary acceptor, X$^-$ (g = 1.8). As shown in Fig. 23, the UQ-depleted RC particles gave no light-induced absorption change or EPR signals. In the UQ-depleted particles, a bacteriochlorophyll triplet signal was observed instead (cf. Dutton *et al.*, 1973a,b). In the reconstituted particles, illumination produced identical absorption changes and EPR signals to those in the unextracted particles.

Quantitative measurements of the low-temperature photochemical activity as a function of the amount of UQ removed from the native particles or the amount of UQ re-added to the UQ-depleted particles showed that full activity was maintained as long as one UQ still remains in one RC (Fig. 24, top), whereas full activity was restored when only one UQ was added back to each RC of the UQ-depleted particles (Fig. 24, bottom).

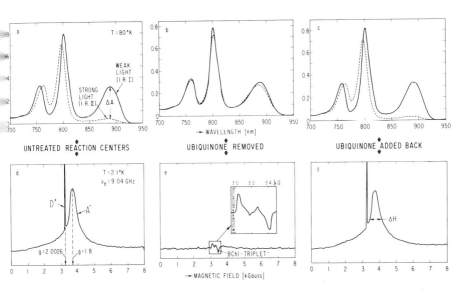

FIG. 23. Light-induced absorption changes (upper row) and electron paramagnetic resonance (EPR) signals (lower row) of reaction centers of *Rhodopseudomonas spheroides* R-26 with (first column) and without (second column) ubiquinone. Untreated reaction centers (d) and those to which ubiquinone was added back after extraction (f) show a narrow signal at g = 2.0026 due to P870$^+$ and a broad g = 1.8 signal due to the reduced primary electron acceptor. Reaction centers from which virtually all the ubiquinone had been extracted show negligible bleaching (b) and a small EPR signal (e; also see inset) attributed to the bacteriochlorophyll triplet. From Okamura *et al.* (1975).

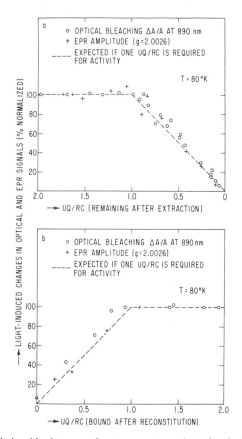

FIG. 24. The relationship between low-temperature photochemical activity and the number of ubiquinones (UQ) per reaction center (RC). (a) Obtained with reaction centers exposed to varying concentrations of LDAO and *o*-phenanthroline. (b) ^{14}C-labeled ubiquinone was added in varying amounts to reaction centers that had less than 0.05 ubiquinone per reaction center. The Fe content (1 Fe/RC) remained unaffected by extraction. From Okamura *et al.* (1975).

The quantitative correlation between the presence and the absence of UQ and the appearance and disappearance of low-temperature absorption changes and EPR signals indicates that in *R. spheroides* UQ plays an obligatory role in the primary photochemical reaction. These results also provide a reconciliation of the earlier extraction studies; most likely, previous extraction conditions were not adequate to remove the tightly bound UQ. In fact, the question of the primary acceptor in *Chromatium* was reexamined, applying the new extraction method to both chromatophore and RC preparations (Okamura *et al.*, 1976). It was found that extracted preparations that still retained photochemical activity at cryogenic temperatures contained one menaquinone per reaction center, but less than 0.1 UQ per RC. The primary electron acceptor in *Chromatium* presumably consists of a menaquinone–iron complex. These photochemically active preparations also exhibit the broad $g = 1.8$ EPR signals similar to that found in *R. spheroides*.

The decay rate of the EPR signals at 80°K at both $g = 2.0026$ and 1.8 were found to be identical. The unextracted and reconstituted RCs had the same decay time of 28 msec. These results are consistent with the recombination time between the photooxidized primary donor, P870$^+$, with the photoreduced primary acceptor. Since the linewidth of the EPR signal and its decay time are characteristic features of the environment of the primary donor and acceptor, the quantitative reconstitution indicates that UQ extraction did not alter the conformation of the active site.

It was also found that UQ could be replaced with a number of other quinones in the reconstitution of the UQ-depleted preparations. However, the linewidth and the decay rate of the EPR signal were different. The decay rate constant is considered to be a sensitive parameter dependent on the shape and height of the potential barrier and the redox (ionization) potential of the primary electron carriers, in relation to the electron-tunneling process. The fact that replacement of UQ by other quinones affects the signal linewidth and decay time, whereas replacement of iron by manganese does not, provides an argument in favor of assigning the major role to UQ in the primary photochemical reaction. Currently available evidence suggests that iron may serve as an electron-transport pathway linking the primary and secondary UQ molecules.

B. OXIDATION-REDUCTION POTENTIAL OF THE PRIMARY ELECTRON ACCEPTOR

In principle, the redox potential of the primary electron acceptor of a photosynthetic bacterium may be determined by chemical reduction and monitoring an appropriate signal representative of the reduced acceptor as a function of the extent of reduction. It may also be determined by monitoring the loss of photochemical activity as a function of the extent of reduction of the primary acceptor. These two titration curves should bear an inverse relationship. Because of the lack of knowledge on the identity of the primary electron acceptor until recently, most determinations have been carried out using the indirect method. However, Dutton *et al.* (1973b) performed a direct titration by monitoring the appearance of reduced iron–ubiquinone complex (Fe·UQ) with an EPR spectrum at $g = 1.82$. Redox-titration data for the primary electron acceptor of seven photosynthetic bacteria have been reviewed by Parson and Cogdell (1975), and all data appearing between 1966 and 1974 were summarized.

The midpoint potential values for the seven photosynthetic bacteria range from -20 to -160 mV, mostly measured in the pH range 7–8. In most cases where the pH effect was reported, a $\Delta E/\Delta$pH dependence of -60 mV was found, the RC preparations being exceptional.

Prince and Dutton (1976) recently measured the redox potential of the primary electron acceptor over a wide pH range (4.7 to 11.0). The midpoint potential of the primary electron acceptor was measured indirectly by monitoring the change in the photochemical activity of the chromatophores of *Rhodopseudomonas spheroides, Chromatium vinosum*, and *Rhodospirillum rubrum*. The photochemical activity was assayed by measuring the light-induced absorption changes associated with the photooxidation of P870 (and in the case of *R. spheroides* also the photooxidation of cytochrome c_{552} or the carotenoid band shift). The midpoint potentials for four species plotted vs. pH are reproduced in Fig. 25. All plots also included data reported previously by other workers. The plots show that the redox potential of the primary electron acceptor at physiological pH involves one electron and one proton ($\Delta E/\Delta$pH $= -60$ mV). At higher pH, a break is seen in the plot, which represents the apparent pK of the reduced form of the primary acceptor. Above the pK, the redox potential was essentially independent of pH. The good agreement in the composite plot of data from such diverse sources indicates an excellent consistency of the results. Interestingly, the redox potential and the pK values of all four organisms lie in a relatively narrow range of -175 ± 25 mV and 9 ± 1, respectively.

o-Phenanthroline is known to be an inhibitor for the electron transport between the primary and secondary acceptors. It has also been reported that the presence of o-phenanthroline shifts the equilibrium redox potential of the primary acceptor toward a more positive value (Jackson et al., 1973; Dutton et al., 1973c). As shown in Fig. 25, the redox reaction still involves one electron and one proton at physiological pH despite the shift, and that although the pK is shifted toward a higher pH value, the midpoint potential of the primary acceptor above the pK was unchanged.

The results in Fig. 25 indicate that at physiological pH, the equilibrium reduction of the primary acceptor involves both an electron and a proton (X/XH), whereas above the pK, the reduction involves only an electron (X/X$^-$). Prince and Dutton (1976) pointed out that the kinetically important redox couple is X/X$^-$ instead of X/XH at equilibrium, which suggests that the kinetically operational midpoint potential is the equilibrium value measured above the pK of the reduced form of the primary acceptor. The more reducing midpoint potential would enhance the likelihood of the primary acceptor being able to reduce NAD$^+$ and other low-potential dyes under some nonphysiological conditions as occasion-

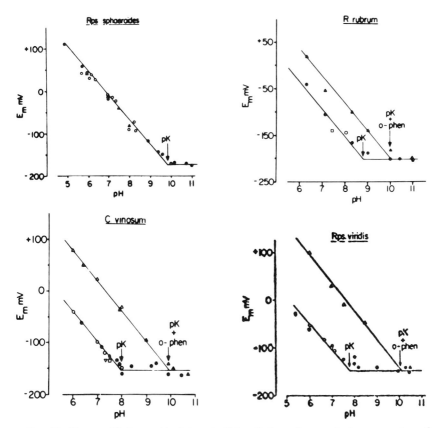

FIG. 25. The equilibrium midpoint potentials of the primary electron acceptor of *Rhodopseudomonas spheroides, Chromatium vinosum, Rhodospirillum rubrum,* and *Rhodopseudomonas viridis* as a function of pH. Each of the plots consists of original data as well as data points reported previously by other workers. From Prince and Dutton (1976) and Prince *et al.* (1976); consult these references for details.

ally claimed in the literature. These results also readily reconcile some anomalies previously reported on the pH-dependence of the redox potential in *Chromatium* (Case and Parson, 1971) and *Rhodopseudomonas viridis* (Cogdell and Crofts, 1972).

Although much has been studied on the primary electron acceptor of purple sulfur and nonsulfur bacteria, relatively little is known about the primary electron acceptor in the green bacteria. Probably this has been due to a lack of purified RC preparations from this bacterium. The currently available bacteriochlorophyll RC complex contains 80–100 bacteriochlorophyll molecules per RC. Nevertheless, Prince and Olson (1976) recently titrated the primary electron acceptor in a bacteriochlorophyll RC complex of *Chlorobium limicola* f. *thiosulfatophilum*, strain Tassajara, by measuring flash-induced oxidation of cytochrome c_{553} in the sample poised at low potentials. It was found that neither the magnitude nor the kinetics of cytochrome c_{553} photooxidation was affected down to -450 mV, the lowest potential that could be obtained at pH 6.8. Subsequent titrations at higher pH extended the potential range, and the midpoint potential of the primary electron-acceptor was estimated to be -540 mV.

In the meantime, Knaff and Malkin (1976) carried out a characterization of the membrane-bound iron–sulfur proteins in *Chlorobium* by redox titration monitored by EPR spectroscopy. Four membrane-bound iron–sulfur proteins were detected. One is a "Rieske" type iron–sulfur protein with a g value of 1.90 and a midpoint potential of $+160$ mV at pH 7. Three iron–sulfur proteins have g value of 1.94 and midpoint potentials of -25, -175, and -550 mV. Because of the similarity in the midpoint potential of the latter protein and the value obtained from cytochrome photooxidation, it was suggested that the latter protein could function as the primary electron acceptor in *Chlorobium*. These findings are consistent with the previous demonstration of a direct ferredoxin-dependent photoreduction of NAD by this organism (Evans and Buchanan, 1965). It is also of interest to note that the redox potential of the primary electron acceptor in the green bacterium *Chlorobium* is nearly of the same value as the primary electron acceptor in PS I of green plants.

V. The Intermediary Electron Acceptor and Picosecond Spectroscopy

The most important development during the past two years in bacterial photosynthesis is probably the discovery of an intermediary electron acceptor in the primary photochemical reaction. Picosecond spectroscopy has been an indispensable tool for making these discoveries (see the Seibert article in this volume, devoted entirely to this subject). As mentioned earlier, Dutton *et al.* (1973a) reported that flash excitation of reaction centers in which the primary electron acceptor has been prereduced generates an EPR signal of the bacteriochlorophyll triplet. By measuring flash-induced absorption changes under similar conditions, Parson *et al.* (1975) observed the formation of two different transient states in RCs poised at low potentials. The two states, designated as P^F and P^R, differ spectrally and kinetically. The difference spectrum of the P^F state resembles that of bacteriochlorophyll triplet *in vitro*, but the spectrum also consists of additional changes in the BPh absorption region.

A. INVOLVEMENT OF BACTERIOPHEOPHYTIN

In the meantime, it was found that *Rhodopseudomonas spheroides* reaction centers excited with picosecond flashes produced a transient state whose spectrum is similar to that of the P^F state detected in reaction centers where the primary acceptor was prereduced (Kaufmann *et al.*, 1975, 1976; Rockely *et al.*, 1975). These results provide additional evidence that both BChl and BPh play a role in the excited state. The transient state relaxes in 100–200 psec. This decay time has been interpreted as the reduction time of the primary electron acceptor, X or Fe·UQ. Dutton *et al.* (1975) used picosecond excitation and monitored the absorption change at 1250 nm, the wavelength that is unique for the oxidized state of $(BChl)_2^{·+}$, and found that the oxidized dimer was formed in $\leq$ 10 psec. Neither prior reduction of X(=Fe·UQ) nor extraction of UQ affects this risetime. These results indicate that $(BChl)_2^{·+}$ is an integral part of the transient state.

Based on information derived from combined electrochemical and spectral studies of bacteriochlorophyll and bacteriopheophytin in solution, Fajer *et al.* (1975) were able to equate the absorption change of the transient state P^F with the primary photochemical reaction involving P870 photooxidation and the reduction of the transient electron acceptor which is bacteriopheophytin. Figure 26. presents the experimentally

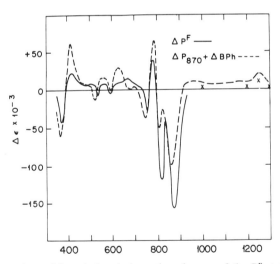

FIG. 26. Comparison of laser-induced absorption changes of the P^F state (solid curve and x), with changes calculated (– – –) on the assumption $\Delta P^F = (P870^+ - P870) + (BPh^- - BPh)$, i.e., changes caused by photooxidation of P870 and photoreduction of BPh. From Fajer *et al.* (1975).

measured absorption change of the P^F state and the absorption changes near 1250 nm measured by Dutton *et al.* (1975), compared with the calculated sum of two difference spectra, one for the photooxidation of P870, i.e., (P870$^+$ − P870), and the other for the photoreduction of a BPh to an anion, i.e., (BPh$^-$ − BPh). There is a good agreement between practically all the maxima and minima in the experimental and calculated spectra over the entire 350–1300 nm region. The experimentally measured absorption decrease at 760 nm and the broad absorption increase between 600 and 680 nm also agree with that observed earlier by Krasnovsky and Vojnovskaya (1951) in the photoreduction of bacteriopheophytin by sodium sulfide in pyridine.

Now that the individual components involved in the primary photochemical reaction in the bacterial reaction center have been identified, the remainder of this section may more conveniently continue the discussion by using the abbreviated formula for the reaction center, $C[B_2 \cdot I]X$, where X stands for the primary electron acceptor, or the iron-ubiquinone complex (Fe·UQ, ferroquinone); I stands for the intermediary electron acceptor, a bacteriopheophytin, BPh; B_2 for the bacterio-chlorophyll dimer $(BChl)_2$; and C for a functional c-type cytochrome, such as cytochrome c_{552} in *Chromatium* chromatophores, present either in the reduced state, Cyt^{2+}, or the oxidized state, Cyt^{3+}.

Thus, the *Rhodopseudomonas spheroides* RC may be represented simply by $[B_2 \cdot I]X$, the biradical state $[B_2 \cdot^+ \cdot I^-]X$ can be formed in ≤ 10 psec. In the absence of the secondary donor Cyt^{2+}, or when the cytochrome is preoxidized chemically, the primary electron acceptor becomes reduced to $[B_2 \cdot^+ \cdot I]X^-$ in 100–200 psec at 295°K. The return of the electron from X^- to $B_2 \cdot^+$ takes ~20 msec.

Okamura *et al.* (1975) showed that removal of UQ from the reaction center also results in the formation of BChl triplet (cf. Fig. 23). By examination of the effect of UQ extraction and reconstitution by picosecond spectroscopy, Kaufmann *et al.* (1976) showed that, in the extracted RC preparations, the $[B_2 \cdot^+ \cdot I^-]$ state is formed in ≤ 10 psec and decays in more than 1 nsec. Readditions of UQ-10 reconstitute the rapid electron transport rate between I^- and X at 100–200 psec.

B. Trapping of the Intermediary Electron Acceptor in the Reduced State; Spectral and EPR Properties

Two groups of workers have recently investigated the trapping of the intermediary electron acceptor in the reduced state and measured its spectral and EPR properties. Both groups exploited the properties of *Chromatium*, in which two photochemically active cytochromes remain intimately associated with the RC of the isolated chromatophores or RC preparations. One of the two cytochromes, the low-potential cytochrome c_{552} is capable of donating an electron irreversibly to the photooxidized P870 even down to liquid helium temperatures (DeVault and Chance, 1966).

Thus, starting with the RC containing a functional cytochrome and poised at a highly reducing potential to maintain the primary electron acceptor as well as the cytochrome in the reduced state, $C^{2+}[B_2 \cdot I]X^-$, the transient state $C^{2+}[B_2^{\cdot +} \cdot I^-]X^-$ formed in $\leqq 10$ psec has two alternative routes for relaxation: the recombination between $B_2^{\cdot +}$ and I^-, which takes 10 nsec, or the donation of an electron from C^{2+} to $B_2^{\cdot +}$ in 1 μsec. Evidently, under a highly reducing condition (-250 to -530 mV), which can maintain the cytochrome in the reduced state, the state $C^{2+}[B_2 \cdot I^-]X^-$ may be isolated spectrally as shown in Fig. 27, top (Shuvalov et al., 1976b; Shuvalov and Klimov, 1976). The intermediate electron acceptor may also be trapped in the reduced state by steady illumination of the *Chromatium vinosum* reaction center poised at -440 mV at a lower temperature (200°K). Under this condition, the photooxidized cytochrome remains trapped in the oxidized state, $C^{3+}[B_2 \cdot I^-]X^-$. A spectral

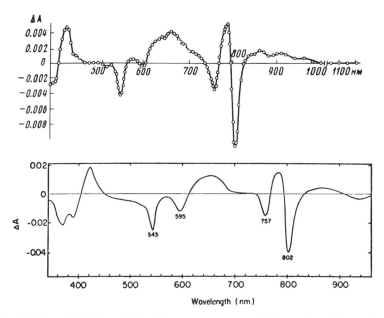

FIG. 27. *Top:* Light-minus-dark difference spectrum of *Chromatium (minutissimum)* reaction centers poised at -450 mV and measured at room temperature. From Shuvalov *et al.* (1976b). *Bottom:* Light-minus-dark difference spectrum of *Chromatium (vinosum)* reaction centers poised at -400 mV and measured at 200°K. Between 360 and 660 nm, the difference spectrum due to cytochrome c_{552} oxidation was computer-subtracted. From Tiede *et al.* (1976).

isolation of I^-, i.e., $C^{3+}[B_2 \cdot I^-]X^-$ minus $C^{2+}[B_2 \cdot I]X^-$, would require an additional subtraction of the separately measured difference spectrum due to cytochrome oxidation. By ingenious instrumental manipulations, a net difference spectrum of the reduced intermediary electron acceptor was obtained (Tiede et al., 1976), as shown in Fig. 27, bottom. Note that the two difference spectra, though obtained under different experimental conditions, are in excellent agreement.

Shuvalov and Klimov (1976) also demonstrated trapping of reduced BPh and oxidized cytochrome in a film of subchromatophores dried in the presence of dithionite and o-phenanthroline. A stoichiometric relationship was found for the BPh and cytochrome changes. The kinetics of BPh reduction corresponds to that of cytochrome oxidation, and the kinetics of BPh^- reoxidation corresponds to that of cytochrome reduction at 293°K. The reaction is largely irreversible at 77°K.

The difference spectra reveal many features which are qualitatively similar to the absorption change ascribed to the in vitro bacteriopheophytin photoreduction (Fajer et al., 1975; Krasnovsky and Vijnovskaya, 1951): the double peak at 360 and 390 nm, the broad increase at 430 and 650 nm, and the bleaching at 543 and 760 nm. The spectra show no changes attributable to P890 photooxidation at 883 or 605 nm. However, a blue shift near 800 nm and its companion change at 595 nm are apparent. As the latter changes cannot be accounted for by bacteriopheophytin alone, changes by a bacteriochlorophyll molecule must be responsible (Shuvalov and Klimov, 1976; Tiede et al., 1976). This suggests that the 800 nm absorbing bacteriochlorophyll is most likely a part of the intermediary acceptor I^-. The spectral change may have been brought about by an electrochromic shift or some mechanism of electron sharing between the bacteriopheophytin and the 800-nm absorbing bacteriochlorophyll.

EPR spectra were also measured for the trapped I^-. Shuvalov and Klimov (1976) found a light-induced EPR signal for I^- trapped at room temperature in Chromatium subchromatophores poised at -430 mV. The EPR signal with g at 2.0025 and linewidth of 12.5 Gauss (G) was attributed to the pigment radical anion. The kinetics of the light-induced EPR signal corresponds to the optical change for I^- formation monitored at 680 nm. The linewidth was interpreted to indicate a shift of electron localization to a monomeric pigment, probably the bacteriopheophytin.

Tiede et al. (1976) measured an EPR spectrum in the Chromatium RC complex treated to generate I^- in the way used for obtaining the difference spectrum (cf. Fig. 27, bottom). Signal A in Fig. 28 was obtained at -400 mV, where both cytochrome c_{552} and X were reduced. This light-minus-dark difference EPR spectrum is apparently composed of a free-radical signal at $g = 2.0023$ but split by 60 G to form a doublet

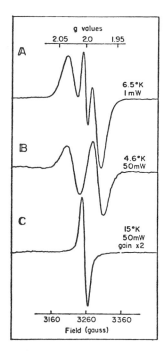

FIG. 28. Electron paramagnetic resonance spectrum associated with the reduced intermediate electron acceptor, I⁻, in the *Chromatium (vinosum)* reaction centers poised at −400 mV and measured at 200°K. Trace A represents difference between the signals taken after 3-minute illumination at 200°K and unilluminated material; B and C are component signals resolved from signal A. See text for details. From Tiede *et al.* (1976).

(g at 2.034 and 1.976) plus a free-radical signal also centered at 2.003 with a linewidth of ∼15 G. The doublet may have an origin in the interaction between I⁻ and the Fe in X. Figures 28B and 28C show separate spectra resolved from spectrum A by varying temperature; the doublet is highly temperature sensitive and is not easily saturated with microwave power, whereas the g = 2.003 is less sensitive to temperature but readily saturated. *Rhodopseudomonas viridis* RCs treated in the manner to trap the intermediate electron acceptor in the reduced state exhibit similar EPR spectra, except the doublet is split by about 100 G with g values at 2.06 and 1.96 (Prince *et al.*, 1976).

When *Chromatium* subchromatophores are poised at +300 mV, a light-induced fluorescence-yield increase related to P890 photooxidation can be observed. By contrast, when the subchromatophores were poised at −450 mV, Klimov *et al.* (1976) found that the "dark" fluorescence (luminescence) was enhanced, and illumination caused a decrease in luminescence yield. The kinetics of the light-induced luminescence change corresponds to that of BPh photoreduction. The luminescence emission most likely originates from bacteriochlorophyll, since the excitation maximum is at 890 nm and the emission lies above 900 nm. The light-induced luminescence decrease was attributed to a drop in the quantum yield of recombination luminescence rather than fluorescence. The variable luminescence has an activation energy of 0.12 eV and a lifetime of about 6 nsec. The latter is consistent with the lifetime of the P^F state under these conditions.

C. TRIPLET FORMATION AND THE REDOX POTENTIAL OF I/I$^-$

When X is prereduced, or when the UQ part of Fe·UQ is removed by extraction, $[B_2{}^{\cdot+}\cdot I^-]$ can still form in ≤ 10 psec. However, since electron transport from I$^-$ to X is prevented, the return of the electron from I$^-$ to $B_2{}^{\cdot+}$ takes place instead, in 10–30 nsec (Parson et al., 1975). At lower temperatures (say, 8°K), it decays via the spin polarized triplet or biradical state $(B^{\cdot+}B^{\cdot-})$ as the electron from I$^-$ returns to spin couple with the unpaired electron of $B_2{}^{\cdot+}$ (Dutton et al., 1975). In RCs where carotenoids are present, the bacteriochlorophyll triplet decays rapidly by transferring energy to the carotenoids, promoting the latter to a triplet state. These have been demonstrated by Cogdell et al. (1975) with purified RCs from three strains of *Rhodopseudomonas spheroides* and two strains of *Rhodospirillum rubrum*, and by Shuvalov and Klimov (1976) with subchromatophores of *Chromatium minutissimum*.

When the state $C^{3+}[B_2\cdot I^-]X^-$ is formed from $C^{2+}[B_2\cdot I]X^-$ by illumination, triplet-state formation should be prevented. This was indeed demonstrated by Tiede et al. (1976) to be the case. However, the mechanism of dissipation of the energy absorbed under these conditions is not known. Prince et al. (1976) further reasoned that this relationship may be utilized for estimating the redox potential of the I/I$^-$ couple. By poising the RCs at known potentials in the range that "I" can be reduced to I$^-$ and by measuring the light-induced triplet formation, the magnitude of triplet formation vs. the redox potential would reflect the redox-titration course of the I/I$^-$ couple. Considering the fact that the light-generated redox span in the bacteriochlorophyll-*b*-containing *Rhodopseudomonas viridis* should be less than that in *R. spheroides*, the redox potential of the intermediate electron acceptor should also be more positive, thus bringing the latter within experimental reach. Prince et al. chose *R. viridis* for this determination. The midpoint potential of the I/I$^-$ couple in *R. viridis* was found to be -400 ± 25 mV.

Similarly, Klimov et al. (1976) examined the photoreduction of "I" in *Chromatium minutissimum* by monitoring the absorption increase at 680 nm directly and found it was unaffected between -200 to -620 mV.

These results suggest that the midpoint potential of I/I^- would be $\leqq$ -620 mV.

VI. Concluding Remarks

Since photosynthesis is the process that converts light energy into chemical energy, a consideration of the efficiency of photosynthesis, namely, the amount of light energy absorbed relative to the amount of chemical energy fixed, is of both theoretical and practical importance. In the simplest term, one may consider the net reaction of photosynthesis as CO_2 + H_2O → carbohydrate + O_2, a reaction requiring 120 kcal of (Gibbs) free energy. As each red photon absorbed by chlorophyll (at 680 nm) has an energy content equivalent to 41 kcal, the reaction would require 3 quanta if the conversion of light energy were 100% efficient. Duysens (1958) considered the thermodynamic limitations of light-energy conversion in photosynthesis and concluded that the maximum attainable efficiency is about 70%. It follows that the reaction would require a minimum of 4 quanta. In the simple mechanistic viewpoint, i.e., in terms of Einstein's law of photochemical equivalence, one would also arrive at the same conclusion. In light of the two photochemical systems acting in coordination in green-plant photosynthesis, 8 quanta would be required. It is now recognized that in the reduction of one CO_2 to the level of carbohydrate, numerous other biochemical processes are involved which also demand energy supply and which are not reflected in the simple formulation. Most experimental measurements with isolated chloroplasts or whole algae yield minimum requirements of the order of 2–3 quanta per equivalent and 5–7 quanta per ATP (Kok, 1965).

The efficiency of photosynthesis may be more advantageously expressed as quantum efficiency or quantum yield. This expression is based on Einstein's law of photochemical equivalence, i.e., each molecule taking part in a chemical reaction induced by exposure to light absorbs one quantum of the radiation that causes the reaction. Obviously this law is more applicable to reactions taking place in the photosynthetic reaction center or reactions very closely linked to the primary photochemical reaction. Thus, the quantum requirements for $NADP^+$ reduction and cytochrome c oxidation have been reported to approach the theoretical limit of one quantum per equivalent.

With the nature of the primary photochemical reaction in the photosystems of green plants and photosynthetic bacteria better understood and the primary reactants better identified, a consideration of the quantum efficiency has gained renewed interest (Fajer *et al.*, 1975; Dutton *et al.*, 1976). The energy of the quantum absorbed by the respective photosynthetic reaction centers and "stored" as free energy may be considered in terms of the thermodynamic potential generated in the reaction center, namely, the difference in potential between the photooxidant and photoreductant formed in the primary charge-separation process.

In the case of photosynthetic bacteria, using 70% as the maximum attainable efficiency (Duysens, 1958; Ross and Calvin, 1967), the quantum absorbed at 870 nm represents an energy of 1.43 eV, and therefore $\sim$1 V would be available to be distributed among the primary photooxidant, $P870^+$ at +0.45 V and the primary photoreductant at -0.55 V (See Table I). The latter potential value has been obtained as the halfwave potential of the BPh anion in various organic solvents by electrochemical reduction (Fajer et al., 1975).

For the bacteriochlorophyll b-containing reaction center of *Rhodopseudomonas viridis*, for which the redox potential of the intermediate electron acceptor has been experimentally determined to be -0.4 V (Prince et al., 1976), the potential span of 0.9 V = +0.5 $-$ (-0.4) also turned out to be approximately 70% of the energy of the 985 nm quantum (1.26 eV). The reduced intermediary electron acceptor I^- transfers an electron to the primary acceptor in 100–200 psec, forming $[B_2{}^{+}\cdot I]X^-$. As noted by Prince et al. (1976), interestingly the redox span between the primary donor and the stable primary acceptor is approximately 0.65 V for all four species, despite the fact that the photonic energy absorbed by *R. viridis* is only about 90% of that absorbed by the bacteriochlorophyll a-containing species (Table I).

As recent electrochemical titrations indicated a midpoint potential for the intermediate electron acceptor of PS I to be -0.73 V, a tentative thermodynamic account may also be made. As seen from Table I, currently available data are in close agreement with a 70% conversion of the absorbed photonic energy in photosystem I. It should be noted that literature values on the midpoint potential of P700 show a wide variation, ranging from +0.375 V (Evans et al., 1977) to +0.525 V (Knaff and Malkin, 1973).

More recently, Mohanty and Ke (1977, unpublished) examined the effect of pH on the redox transitions of fluorescence yield in the PS II particles and obtained an apparent pK value of about 8.9. Since the kinetically important redox couple of the primary electron acceptor in PS II is Q/Q^- instead of Q/QH at equilibrium (Stiehl and Witt, 1969; Van Gorkom, 1974; Pulles et al., 1974), the operational midpoint potential of the primary acceptor would be the equilibrium value measured above the pK of the reduced form of the acceptor, which was estimated to be -450 mV. Taking the midpoint potential of the PS II primary electron donor as -0.82 V, the electrochemical potential span created in PS II by

TABLE I

THERMODYNAMIC ACCOUNT FOR THE PHOTOSYNTHETIC REACTION CENTERS

	Rhodopseudomonas spheroides	Chromatium	Rhodopseudomonas viridis	PS I	PS II
Photon, nm	870	890	985	700	675
Photonic energy, eV	1.43	1.39	1.26	1.77	1.84
Max. free energy, eV[a]	1.0	0.97	0.88	1.24	1.29
$E(P/P^+)$,[b] V	+0.45[c]	+0.49[d]	+0.50[e]	+0.47[i]	~+0.82
$E(I/I^-)$, V	−0.55[f]	−0.62[g]	−0.40[e]	−0.73[j]	
$E(X/X^-)$, V	−0.18[h]	−0.16[h]	−0.15[e]	−0.53[k]	−0.45[l,m]
ΔE (P/P^+) − (I/I^-), V	1.0	1.1	0.9	1.2	1.27
ΔE (P/P^+) − (X/X^-), V	0.63	0.65	0.65	1.0	—

[a] Based on 70% of the absorbed photonic energy.
[b] P/P^+ is used in place of $B_2/B_2^{\cdot+}$.
[c] Dutton and Jackson (1972).
[d] Cusanovich et al. (1968); Dutton (1971).
[e] Prince et al. (1976).
[f] Fajer et al. (1975).
[g] Klimov et al. (1976).
[h] Prince and Dutton (1976).
[i] Ke et al. (1975).
[j] Ke et al. (1977).
[k] Ke (1974).
[l] Mohanty and Ke (1977, unpublished).
[m] Ke et al. (1976a).

a photon would be 1.27 V, which also amounts to 70% of the photonic energy absorbed by PS II.

ACKNOWLEDGMENTS

I wish to thank J. Breton, R. K. Clayton, P. L. Dutton, J. Fajer, G. Feher, W. Junge, V. A. Shuvalov, and A. Vermeglio for permission to use their illustrations in this review, and to express my gratitude to H. Beinert for his collaboration in much of the EPR work on photosystem I.

Our own work reported here was supported in part by a National Science Foundation grant, No. 29161.

REFERENCES

Arnon, D. I. (1965). *Science* **149**, 1460–1469.
Bales, H., and Vernon, L. P. (1963). *In* "Bacterial Photosynthesis" (H. Gest *et al.*, eds.), pp. 269–274. Antioch Press, Yellow Springs, Ohio.
Bearden, A. J., and Malkin, R. (1972a). *Biochem. Biophys. Res. Commun.* **46**, 1299–1305.
Bearden, A. J., and Malkin, R. (1972b). *Biochim. Biophys. Acta* **283**, 456–468.
Bearden, A. J., and Malkin, R. (1975). *Q. Rev. Biophys.* **7**, 131–177.
Benssason, R., and Land, E. J. (1973). *Biochim. Biophys. Acta* **325**, 175–181.
Black, C. C. (1966). *Biochim. Biophys. Acta* **120**, 332–340.
Blankenship, R., McGuire, A., and Sauer, K. (1975). *Proc. Natl. Acad. Sci. U.S.A.* **72**, 4943–4947.
Bolton, J. R., and Cost, K. (1973). *Photochem. Photobiol.* **18**, 417–421.
Borisov, A. Yu., and Il'ina, M. D. (1973). *Biochim. Biophys. Acta* **305**, 368–371.
Bray, R. C., Palmer, G., and Beinert, H. (1964). *In* "Rapid Mixing and Sampling Techniques in Biochemistry" (B. Chance *et al.*, eds.), pp. 195–203. Academic Press, New York.
Breton, J., and Roux, E. (1971). *Biochem. Biophys. Res. Commun.* **45**, 557–563.
Breton, J., Roux, E., and Whitmarsh, J. (1975). *Biochem. Biophys. Res. Commun.* **64**, 1274–1277.
Butler, W. L. (1973). *Acc. Chem. Res.* **6**, 177–184.
Cammack, R. (1973). *Biochem. Biophys. Res. Commun.* **54**, 548–550.
Cammack, R., and Evans, M. C. W. (1975). *Biochem. Biophys. Res. Commun.* **67**, 544–549.
Case, G. D., and Parson, W. W. (1971). *Biochim. Biophys. Acta* **253**, 187–202.
Clayton, R. K. (1963). *Biochim. Biophys. Acta* **75**, 312–323.
Clayton, R. K., and Straley, S. C. (1970). *Biochem. Biophys. Res. Commun.* **38**, 1114–1118.
Clayton, R. K., and Straley, S. C. (1972). *Biophys. J.* **12**, 1221–1234.
Clayton, R. K., and Wang, R. T. (1971). *In* "Methods in Enzymology" (A. San Pietro, ed.), Vol. 23, pp. 696–704. Academic Press, New York.
Cogdell, R. J., and Crofts, A. R. (1972). *FEBS Lett.* **27**, 176–178.
Cogdell, R. J., Brune, D. C., and Clayton, R. K. (1974). *FEBS Lett.* **45**, 344–347.
Cogdell, R. J., Monger, T. G., and Parson, W. W. (1975). *Biochim. Biophys. Acta* **408**, 189–199.
Cusanovich, M. A., Bartsch, R. G., and Kamen, M. D. (1968). *Biochim. Biophys. Acta* **153**, 397–417.
DeVault, D. C., and Chance, B. (1966). *Biophys. J.* **6**, 825–847.

Döring, G., Bailey, J. L., Kreutz, W., Weikard, J., and Witt, H. T. (1968). *Naturwissenschaften* 55, 219–224.

Dratz, E. A., Schultz, A. J., and Sauer, K. (1967). *Brookhaven Symp. Biol.* 19, 303–318.

Dutton, P. L. (1971). *Biochim. Biophys. Acta* 226, 63–80.

Dutton, P. L., and Jackson, J. B. (1972). *Eur. J. Biochem.* 30, 495–510.

Dutton, P. L., Leigh, J. S., and Seibert, M. (1973a). *Biochem. Biophys. Res. Commun.* 46, 406–413.

Dutton, P. L., Leigh, J. S., and Reed, D. W. (1973b). *Biochim. Biophys. Acta* 292, 654–664.

Dutton, P. L., Leigh, J. S., and Wraight, C. A. (1973c). *FEBS Lett.* 36, 169–173.

Dutton, P. L., Kaufmann, K. J., Chance, B., and Rentzepis, P. M. (1975), *FEBS Lett.* 60, 275–280.

Dutton, P. L., Prince, R. C., Tiede, D. M., Petty, K., Cobley, U. T., Kaufmann, K. J., Netzel, T. L., and Rentzepis, P. M. (1977). *Brookhaven Symp. Biol.* 28, 213–237.

Duysens, L. N. M. (1954). *Nature (London)* 173, 692–693.

Duysens, L. N. M. (1958). *Brookhaven Symp. Biol.* 11, 10–23.

Duysens, L. N. M., and Sweers, H. E. (1963). *In* "Studies on Microalgae and Photosynthetic Bacteria" (Jpn. Soc. Plant Physiol., ed.), pp. 353–372. Univ. of Tokyo Press, Tokyo).

Evans, M. C. W., and Buchanan, B. B. (1965). *Proc. Natl. Acad. Sci. U.S.A.* 53, 1420–1425.

Evans, M. C. W., Telfer, A., and Lord, A. V. (1972). *Biochim. Biophys. Acta* 267, 530–537.

Evans, M. C. W., Reeves, S. G., and Cammack, R. (1974). *FEBS Lett.* 49, 111–114.

Evans, M. C. W., Sihra, C. K., Bolton, J. R., and Cammack, R. (1975). *Nature (London)* 256, 668–670.

Evans, M. C. W., Sihra, C. K., and Slabs, A. R. (1977). *Biochem. J.* 162, 75–85.

Fajer, J., Brune, D. C., Davis, M. S., Forman, A., and Spaulding, C. D. (1975). *Proc. Natl. Acad. Sci. U.S.A.* 72, 4956–4960.

Feher, G. (1971). *Photochem. Photobiol.* 14, 373–387.

Feher, G., Okamura, M. Y., and McElroy, J. D. (1972). *Biochim. Biophys. Acta* 267, 222–226.

Feher, G., Isaacson, R. A., McElroy, J. D., Ackerson, L. C., and Okamura, M. Y. (1974). *Biochim. Biophys. Acta* 368, 135–139.

Feher, G., Hoff, A. J., Isaacson, R. A., and Ackerson, L. C. (1975). *Ann. N.Y. Acad. Sci.* 244, 239–259.

Fuller, R. C., and Nugent, N. A. (1969). *Proc. Natl. Acad. Sci. U.S.A.* 63, 1311–1318.

Geacintov, N. E., Van Nostrand, F., Pope, M., and Tinkel, J. L. (1971). *Biochim. Biophys. Acta* 226, 486–491.

Golbeck, J. H., Lien, S., and San Pietro, A. (1976). *Biochim. Biophys. Res. Commun.* 71, 452–458.

Haehnel, W. (1973). *Biochim. Biophys. Acta* 305, 618–631.

Haehnel, W. (1976). *Biochim. Biophys. Acta* 423, 499–509.

Halsey, Y. D., and Parson, W. W. (1974). *Biochim. Biophys. Acta* 347, 404–416.

Hawkridge, F. M., and Ke, B. (1977). *Anal. Biochem.* 78, 76–85.

Hiyama, T., and Ke, B. (1971a). *Proc. Natl. Acad. Sci. U.S.A.* 68, 1010–1013.

Hiyama, T., and Ke, B. (1971b). *Arch. Biochem. Biophys.* 147, 99–108.

Hiyama, T., and Ke, B. (1972a). *Photosynth., Two Centuries Its Discovery Joseph Priestley, Proc. Int. Congr. Photosynth. Res., 2nd, 1971* Vol. 1, pp. 491–497.

Hiyama, T., and Ke, B. (1972b). *Biochim. Biophys. Acta* 267, 160–171.

Hochstrasser, R. M., and Kasha, M. (1964). *Photochem. Photobiol.* 3, 317–331.

136 BACON KE

Jackson, J. B., Cogdell, R. J., and Crofts, A. R. (1973). *Biochim. Biophys. Acta* **292**, 218–225.

Junge, W. (1974). *Proc. Int. Congr. Photosynth. Res., 3rd,* pp. 229–234.

Junge, W., and Eckhof, A. (1973). *FEBS Lett.* **36**, 207–212.

Junge, W., and Eckhof, A. (1974). *Biochim. Biophys. Acta* **357**, 103–117.

Kamen, M. D. (1961). *In* "Light and Life" (W. D. McElroy and B. Glass, eds.), pp. 483–488. Johns Hopkins Univ. Press, Baltimore, Maryland.

Kamen, M. D. (1963). "Primary Processes in Photosynthesis," p. 154. Academic Press, New York.

Karapetyan, N. V., and Klimov, V. V. (1970). *Plant Physiol. (USSR)* **18**, 223–228.

Karapetyan, N. V., Klimov, V. V., and Krasnovsky, A. A. (1973). *Dokl. Akad. Nauk SSSR* **211**, 729–732.

Kasha, M. (1963). *Radiat. Res.* **20**, 55–71.

Kassner, R. J., and Kamen, M. D. (1967). *Proc. Natl. Acad. Sci. U.S.A.* **58**, 2445–2450.

Katz, J. J., and Norris, J. R. (1973). *Curr. Top. Bioenerg.* **5**, 41–75.

Katz, J. J., Ballschmitter, K., Garcia-Morin, M., Strain, H. H., and Uphaus, R. A. (1968). *Proc. Natl. Acad. Sci. U.S.A.* **60**, 100–107.

Kaufmann, K. J., Dutton, P. L., Netzel, T. L., Leigh, J. S., and Rentzepis, P. M. (1975). *Science* **188**, 1301–1304.

Kaufmann, K. J., Petty, K. M., Dutton, P. L., and Rentzepis, P. M. (1976). *Biochem. Biophys. Res. Commun.* **70**, 839–845.

Ke, B. (1972a). *Biochim. Biophys. Acta* **267**, 595–599.

Ke, B. (1972b). *Arch. Biochem. Biophys.* **152**, 70–77.

Ke, B. (1973). *Biochim. Biophys. Acta* **301**, 1–33.

Ke, B. (1974). *Proc. Int. Congr. Photosynth. Res., 3rd,* Vol. 1, pp. 373–382.

Ke, B. (1975a). *Bioelectrochem. Bioenerg.* **2**, 93–105.

Ke, B. (1975b). *Photochem. Photobiol.* **20**, 543–546.

Ke, B., and Beinert, H. (1973). *Biochim. Biophys. Acta* **305**, 689–693.

Ke, B., Vernon, L. P., Garcia, A. F., and Ngo, E. (1968). *Biochemistry* **7**, 311–318.

Ke, B., Chaney, T. H., and Reed, D. W. (1970). *Biochim. Biophys. Acta* **216**, 373–383.

Ke, B., Ogawa, T., Hiyama, T., and Vernon, L. P. (1971). *Biochim. Biophys. Acta* **226**, 53–62.

Ke, B., Garcia, A. F., and Vernon, L. P. (1973a). *Biochim. Biophys. Acta* **292**, 226–236.

Ke, B., Hansen, R. E., and Beinert, H. (1973b). *Proc. Natl. Acad. Sci. U.S.A.* **70**, 2941–2945.

Ke, B., Sahu, S., Shaw, E. R., and Beinert, H. (1974a). *Biochim. Biophys. Acta* **347**, 36–48.

Ke, B., Sugahara, K., Shaw, E. L., Hansen, R. E., Hamilton, W. D., and Beinert, H. (1974b). *Biochim. Biophys. Acta* **368**, 401–408.

Ke, B., Sugahara, K., and Shaw, E. R. (1975). *Biochim. Biophys. Acta* **408**, 12–15.

Ke, B., Hawkridge, F. M., and Sahu, S. (1976a). *Proc. Natl. Acad. Sci. U.S.A.* **73**, 2211–2215.

Ke, B., Sugahara, K., and Sahu, S. (1976b). *Biochim. Biophys. Acta* **449**, 84–94.

Ke, B., Dolan, E., Sugahara, K., Hawkridge, F. M., Demeter, S., and Shaw, E. R. (1977). *Plant Cell Physiol.* (special issue on Photosynthetic Organelles) pp. 187–199.

Klimov, V. V., Shuvalov, V. A., Krakhmaleva, I. N., and Krasnovsky, A. A. (1976). *Biochimiya* **201**, 1244–1247.

Knaff, D. B., and Malkin, R. (1973). *Arch. Biochem. Biophys.* **159**, 558–562.

Knaff, D. B., and Malkin, R. (1976). *Biochim. Biophys. Acta* **430**, 244–252.

Kok, B. (1956). *Biochim. Biophys. Acta* **22**, 399–401.

Kok, B. (1965). *In* "Plant Biochemistry" (J. Bonner and J. E. Varner, eds.), 2nd ed., pp. 904–960. Academic Press, New York.

Kok, B., Rurainski, H. J., and Owen, O. V. H. (1965). *Biochim. Biophys. Acta* **109**, 347–356.

Krasnovsky, A. A., and Vojnovskaya, K. K. (1951). *Dokl. Akad. Nauk SSSR* **81**, 879–882.

Leigh, J. S., and Dutton, P. L. (1972). *Biochem. Biophys. Res. Commun.* **46**, 414–421.

Lin, L., and Thornber, J. P. (1975). *Photochem. Photobiol.* **22**, 37–40.

Loach, P. A., and Hall, R. L. (1972). *Proc. Natl. Acad. Sci. U.S.A.* **69**, 786–790.

Lozier, R. H., and Butler, W. L. (1974). *Biochim. Biophys. Acta* **333**, 460–464.

McIntosh, A. R., Chu, M., and Bolton, J. R. (1975). *Biochim. Biophys. Acta* **376**, 308–314.

Malkin, R., and Bearden, A. J. (1971). *Proc. Natl. Acad. Sci. U.S.A.* **68**, 16–19.

Malkin, R., Aparicio, P. J., and Arnon, D. I. (1974). *Proc. Natl. Acad. Sci. U.S.A.* **71**, 2362–2366.

Malkin, R., Bearden, A. J., Hunter, F. A., Alberte, R. S., and Thornber, J. P. (1976). *Biochim. Biophys. Acta* **430**, 389–394.

Murata, N., and Takamiya, A. (1969). *Plant Cell Physiol.* **10**, 193–202.

Nelson, N., Bengis, C., Silver, B. L., Getz, D., and Evans, M. C. W. (1975). *FEBS Lett.* **58**, 363–365.

Netzel, T. L., Rentzepis, P. M., and Leigh, J. S. (1973). *Science* **182**, 238–241.

Nöel, H., Van der Rest, M., and Gingras, G. (1972). *Biochim. Biophys. Acta* **275**, 219–230.

Norris, J. R., Uphaus, R. A., Crespi, H. L., and Katz, J. J. (1971). *Proc. Natl. Acad. Sci. U.S.A.* **68**, 625–628.

Norris, J. R., Scheer, H., Druyan, M. E., and Katz, J. J. (1974). *Proc. Natl. Acad. Sci. U.S.A.* **71**, 4897–4900.

Ohnishi, T., Asakura, T., Wilson, D. F., and Chance, B. (1972). *FEBS Lett.* **21**, 59–62.

Okamura, M. Y., Issacson, R. A., and Feher, G. (1975). *Proc. Natl. Acad. Sci. U.S.A.* **72**, 3491–3495.

Okamura, M. Y., Ackerson, L. C., Isaacson, R. A., Parson, W. W., and Feher, F. (1976). *Biophys. J.* **16**, 223.

Olson, J. M., Prince, R. C., and Brune, D. C. (1976). *Brookhaven Symp. Biol.* **28**, 238–246.

Orme-Johnson, N. R., Orme-Johnson, W. H., Hansen, R. E., Beinert, H., and Hatefi, Y. (1971). *Biochem. Biophys. Res. Commun.* **44**, 446–452.

Parson, W. W., and Case, G. D. (1970). *Biochim. Biophys. Acta* **205**, 233–245.

Parson, W. W., and Cogdell, R. J. (1975). *Biochim. Biophys. Acta* **416**, 105–149.

Parson, W. W., Clayton, R. K., and Cogdell, R. J. (1975). *Biochim. Biophys. Acta* **387**, 265–278.

Penna, F., Reed, D. W., and Ke, B. (1974). *Proc. Int. Congr. Photosynth. Res., 3rd,* Vol. 1, pp. 421–425.

Petering, D. H., Fee, J. A., and Palmer, G. (1971). *J. Biol. Chem.* **246**, 643–653.

Philipson, K. D., and Sauer, K. (1973). *Biochemistry* **12**, 535–539.

Philipson, K. D., Sato, V. L., and Sauer, K. (1972). *Biochemistry* **11**, 4591–4595.

Prince, R. C., and Dutton, P. L. (1976). *Arch. Biochem. Biophys.* **172**, 329–334.

Prince, R. C., and Olson, J. M. (1976). *Biochim. Biophys. Acta* **423**, 357–362.

Prince, R. C., Leigh, J. S., and Dutton, P. L. (1976). *Biochim. Biophys. Acta* **440**, 622–636.

Pulles, M. P. J., Kerhoff, P. L. M., and Amesz, J. (1974). *FEBS Lett.* **47**, 143–145.

Rawlings, J., Siiman, O., and Gray, H. B. (1974). *Proc. Natl. Acad. Sci. U.S.A.* **71**, 125–127.

Reed, D. W. (1969). *J. Biol. Chem.* **244**, 4936–4941.

Reed, D. W., and Clayton, R. K. (1968). *Biochim. Biophys. Res. Commun.* **30**, 471–475.

Reed, D. W., and Ke, B. (1973). *J. Biol. Chem.* **248**, 3041–3045.

Reed, D. W., and Peters, G. A. (1972). *J. Biol. Chem.* **247**, 7148–7152.

Rockley, M. G., Windsor, M. W., Cogdell, R. J., and Parson, W. W. (1975). *Proc. Natl. Acad. Sci. U.S.A.* **72**, 2251–2255.

Ross, R. T., and Calvin, M. (1967). *Biophys. J.* **7**, 595–614.

Rumberg, B. (1964). *Z. Naturforsch. Teil B* **19**, 707–716.

Sauer, K. (1965). *Biophys. J.* **5**, 337–348.

Sauer, K. (1975). *In* "Bioenergetics of Photosynthesis" (Govindjee, ed.), pp. 139–144. Academic Press, New York.

Sauer, K., and Calvin, M. (1962). *J. Mol. Biol.* **4**, 451–466.

Sauer, K., Dratz, E. A., and Coyne, L. (1968). *Proc. Natl. Acad. Sci. U.S.A.* **61**, 17–24.

Shipman, L. L., Cotton, T. M., Norris, J. R., and Katz, J. J. (1976). *Proc. Natl. Acad. Sci. U.S.A.* **73**, 1791–1794.

Shuvalov, V. A. (1976). *Biochim. Biophys. Acta* **430**, 113–121.

Shuvalov, V. A., and Klimov, V. V. (1976). *Biochim. Biophys. Acta* **440**, 587–599.

Shuvalov, V. A., Klimov, V. V., and Krasnovsky, A. A. (1976a). *Mol. Biol.* **10**, 326–337.

Shuvalov, V. A., Klimov, V. V., Krakhmaleva, I. N., Moskalenko, A. A., and Krasnovsky, A. A. (1976b). *Dokl. Akad. Nauk SSSR* **227**, 984–987.

Slooten, L. (1972). *Biochim. Biophys. Acta* **275**, 208–218.

Stiehl, H. H., and Witt, H. T. (1969). *Z. Naturforsch., Teil B* **24**, 1588–1598.

Straley, S. C., Parson, W. W., Mauzerall, D., and Clayton, R. K. (1973). *Biochim. Biophys. Acta* **305**, 597–609.

Takamiya, K.-I., and Takamiya, A. (1970). *Biochim. Biophys. Acta* **205**, 72–85.

Thomas, J. B., VanLierop, J. H., and Ten Han, M. (1967). *Biochim. Biophys. Acta* **143**, 204–220.

Tiede, D. M., Prince, R. C., and Dutton, P. L. (1976). *Biochim. Biophys. Acta* **449**, 447–467.

Tinoco, I., Jr. (1963). *Radiat. Res.* **20**, 133–142.

Trosper, T. L., Benson, D. L., and Thornber, J. P. (1977). *Biochim. Biophys. Acta* **460**, 318–330.

Tsujimoto, H. Y., Chain, R. K., and Arnon, D. I. (1973). *Biochem. Biophys. Res. Commun.* **51**, 917–923.

Van Gorkom, H. M. (1974). *Biochim. Biophys. Acta* **347**, 439–442.

Vermeglio, A., and Clayton, R. K. (1977). *Biochim. Biophys. Acta* **449**, 500–515.

Vermeglio, A., Breton, J., and Mathis, P. (1976). *J. Supramol. Struct.* (in press).

Vernon, L. P., and Ke, B. (1966). *In* "The Chlorophylls" (L. P. Vernon and G. R. Seely, eds.), pp. 569–607. Academic Press, New York.

Vernon, L. P., Yamamoto, H. Y., and Ogawa, T. (1969). *Proc. Natl. Acad. Sci. U.S.A.* **63**, 911–917.

Visser, J. W. M., Rijersberg, K. P., and Amesz, J. (1974). *Biochim. Biophys. Acta* **368**, 235–246.

Wang, J. H. (1970). *Science* **167**, 25–30.

Wraight, C. A., and Clayton, R. K. (1974). *Biochim. Biophys. Acta* **333**, 246–260.

Yocum, C., and San Pietro, A. (1969). *Biochem. Biophys. Res. Commun.* **36**, 614–670.

Zweig, G., and Avron, M. (1965). *Biochem. Biophys. Res. Commun.* **19**, 397–400.

The Primary Reaction of Chloroplast Photosystem II

David B. Knaff[1] and Richard Malkin

Department of Cell Physiology
University of California
Berkeley, California

I. Introduction

The concept that photosynthesis in oxygen-evolving organisms involves the cooperative action of two different photosystems operating in series has been widely accepted. Many of the data supporting this model have recently been reviewed in a monograph (Govindjee, 1975). One of the two photosystems (Photosystem II, PSII) generates in the light a strong oxidant and concomitantly produces a weak reductant. The light-produced oxidant is capable of oxidizing H_2O to O_2. The other photosystem (Photosystem I, PSI) produces in the light a weak oxidant and concomitantly produces a strong reductant that is capable of reducing pyridine nucleotide. Electrons removed from water by PSII are transferred to PSI by a series of dark electron-transfer reactions. This

---

[1] Present address: Department of Chemistry, Texas Tech University, Lubbock, Texas.

pathway showing involvement of some intermediate electron carriers is seen in Fig. 1.

The reactions associated with PSII are some of the most important biochemical events in the biosphere. They provide the eventual source of electrons for all reductive biosynthetic reactions (including carbohydrate and fatty acid synthesis) of algae and higher plants. As a by-product of its ability to utilize water as an electron donor, PSII releases molecular oxygen. There is considerable evidence from geological records that the early atmosphere of the earth contained no free oxygen and that all the oxygen in our present atmosphere is of photosynthetic origin (Berkner and Marshall, 1968). PSII not only plays an essential role in providing food, but also supplies the terminal electron acceptor, oxygen, which we require for respiration.

In this chapter we will consider the nature of the primary reaction of PSII, i.e., the initial light-driven charge separation. Three requirements may be applied to reactions considered to be primary photochemical events: (1) The reaction must be rapid—this is the most direct demonstration of a primary event, since the primary reactants are the first products formed after the absorption of light. To satisfy this kinetic requirement, it is necessary to measure the rate of formation of the primary reactants, but it is now apparent from studies of the primary reaction in photosynthetic bacteria (Parson and Cogdell, 1975) that such

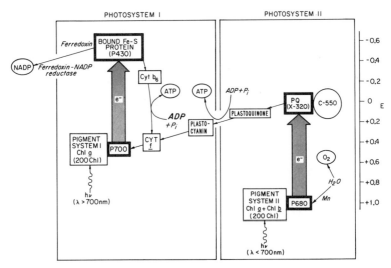

FIG. 1. The chloroplast electron-transport chain, showing the cooperation of two photosystems in the transfer of electrons from water to $NADP^+$.

events occur in the picosecond time domain and are difficult to detect by conventional spectroscopic techniques. (2) The reaction should proceed at cryogenic temperatures—at the temperature of liquid nitrogen (77°K) or lower. At cryogenic temperatures, ordinary chemical reactions requiring molecular diffusion do not commonly occur. Reactions dependent solely upon photon capture are still expected to occur at these temperatures. Temperature-independent reactions are not, however, limited to primary photochemical events, as will be documented by further discussion. (3) The reaction should proceed with a high quantum efficiency—the quantum yield of the primary reaction should be at or close to one. However, this criterion alone is not sufficient for the identification of a primary reactant, because other electron carriers undergo photoreactions with high quantum efficiency.

With this general background, we will proceed to a discussion of the chemical identities and oxidation–reduction properties of the PSII primary reactants. Earlier reviews deal with other aspects of the primary photochemistry in photosynthesis and some cover secondary electron transfer processes in PSII that are beyond the scope of this chapter (Butler, 1973; Van Gorkom and Donze, 1973; Bearden and Malkin, 1975; Sauer, 1975; Radmer and Kok, 1975; Joliot and Kok, 1975).

II.   The Photosystem II Reaction-Center Chlorophyll, P680

The reaction-center chlorophyll of PSII has been designated P680 on the basis of a light-induced absorbance change in the red spectral region (Floyd et al., 1971; Butler, 1972b). Witt and co-workers first observed the spectral change and originally named the component "chlorophyll a_{II}" (Döring et al., 1969), but the designation "P680" (Butler, 1972b) is more consistent with the terminology used for the reaction-center chlorophyll found in other photoreactions, and it therefore will be used in this review.

A.   OPTICAL PROPERTIES IN UNTREATED CHLOROPLASTS

Absorbance changes of P680 have been observed in untreated chloroplasts illuminated with repetitive flashes. As shown in Fig. 2, the absorbance change at 690 nm after a 20-μsec flash displays biphasic decay kinetics: one component exhibits a rapid half-time decay of 200 μsec; a second component has a slower half-time of decay, 20 msec. Spectral analysis of the slowly decaying component led to its association with P700, the reaction-center chlorophyll of PSI. The rapidly decaying component was identified as P680 (Döring et al., 1969; Döring and Witt, 1972), and its light *minus* dark difference spectrum was typical of a

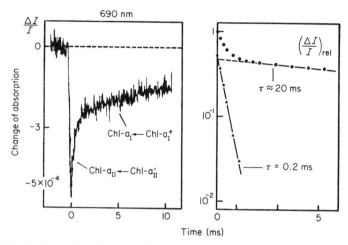

FIG. 2. *Left:* Absorption change at 690 nm in spinach chloroplasts as a function of time after a 20-μsec flash. Chl a_i represents P_{700} while Chl a_π represents P_{680}. *Right:* Relative absorption change at 690 nm as a function of time plotted on a log scale. From Döring *et al.* (1969).

chlorophyll a species, with absorbance maxima in the red and blue spectral regions (see Fig. 3).

More complete characterization of these absorbance changes led to the conclusion that they were associated with chloroplast PSII (Döring and Witt, 1972). The PSII electron-transfer inhibitor 3-(3′,4′-dichlorophenyl)-1,1-dimethylurea (DCMU) inhibits at similar concentrations both oxygen evolution and the P680 change. Long-wavelength illumination (728 nm), which preferentially activates PSI, did not affect the P680 change. Döring and Witt found a similar absorption change in subchloroplast fragments (prepared by digitonin treatment), which are known to contain predominantly PSII.

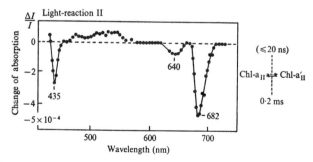

Fig. 3. Absorption change with a half-time of 0.2 msec in Photosystem II subchloroplast fragments. From Döring *et al.* (1969).

The identification of a P680 component exhibiting a 200-μsec decay agreed with a reaction time measured by Kok *et al.* (1970) for an intermediate on the oxidizing side of PSII. It was therefore assumed that the decay at 690 nm represented a reduction of the oxidized reaction-center chlorophyll by an electron released through an intermediate involved in the oxidation of water. However, the magnitude of the absorbance change associated with the P680 change was considerably smaller than expected on the basis of a one-to-one correspondence of PSII and PSI reaction centers in a linear electron-transport chain.

This problem of the stoichiometry of the PSI and II reaction centers led to a reinvestigation on a faster time-scale of the kinetic properties of P680. With a time-resolution of about 1 μsec, a new absorbance change was observed in chloroplasts (Gläser *et al.*, 1974). This component had a decay time of 35 μsec, which is to be compared with about 200 μsec for the slowly decaying P680 component under similar conditions. The absorbance maximum of the rapidly decaying component was 690 nm in the red region, as compared with 687 nm for the slowly decaying component. Spectral analysis led to the conclusion that this rapidly decaying component was associated with the PSII reaction-center chlorophyll. More important in terms of the stoichiometry of PSII and of PSI, when both components were assumed to arise from P680, there was about one P680 per active electron-transport chain, i.e., the ratio of P680 to P700 was near one.

The P680 component decaying in 35 μsec was sensitive to DCMU at a concentration similar to that required to inhibit the 200-μsec component. Another inhibitor, 2,5-dibromo-3-methyl-6-isopropyl-1,4-benzoquinone (DBMIB), which is known to inhibit the plastoquinone pool between the two light reactions (Trebst *et al.*, 1970) [and also known to function as a PSII electron acceptor (Gould and Izawa, 1973)] was found to eliminate PSI transients and to have little effect on either of the P680 kinetic components. Far-red background illumination also had no effect on the 35-μsec component, confirming the involvement of the component in PSII.

The biphasic reduction kinetics of P680 in oxygen-evolving chloroplasts at room temperature led to a tentative model (Wolff *et al.*, 1974) in which these two decay times represent the linear release of electrons from two different electron donors on the oxidizing side of PSII, as shown in Eq. (1).

$$P680 \xleftarrow[35\ \mu sec]{e^-} D \xleftarrow[200\ \mu sec]{e^-} Y \xleftarrow{e^-} H_2O \tag{1}$$

The relative proportion of the component decaying at 35 μsec or 200 μsec would be governed, according to this model, by the distribution of

D and Y in the oxidized state. Wolff *et al.* (1974) suggested that the two electron donors, D and Y, could be components of the water-splitting system that functions in PSII and hence may be related to the S− states proposed by Kok *et al.* (1970) as intermediates in oxygen evolution.

Another approach to the study of P680 in untreated chloroplasts has been described by Haveman and Mathis (1976). It was demonstrated previously that chlorophyll cation radicals exhibit a broad absorption band in the near-infrared region (about 820 nm) with an extinction coefficient of about 7000 M^{-1} cm^{-1} (Borg *et al.*, 1970; Seki *et al.*, 1973). Spectral changes in this region offer the advantage of being free from overlapping signals associated with other electron carriers and free from fluorescence changes known to occur in the red spectral region. Haveman and Mathis (1976) attempted to follow spectrally the oxidation-reduction changes of P680 at 825 nm after single-flash activation of chloroplast samples under physiological conditions; they could not detect any P680 light-induced changes with a time-resolution of 50 μsec. In subsequent work, the time-resolution of the kinetic studies was increased such that absorbance changes in about 5-μsec would have been observed (Mathis *et al.*, 1976). However, even on this fast time-scale, no PSII absorbance changes associated with P680 were detected; it was therefore concluded that, under physiological conditions, P680^{+} must be reduced by electrons from water in less than 5 μsec. It was shown that P680 absorbance changes at 825 nm could be observed after inhibition of water oxidation by Tris treatment or low pH (see below), an indication that the PSII reaction-center chlorophyll could indeed be monitored in this spectral region.

The discrepancy between the two groups who have reported kinetic measurements of P680 under physiological conditions is of some concern in attempts to describe the reactions of the PSII reaction center. Witt (1975) and co-workers have used a repetitive-flash technique and thereby created an equilibrium condition of the S− states on the oxidizing side of PSII during their measurements. It is possible that decay times are influenced by S− states. It is not obvious, however, that an equilibrium of S− states is the cause of the difference in results—Mathis *et al.* (1976) have been unable to detect in their single-flash experiments any dependence on S− state. Another possibility not yet adequately explored is that the absorption properties of P680^{+} are altered under physiological experimental conditions (possibly by oxidized species between water and P680), making the 820-nm band difficult to detect.

B. Optical Properties in Modified Chloroplast Preparations

Döring (1975) reported investigations of P680 in several different chloroplast preparations known to have little or no oxygen evolution. In all these preparations, a P680 component with a decay time of about 200 μsec was observed, and in general the absorbance changes of treated preparations remained as sensitive to DCMU as they were in untreated chloroplasts. Döring proposed that the linear electron flow in PSII in untreated chloroplasts is replaced in the inhibited system by a cyclic electron flow that is DCMU-sensitive. This cyclic flow is thought to involve carriers on the oxidizing side of PSII in addition to P680 and the primary electron acceptor. One exception to this behavior was noted with chloroplasts treated with the proteolytic enzyme, trypsin; in the presence of ferricyanide, the 200-μsec decaying component was insensitive to DCMU, a finding that indicates the operation of another type of cyclic pathway. The nature of these reactions is not yet clear.

Absorbance changes of P680 around 820 nm after single-flash activation have been reported for chloroplast fragments inhibited by low pH or by treatment with 0.8 M Tris buffer (pH 8.0) [the latter treatment specifically inactivates oxygen evolution; see Yamashita and Butler, 1968] (Haveman and Mathis, 1976). Under these conditions, the decay time of P680 was found to be about 150 μsec and the decay was thought to arise from a reaction of P680$^+$ with the reduced PSII primary electron acceptor. This conclusion is consistent with double-flash experiments in which the second flash was found to be as effective as the first in photooxidizing P680, even in the presence of DCMU. In addition, the decay time of about 150 μsec is similar to the decay time for strong luminescence observed by other workers under similar conditions (Haveman and Lavorell, 1975; Van Gorkom et al., 1976); similar decay times have been detected by following absorbance changes of the primary electron acceptor in Tris-treated chloroplasts (Renger and Wolff, 1976).

P680 absorbance changes have also been observed in deoxycholate-treated PSII subchloroplast fragments that have lost the ability to evolve oxygen (Van Gorkom et al., 1974). The spectrum of this change had absorbance maxima at 680 nm and 434 nm, similar to the absorbance changes for P680 measured by Döring et al. (1969) in untreated chloroplasts. These absorbance changes were, however, observed only in the presence of ferricyanide during continuous illumination. This finding suggests that ferricyanide oxidizes an as yet unknown electron donor which usually functions by rereducing P680$^+$. The P680 changes were reversible after the cessation of illumination and could be inhibited by heat, but not by DCMU. In these fragments, a back-reaction may occur between P680$^+$ and the primary electron acceptor.

In a PSII reaction-center complex, absorbance changes of P680 were also detected in the presence of ferricyanide (Van Gorkom et al., 1975). The changes were similar to those observed in the deoxycholate preparation described above in that they were insensitive to DCMU. As in other preparations in which oxygen evolution is impaired, there may be a cyclic reaction around PSII or a back-reaction between the reduced primary electron acceptor and $P680^+$.

Although the absorbance changes measured in these non-oxygen-evolving preparations during continuous illumination clearly represent the photooxidation of a chlorophyll species, it is possible that they arise from photooxidized bulk chlorophyll, not from P680. Yamashita and Butler (1969) observed in chloroplasts in which oxygen evolution is inhibited by Tris treatment that strong continuous illumination results in the photooxidation of about 10% of the total chlorophyll and that photooxidation can be decreased by the addition of reagents that function as PSII electron donors. Therefore, the observation of an oxidized chlorophyll molecule in a preparation in which there is no oxygen evolution is not sufficient evidence for the assignment of this chlorophyll as $P680^+$. If changes were obtained with single-turnover flashes, they would provide more conclusive evidence to this point. In this regard, correlations under various conditions of the presumed P680 absorbance changes with those of the PSII primary electron acceptor (see Section III) would be particularly useful.

C. STUDIES OF P680 AT CRYOGENIC TEMPERATURE

Optical changes of P680 were first reported for leaves and chloroplasts at liquid-nitrogen temperatures by Floyd et al. (1971). After flash-activation, it was found that $P680^+$ decayed with biphasic kinetics, one reductive phase having a half-time of 30 μsec and the second having a half-time of 4.5 msec. As discussed by Butler (1972b), control experiments necessary to eliminate fluorescence transients originating from PSII were not considered in these studies; it is therefore not clear how much the observations were affected by artifacts. It does seem clear that the absorbance changes in the red region observed by Floyd et al. (1971) after a series of flashes can be adequately explained as arising solely from P700.

An electron paramagnetic resonance (EPR) free-radical signal attributed to $P680^+$ has been observed in chloroplasts and PSII subchloroplast fragments at $35°K$ (Malkin and Bearden, 1975). It was seen after flash-activation with red light, but not with far-red light, and it showed a decay at low temperature with a half-time of about 5 msec. The EPR spectrum of the reversible signal was similar to that of chlorophyll cation free radicals ($g = 2.002$; linewidth about 8 gauss) and was considered to originate from $P680^+$. This signal was assigned to the PSII reaction center on the basis of its response to monochromatic illumination, the effects of inhibitors such as DCMU, and the effect of ambient oxidation–reduction potential. The millisecond decay of this component after flash-activation is caused by a back reaction between P680$^+$ and the reduced primary electron acceptor of PSII, a reaction that occurs at cryogenic temperatures (see below).

Optical changes of P680 have also been observed in the near-infrared region at 825 nm in chloroplasts at liquid-nitrogen temperature (Mathis and Vermeglio, 1975). These absorbance changes were reversible following flash-activation, and they decayed with a half-time of about 3 msec. Because this decay time is similar to that previously described for the P680$^+$ component detected by EPR spectroscopy, it is most likely that the near-infrared absorbance changes at cryogenic temperature also monitor the PSII reaction-center chlorophyll.

In contrast to these reports of reversible P680 reactions at cryogenic temperatures, Ke *et al.* (1974) have described irreversible absorbance changes, which they attributed to P680, in a PSII reaction-center complex preparation. A photobleached chlorophyll with an absorption maximum at 680 nm was observed after illumination at 77°K of a sample pretreated with ferricyanide. Under similar conditions, an irreversibly formed free-radical EPR signal was photoinduced at cryogenic temperatures. Although these authors assign both of these components to P680$^+$, an equally likely possibility is that they arise from another chlorophyll molecule that functions as a secondary electron donor to P680$^+$. It is now widely accepted that P680$^+$ does not accumulate at cryogenic temperatures—it either is reduced via the primary electron acceptor or is reduced via a secondary electron donor. It may be that in this particular preparation these routes of reduction are absent or inhibited, but conclusive assignment of this optically and EPR spectroscopically detected component to P680 can be made only after study of its flash-induced properties.

III.   The Photosystem II Primary Electron Acceptor

A.   THE FLUORESCENCE QUENCHER, Q

The first experimental parameter that was available for monitoring the oxidation-reduction state of the PSII primary electron acceptor was the fluorescence yield of chlorophyll *a*. Duysens and Sweers (1963) found that chlorophyll fluorescence yield increased when the primary electron acceptor of PSII was reduced and decreased as the acceptor was oxidized. To explain these observations, they proposed that the primary electron acceptor (which they called Q, for quencher) quenched chlorophyll fluorescence in its oxidized but not in its reduced state. The relationship of this fluorescence yield increase to primary photochemistry was strengthened by the finding that the increase in chlorophyll fluorescence yield also occurred on illumination at cryogenic temperatures (Kok *et al.*, 1963).

The increase in chlorophyll fluorescence could also be obtained by the addition of a reductant in the dark, a technique that permitted the determination of the midpoint oxidation-reduction potential(s) of the component(s) involved in the quenching process. Titrations with untreated chloroplasts, shown in Fig. 4 (Cramer and Butler, 1969; see also Ke *et al.*, 1976), indicated the presence of two components, both one-electron carriers that quench in the oxidized form, with midpoint potentials of approximately −35 mV and −270 mV (at pH 7.0). Oxidation-reduction titrations of the photochemical activity of PSII at cryogenic temperature showed that PSII activity was lost as a component with a midpoint potential near 0 V became reduced (Erixon and Butler, 1971b; Knaff, 1975a,b). Presumably, the loss in activity occurred because the primary acceptor of PSII was reduced prior to illumination. Concurrent measurements of light-induced fluorescence increase at cryogenic temperature showed that the component controlling fluorescence has the same midpoint potential as that of the PSII primary acceptor (Erixon and Butler, 1971b). This value corresponds within experimental error to the more electropositive quenching component observed at room temperature.

More recently, the relationship of the fluorescence yield change to the PSII primary acceptor has been questioned. Measurements by Mauzerall (1972) of the time-course of the fluorescence yield of *Chlorella* cells after flash-activation led to the conclusion that the fluorescence changes were too slow to be associated with the primary reaction of PSII. Furthermore, Mauzerall's experimental results required a minimum of six different fluorescence states, a situation that was incompatible with a

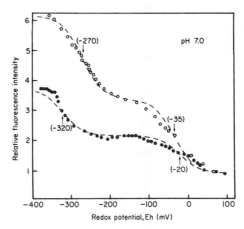

FIG. 4. Relative fluorescence intensity of spinach chloroplasts at pH 7.0 as a function of oxidation–reduction potential. From Cramer and Butler (1969).

single component (the primary acceptor) controlling fluorescence yield. Butler (1972a), Okayama and Butler (1972a), and Butler et al. (1973a) found that the fluorescence yield at cryogenic temperature depended on the oxidation state of cytochrome b_{559} as well as on the state of the acceptor and that the rate of the light-induced fluorescence increase is slower than the rate of the primary reaction of PSII. The effect of cytochrome b_{559} on fluorescence yield was confirmed in our laboratory (Malkin et al., 1974), and, in addition, it was shown that a component with a midpoint potential of $+475$ mV affected the fluorescence yield at cryogenic temperature. These findings clearly indicate that fluorescence yield is not controlled by the oxidation state of a single component (the primary acceptor of PSII), but is controlled by the oxidation state of several components associated with PSII. Butler (1973) proposed that P680$^+$ quenches fluorescence and that many of the above-mentioned effects result from the rates at which secondary electron donors reduce P680$^+$ to its nonquenching reduced state. The proposal that P680$^+$ is a quencher has gained recent support from measurements of the effect of treatments that inhibit PSII on chlorophyll fluorescence at room temperature in the time range from 0.5 μsec to 100 msec (Etienne, 1974; Den Haan et al., 1974; Van Gorkom et al., 1976).

It seems clear that the fluorescence yield of PSII depends on several parameters, one of which is the oxidation state of the PSII primary electron acceptor. However, the fluorescence yield changes that can accompany the reduction of the primary electron acceptor do not appear to be a necessary consequence of the PSII primary reaction. For example, several investigators (Erixon and Butler, 1971c; Katoh and Kimimura, 1974; Okada et al., 1976) demonstrated in chloroplasts irradiated with ultraviolet light that the loss of the fluorescence-yield change phenomenon (variable fluorescence) was much more pronounced than was the loss of photochemical activity of PSII.

The demonstration that UV-irradiated chloroplasts could retain photochemical activity at low temperature with little or no change in light-induced fluorescence yield was supported by work in our laboratory where PSII fragments were treated with the strong oxidant K_2IrCl_6 (Malkin and Knaff, 1973; Knaff and Malkin, 1976). These oxidized fragments retained full PSII activity, as measured by either the rate of DCMU-sensitive photoreduction of electron acceptors at room temperature or, more important, by the magnitude of P680 photooxidation after a single-turnover flash at cryogenic temperature (Knaff and Malkin, 1976). The fragments show no light-induced increase in fluorescence yield, either at room temperature or at 77°K.

Conclusions concerning fluorescence yield have been placed on a firmer theoretical footing by the recent elegant analysis of PSII fluorescence at cryogenic temperatures by Butler and co-workers (Butler and

Kitajima, 1974, 1975). (Fluorescence at room temperature is a more complicated subject, and a thorough discussion of it would be beyond the scope of this review.) This analysis has as one of its central theses the existence of two different quenching mechanisms: (1) quenching in the bulk chlorophyll and (2) quenching at the reaction center itself. Because of the low concentration of PSII reaction centers, compared to bulk chlorophyll, it is assumed that all of the fluorescence comes from the bulk chlorophyll. Quenching at the reaction center would be expected to decrease the variable fluorescence (the increase in fluorescence yield observed on reduction of the primary acceptor). The proposed mechanism for this phenomenon involves fluorescence quenching by increasing the probability of nonradiative decay processes of the excited reaction-center chlorophyll. Because the rate-constant for photochemical reactions is very much greater than that for fluorescence or for nonradiative processes, a substantial increase in the rate-constant for nonradiative decay can occur with little effect on the photochemical yield. It therefore is understandable that treatment of chloroplasts (with UV light or with K_2IrCl_6) could create quenching centers that preferentially increase the probability of nonradiative decay at the reaction center, thereby eliminating variable fluorescence while maintaining high photochemical yields.

B. C-550

In 1969, a new, light-induced chloroplast absorbance change was discovered (Knaff and Arnon, 1969b). The chloroplast component responsible for this change was named C-550 because the major absorbance feature at room temperature was a light-induced bleaching centered at 550 nm. C-550 was clearly associated with PSII, rather than with PSI, since the absorbance change was detected in PSII subchloroplast fragments but not in PSI fragments and since it was found that the change could not be elicited with long-wavelength light ($\lambda > 700$ nm) that activates PSI (Knaff and Arnon, 1969b). In addition, the action spectrum for production of the C-550 change showed the sharp decrease in quantum efficiency near 680 nm that is characteristic of PSII reactions (Knaff and McSwain, 1971).

A possible role for C-550 in the primary photochemistry of PSII was suggested on the basis of the observation that the absorbance change occurred equally well at 77°K and at room temperature (Knaff and Arnon, 1969b). At cryogenic temperature, as may be seen in Fig. 5, the maximum bleaching shifts slightly to the blue and better resolution reveals a positive peak. Also shown in the figure is the finding that the photoreduction of C-550 is activated only by PSII illumination, even at 77°K.

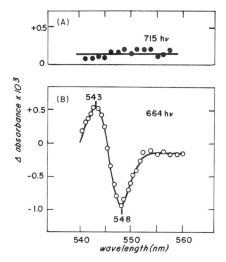

FIG. 5. Photoreduction of C-550 at 77°K in spinach chloroplasts. (A) Photosystem I (715 nm) actinic illumination; (B) Photosystem II (664 nm) actinic illumination.

The connection between C-550 and the primary acceptor was established when it was detected that the absorbance change resulted from a photoreduction rather than from a photooxidation (Erixon and Butler, 1971a,b; Knaff and Arnon, 1969c). This is illustrated in Fig. 6, which shows that the absorbance changes produced in the dark by the addition of a reductant (sodium dithionite) are identical to those produced by illumination.

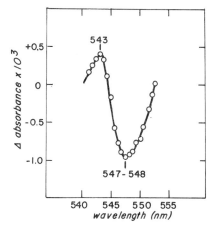

FIG. 6. Reduction of C-550 by dithionite addition in the dark. Reduced minus oxidized spectrum recorded at 77°K.

Not only was it found that the C-550 absorbance change was caused by a photoreduction, but it was also found that C-550 had the same midpoint potential as did the more positive fluorescence component (see Section III,A). Furthermore, the C-550 absorbance change that appeared as the oxidation–reduction potential became more negative had the same midpoint potential as did the component responsible for the PSII activity (Erixon and Butler, 1971b). As discussed earlier (see Section III,A), this behavior would be expected if C-550 were the primary acceptor of PSII.

The nature of the C-550 absorbance change at cryogenic temperature was clarified by a series of observations after chloroplasts were treated with lipase (Butler and Okayama, 1971). This treatment makes it possible to observe the absolute spectra of C-550 in both the oxidized and the reduced forms rather than the reduced minus oxidized difference spectrum shown in Figs. 5 and 6. These experiments showed that oxidized C-550 has an absorption band at 546 nm (at 77°K) and that the absorption band of reduced C-550 (equal in area to that of oxidized C-550) shifts to 544 nm on reduction. The net C-550 absorbance change thus results from a small band-shift to the blue of a component with little or no change in absorbance probability.

These observations on the nature of the C-550 band-shift raised the possibility that C-550 was not the actual primary electron acceptor. It seemed possible that C-550 was a membrane-bound chromophore spatially oriented so that changes in the oxidation–reduction state of the primary electron acceptor altered the electric field near the C-550 chromophore causing a spectral shift. Such electrochromic shifts in carotenoid spectra have been well documented in purple photosynthetic bacteria, where carotenoids play no direct role in primary photochemistry (Arnold and Clayton, 1960; Dutton, 1971).

The possibility that the C-550 band-shift may be simply a convenient indicator of the redox state of the primary acceptor, rather than an absorbance property of the acceptor itself, was supported by experiments with chloroplasts extracted with nonpolar solvents, such as hexane. In these experiments, it was found that the C-550 band-shift required the addition of β-carotene (Okayama and Butler, 1972b; Cox and Bendall, 1974; Knaff et al., 1977). Consideration of this requirement for β-carotene, the existence of carotenoid electrochromic band-shifts in photosynthetic bacteria, and the absence of any evidence to support a direct role for β-carotene in PSII primary photochemistry made it seem even more likely that the C-550 absorbance change resulted from an electrochromic band-shift rather than from the primary acceptor itself. Most important is that recent experiments (Knaff et al., 1977) have shown that extracted chloroplasts reactivated with plastoquinone in the absence of β-carotene have considerable PSII activity (P680 photooxidation at cryogenic temperature) but no C-550 absorbance change. These experiments indicate that the C-550 band-shift is an indirect consequence of the primary reaction, rather than being associated with the primary acceptor.

Further evidence that C-550 is not the PSII primary electron acceptor was obtained with the K_2IrCl_6-treated fragments discussed above. These fragments contain a functional PSII reaction center that exhibits full P680 photooxidation at 15°K, yet they show no C-550 absorbance change, either during illumination (Malkin and Knaff, 1973) or after chemical reduction in the dark (Knaff and Malkin, 1976). An analysis of the kinetics of the back reaction between $P680^+$ and A^- (see Section IV) suggested that the reaction center of PSII probably is unaffected by K_2IrCl_6 treatment. Because the fragments are devoid of C-550, it appears that C-550 is not a necessary component of the PSII reaction center.

C. X-320

Light-induced absorbance changes in the spectral region near 320 nm were first observed by Stiehl and Witt in chloroplasts (1968, 1969) and were interpreted as arising from the primary acceptor of PSII. The evidence supporting the hypothesis that these absorbance changes (attributed to a component called X-320) were associated with the PSII primary acceptor includes (i) the fact that it can be observed in a subchloroplast fragment enriched in PSII and devoid of PSI (Van Gorkom et al., 1975); (ii) the rapid rise-time observed for the appearance of the X-320 absorbance change (less than 30 μsec, limited by instrumental response time); (iii) the fact that the X-320 change can be observed at cryogenic temperatures (Witt, 1973; Haveman et al., 1975; Pulles et al., 1974); (iv) the agreement of the kinetics of X-320 decay ($t_{1/2}$ = 0.6 msec) at room temperature after a short light flash with that expected for the reoxidation of the photoreduced primary acceptor (Döring et al., 1967; Vater et al., 1967); (v) the observation that DCMU, an inhibitor known to block reoxidation of the photoreduced primary acceptor (Duysens and Sweers, 1963; Knaff and Arnon, 1969b,c) had no effect on the light-induced X-320 change but did inhibit its dark decay after a flash (Witt, 1973); (vi) the agreement of the decay kinetics of X-320 at liquid-nitrogen temperature with those expected for the reduced primary acceptor being reoxidized via a back reaction with $P680^+$ (Haveman et al., 1975); and (vii) the agreement of X-320 decay kinetics at room temperature (in chloroplasts treated at low pH to inhibit O_2 evolution) with the kinetics observed for the back reaction under these conditions (Pulles et al., 1976; Mathis et al., 1976; Renger and Wolff, 1976).

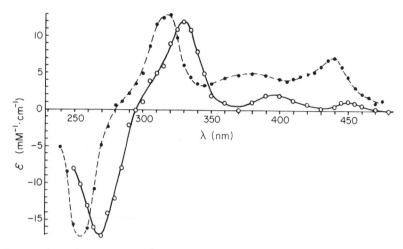

Fig. 7. Light minus dark difference spectrum of X-320 in deoxycholate-treated chloroplast fragments (solid line) and the difference spectrum of plastosemiquinone anion minus plastoquinone. From Van Gorkom (1974).

X-320 thus meets many of the criteria expected for a species that functions as the primary acceptor of PSII. There has as yet been no direct demonstration that the X-320 absorbance change is caused by a reduction rather than by an oxidation, largely because of interfering ultraviolet absorbance of the reducing agents known to be capable of reducing the primary acceptor in the dark. However, the effect of DCMU on the kinetics of X-320 can most readily be interpreted by assuming that X-320 is indeed the PSII primary acceptor. Furthermore, the light *minus* dark difference spectrum of X-320 (Fig. 7) shows the net appearance and disappearance of absorption bands expected for an oxidation–reduction reaction, in contrast to the band-shift type of difference spectrum seen in the C-550 region and a similar band-shift in the chlorophyll region (Van Gorkom, 1974; Lozier and Butler, 1974), both of which appear to be from indirect electrochromic shifts.

D.   CHEMICAL IDENTITY OF THE PRIMARY ELECTRON ACCEPTOR

Because of evidence accumulating to support the hypothesis that X-320 is the primary acceptor of PSII, a logical next step was to deduce its possible chemical identity from the light *minus* dark (reduced *minus* oxidized) difference spectrum. The broken line spectrum seen in Fig. 7 is the difference spectrum observed by Bensasson and Land (1973) for the plastosemiquinone anion (chemically reduced) *minus* fully oxidized plastoquinone. There is good agreement between the X-320 light *minus* dark difference spectrum determined by Van Gorkom (1974) and the plastosemiquinone anion *minus* plastoquinone difference spectrum. Both contain the same major features (a large decrease in absorbance in the 250–270 nm region, a large increase in absorbance in the 315–330 nm region, and two smaller absorbance increases in the region between 370 and 450 nm), but they differ in the exact wavelength positions of the maxima and minima and in the relative extinction coefficients of absorbance. The striking similarities between the two spectra shown in Fig. 7 made it likely that the primary acceptor is a specialized membrane-bound plastoquinone that functions between the fully oxidized and one-electron-reduced (semiquinone) state. With a reduced *minus* oxidized extinction coefficient of 13 mM^{-1} cm^{-1} taken from the *in vitro* studies (Bensasson and Land, 1973), the amount of primary acceptor photoreduced equals one acceptor per 300 chlorophyll molecules (Pulles *et al.*, 1974; Van Gorkom, 1974). This number is in reasonable agreement with the known size of the PSII photosynthetic unit.

The X-320 difference spectrum does not agree with that of the fully (two-electron) reduced *minus* fully oxidized spectrum (Bensasson and Land, 1973; Redfearn and Friend, 1962) of plastoquinone, in agreement with evidence that the primary reaction of PSII involves the transfer of a single electron and produces the P680$^+$ cation radical (Malkin and Bearden, 1975; Mathis and Vermeglio, 1975). Furthermore, the observation that plastoquinone appears to be reduced only to the semiquinone state during the primary reaction would seem to rule out proposals in which a single quinone serves as the acceptor for two P680 molecules (see Witt, 1975). Not only do the *in vitro* studies point to plastosemiquinone as the reduced form of the PSII primary acceptor, but the spectra seem to preclude the protonated semiquinone as the reduced species, making a much better fit for the unprotonated semiquinone anion (Bensasson and Land, 1973).

Another approach to establishment of the chemical identity of the PSII primary acceptor involves extraction of chloroplasts with nonpolar solvents (see discussion above in Section III,B). The first studies reported concerning the effects of such extraction on the PSII primary photochemical reaction produced conflicting results about the roles of plastoquinone and β-carotene (Okayama and Butler, 1972b; Cox and Bendall, 1974). A recent investigation with a similar approach gave direct evidence for the role of plastoquinone as the PSII primary electron acceptor. Figure 8 shows the effects of extraction (with 0.2% methanol in hexane) and reconstitution with pure plastoquinone on P680 photooxidation. P680 photooxidation can be followed by the reversible EPR signal of the P680$^+$ cation radical (see Section II). (The irreversible EPR signals seen at this *g* value are caused by a species other than

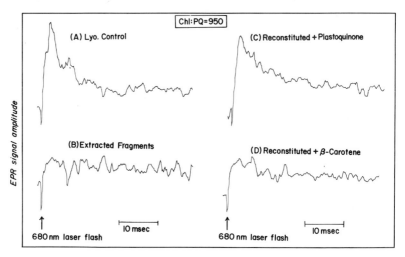

FIG. 8. Effect of extraction and reconstitution on P680 photooxidation at 15°K in Photosystem II subchloroplast fragments. From Knaff *et al.* (1977).

P680$^+$.) Removal of plastoquinone below a level of one plastoquinone per 750 chlorophyll molecules resulted in the loss of P680 photooxidation at 15°K (Knaff *et al.*, 1977). Reconstitution with β-carotene did not restore P680 photooxidation to any measurable extent but reconstitution with pure plastoquinone resulted in an approximately 60% reconstitution (Knaff *et al.*, 1977). These results indicate that plastoquinone, but not β-carotene, is required for P680 photooxidation.

Partial extraction of plastoquinone (plastoquinone to chlorophyll ratios between 1:400 and 1:600) resulted in only partial loss of P680 photooxidation. Brief extractions that left the plastoquinone-to-chlorophyll ratio above 1:400 had no effect on P680 photooxidation. It therefore appears that the plastoquinone compliment of PSII consists of two pools: a tightly bound pool (present at a level equimolar with that of P680) that functions as the primary electron acceptor and a less tightly bound pool (plastoquinone:P680 = 4:1) that is not involved in the primary reaction and presumably functions as a secondary electron acceptor (see Amesz, 1973). The reconstitution of P680 photooxidation at cryogenic temperature by plastoquinone in exhaustively extracted chloroplasts is the most important evidence for plastoquinone functioning as the primary electron acceptor of PSII. This evidence is rendered more convincing by the observation that the half-time for the back-reaction between P680$^+$ and A$^-$ at cryogenic temperatures (a measure of the intactness of the reaction center) is identical in control and reconstituted preparations (Knaff *et al.*, 1977).

β-Carotene appears to play no direct role in the primary photochemistry of PSII. Extraction of chloroplasts with organic solvents is known to remove 99% of the β-carotene (Cox and Bendall, 1974), yet P680 photooxidation can be restored in such extracted preparations by the addition of plastoquinone alone (Knaff et al., 1977). Although β-carotene is not needed for the primary photochemical reaction of PSII, it is required in some as yet undefined manner for the C-550 absorbance change (Okayama and Butler, 1972b; Cox and Bendall, 1974; Knaff et al., 1977).

If, as indicated by the evidence cited above, a specialized plastoquinone molecule is reduced to the semiquinone during the primary reaction of PSII, an EPR signal from this reduced species should be observed. To date, no EPR signal from the PSII primary electron acceptor has been detected under conditions where the acceptor is known to be reduced, an indication that the reduced acceptor is not a simple plastosemiquinone but that a more complicated situation exists. Similar results have been obtained in studies of the primary photochemistry in reaction centers from photosynthetic bacteria (for a recent review, see Parson and Cogdell, 1975). In that case, it was found that a ubiquinone–iron complex functions as the primary electron acceptor and that the nonheme iron alters the environment of the quinone so that the ubisemiquinone EPR signal is not observed and a broad, highly temperature-dependent signal is associated with the acceptor. After removal of the iron, the EPR signal of the ubiquinone radical is found when the acceptor is reduced. It is possible that a similar situation exists in the case of chloroplast PSII, particularly in light of recent observations in our laboratory that PSII subchloroplast fragments contain significant amounts of firmly bound nonheme iron (Knaff et al., 1977). The possible role of this iron and its relationship to the primary electron acceptor remain areas for further investigation.

E. OXIDATION–REDUCTION PROPERTIES

As discussed above, oxidation–reduction titrations of loss of PSII activity (Erixon and Butler, 1971b; Knaff, 1975a), variable fluorescence (Cramer and Butler, 1969; Ke et al., 1976), and the C-550 absorbance change (Erixon and Butler, 1971b; Knaff, 1975a) established that the PSII primary acceptor has a midpoint potential near 0.0 V at pH 7. Fluorescence titrations (Cramer and Butler, 1969; Ke et al., 1976) established that this midpoint potential was pH-dependent, with a -60 mV/pH unit dependence indicating that the primary acceptor was capable of taking up one H^+ ion per electron on reduction. This pH-dependence has been confirmed in titrations of both the C-550 absorbance change (Knaff, 1975b) and the loss of PSII activity (Knaff, 1975a).

These titrations were performed in experiments where long times were allowed for the membrane-bound carriers to come to equilibrium with the oxidation–reduction mediators and, as a result, sufficient time is allowed for proton equilibration. Under these conditions, the reduction of the primary acceptor (A) can be represented by the equation:

$$A + H^+ + e^- \rightarrow AH \qquad (2)$$

However, kinetic results (Ausländer and Junge, 1974) and the identification of the reduced acceptor as the plastosemiquinone anion (Van Gorkom, 1974) indicate that the photoreduced primary acceptor is reoxidized before it can be protonated. Therefore, on the time scale of photosynthetic electron transport, the reduction of the primary acceptor is more accurately represented by the equation:

$$A + e^- \rightarrow A^- \qquad (3)$$

Prince and Dutton (1976) have pointed out for the analogous case of the primary electron acceptors in photosynthetic bacteria that the operative midpoint potential of the primary electron acceptor corresponds to the midpoint potential at a pH greater than 7 of the pK of the reduced form of the acceptor. At this pH, the reduction of the acceptor

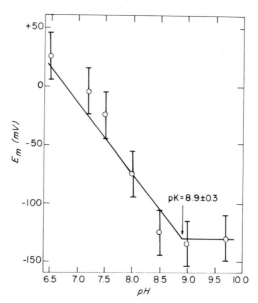

FIG. 9. Effect of pH on the midpoint oxidation–reduction potential of the Photosystem II primary electron acceptor. From Knaff (1975a).

is expressed by Eq. (3) rather than by Eq. (2). Figure 9 shows a determination of the pK for the reduced PSII primary acceptor (Knaff, 1957a). The pK value of 8.9 corresponds to an "effective" E_m of -130 mV, considerably more negative than the usually cited values near 0.0 V obtained at pH 7.

The midpoint potential of the PSII primary acceptor is affected not only by pH: detergents also appear to shift the E_m to more positive values (Knaff, 1975b). Another recent finding was the observation that o-phenanthroline [known to inhibit PSII reactions by blocking the reoxidation of the photoreduced primary acceptor (Knaff and Arnon, 1969c)] shifted the midpoint potential of the primary acceptor by $+70$ mV. The exact mechanism of this shift is not known, but the phenomenon has also been observed in titrations of the primary electron acceptor in purple photosynthetic bacteria (Jackson et al., 1973; Dutton et al., 1973; Evans et al., 1974).

IV.   Electron Donors to P680

A.   DONORS AT CRYOGENIC TEMPERATURE

The following photochemical model for the PSII reaction center (Butler et al., 1973a,b) at cryogenic temperature has been supported by a number of direct and indirect measurements:

$$D \cdot P \cdot A \underset{k_{-1}}{\overset{k_1}{\rightleftharpoons}} D \cdot P^+ \cdot A^- \overset{k_2}{\longrightarrow} D^+ \cdot P \cdot A^- \qquad (4)$$

where A is the primary electron acceptor, P is the reaction-center chlorophyll, P680, and D is an electron donor capable of reducing $P680^+$ at low temperatures. According to this model, the rate constant, k_1, is a function of light intensity and the PSII charge separation results in the formation of P^+ and A^-. Two possible reaction routes are then available for P^+: (1) P^+ can be reduced in a back reaction with A^- or (2) P^+ can be reduced by an electron donor, D. It should be obvious, however, that the reversal of the charge separation annihilates charge stabilization and therefore is a reaction that is wasteful in terms of the overall process of chloroplast photosynthesis.

1. The Photosystem II Back Reaction

Early evidence regarding the presence of a back-reaction between the PSII primary electron acceptor and the primary donor was obtained by indirect methods. In the studies, various components related to the primary electron acceptor (C-550 and Q) were observed before and after saturating flashes of light (Butler et al., 1973b; Vermeglio and Mathis,

1973; Lozier and Butler, 1974; Murata et al., 1973). In some of this work it was shown that single saturating flashes given to the PSII reaction center at cryogenic temperature resulted in about 20–30% of the acceptor being transformed to the reduced state after the flash; such flashes were totally effective in reducing the primary acceptor at physiological temperatures. A second flash was generally less effective, and after a series of about 10 flashes 50–70% of the acceptor was in the reduced state.

The explanation of the relative ineffectiveness of saturating flashes in the production of a stabilized reduced primary acceptor at cryogenic temperature was first put forward by Butler et al. (1973a,b), according to Eq. (4) above. Stabilization of A^- requires reduction of P^+ via an electron donor (characterized by rate k_2) and, if the rate of the back reaction is faster than that for the secondary reaction, the back reaction will occur to a significant extent. At cryogenic temperatures, charge separation would be expected to occur after flash-activation, but most of the A^- produced would react with P^+. Only to the extent that P^+ is reduced by a secondary donor, D, would A^- be stabilized in the reduced state in the dark following flash-activation.

Direct evidence for a back-reaction in the PSII reaction center at cryogenic temperature has been obtained by monitoring reaction-center components during flash activation. Mathis and Vermeglio (1974) reported that C-550 was fully reduced by a laser flash at 77°K and was then reoxidized in a reaction with a half-time of about 4 msec. This report was followed by measurements of the $P680^+$ free-radical EPR signal at temperatures from 10° to 77°K by Malkin and Bearden (1975) and by Visser (1975). $P680^+$ was found to be formed rapidly after flash-activation and to exhibit a decay with a half-time of about 5.0 msec. Mathis and Vermeglio (1975) also detected the reduction of $P680^+$ by observing absorbance changes in the 820-nm region at about 77°K, and a similar decay time was observed following flash-activation. More recently, Haveman et al. (1975) reported that the X-320 absorbance change shows a similar kinetic response after flash-activation at about 77°K.

In the above-described studies of PSII components, numerous control experiments were done to confirm the association of the back reaction with the PSII reaction center. In the presence of DCMU, hydroxylamine, and with preillumination at 300°K (with the primary electron acceptor trapped in the reduced state), no back reaction was observed after subsequent flash-activation at cryogenic temperature. Similarly, chemical reduction of A at 300°K prior to flash-activation at cryogenic temperature resulted in no further photoreduction. Preillumination of samples at cryogenic temperature with continuous light, which activates

PSII, is also effective in eliminating the flash-induced changes since D^+ and A^- accumulate prior to the flash-activation. Although the evidence for the occurrence of the PSII back-reaction at cryogenic temperature is now clear, changes in fluorescence yield under similar conditions failed to reflect any evidence for a millisecond reaction (Den Haan et al., 1973). Although this result has not been fully explained in terms of the above model, the absence of a millisecond decay component may be related to recent results that indicate the components other than the primary electron acceptor, Q, can function as fluorescence quenchers (see Butler, 1972a; Butler et al., 1973a).

2. Photooxidation of Chloroplast Cytochrome b_{559}

Although there were early reports of the photooxidation of chloroplast cytochrome f at cryogenic temperatures (Chance and Bonner, 1963; Witt et al., 1961), recent studies have demonstrated that this assignment was incorrect and that the only chloroplast cytochrome which undergoes photooxidation at cryogenic temperature is cytochrome b_{559}. Knaff and Arnon (1969a) first unequivocally demonstrated the photoreaction of this cytochrome at liquid-nitrogen temperature and, in addition, reported the key finding that this photooxidation is mediated by PSII.

Some of the early confusion regarding the identity of the cytochrome undergoing low-temperature photooxidation arose from the shift in the α-band maximum of cytochrome b_{559} at cryogenic temperatures. Thus, the cytochrome band at 77°K is at 556 nm and at 559 nm at 300°K. This low-temperature band is close to that of cytochrome f (α-band at 554 nm at 300°K and at 552 nm at 77°K), which caused some confusion. Most groups now agree that little, if any, cytochrome f undergoes low-temperature photooxidation.

The PSII nature of the cytochrome b_{559} photooxidation reaction was confirmed in studies by Erixon and Butler (1971a), who found similar reactions in chloroplasts and in PSII subchloroplast preparations. Subsequent work with various other PSII preparations, including highly enriched reaction-center complexes, further correlated the involvement of PSII in cytochrome b_{559} photooxidation at cryogenic temperatures (Boardman, 1972; Kitajima and Butler, 1973; Wessels et al., 1973).

In further characterization of PSII reactions in chloroplasts at liquid-nitrogen temperature, Erixon and Butler (1971b) found that the photooxidation of cytochrome b_{559} was linked to the photoreduction of the PSII primary electron acceptor (monitored by the C-550 absorbance change). It was demonstrated that reduction of the PSII primary acceptor by reductive titration prior to freezing or photochemically in the presence of DCMU and hydroxylamine prior to freezing led to a loss of photooxidation of cytochrome b_{559} at 77°K during a subsequent illumination period. These results led to the conclusion that cytochrome b_{559} can function as an electron donor to the PSII reaction center.

Although cytochrome b_{559} undergoes photooxidation at 77°K, it has been shown that at higher temperatures the extent of cytochrome oxidation decreases. Butler *et al.* (1973b) found little cytochrome oxidation at about 170°K after continuous illumination, even though the PSII primary electron acceptor became fully reduced (as indicated by the extent of the C-550 change). Similarly, Vermeglio and Mathis (1973) reported little cytochrome oxidation at 220°K after continuous illumination, although the PSII primary acceptor again became fully reduced under comparable conditions. In the latter studies, it was shown, however, that complete cytochrome oxidation could occur at 220°K if the chloroplasts received one saturating flash at 273°K prior to measurements at the lower temperature. It therefore does not appear that any inactivation of the photochemical system occurred; it may be that alternate electron donors to P680 may be present and functioning at different temperatures.

In the early studies of cytochrome b_{559} it was demonstrated that the form of the cytochrome which underwent low-temperature photooxidation in untreated chloroplasts was the "high-potential" form. The oxidation–reduction potential of this form is about +350 mV (Knaff and Arnon, 1971; Boardman *et al.*, 1971; Okayama and Butler, 1972a). However, it has also been shown that other forms of cytochrome b_{559}, with lower midpoint potentials, can be photooxidized at 77°K. After washing with Tris buffer, lower potential forms of cytochrome b_{559} are found, and they can be photooxidized at 77°K (Erixon *et al.*, 1972). In addition, in PSII reaction-center complex preparations, cytochrome b_{559} is in a form with a midpoint potential of about +50 mV, and this form also shows low-temperature photoreactions (Wessels *et al.*, 1973).

3. Other Secondary Electron Donors

At temperatures of about 150°K, it has been demonstrated that C-550 photoreduction occurs during continuous illumination and that little or no cytochrome b_{559} is photooxidized (Butler *et al.*, 1973b; Vermeglio and Mathis, 1973). However, preillumination at 300°K makes possible subsequent cytochrome photooxidation at about 150°K (Vermeglio and Mathis, 1973). These findings have been interpreted as indicative of an electron donor other than cytochrome b_{559} functioning to reduce P680$^+$. Little information is available on the chemical identity of this secondary donor.

It has become apparent from a series of extensive studies that electron donors other than cytochrome b_{559} are also capable of reducing oxidized P680 at temperatures below 77°K. The best-documented alternate electron donor appears to be a chlorophyll molecule near the reaction center. The EPR properties of this chlorophyll were first described by Malkin and Bearden (1973), and the component was initially assigned to P680. However, after studies with flash-activation of the PSII reaction center, the above-described model evolved, and it was noted that $P680^+$ is not stable after illumination but always undergoes a subsequent reduction. The EPR component reported by Malkin and Bearden was formed irreversibly and therefore could not originate from $P680^+$ (see also Visser, 1975).

The free-radical signal observed after illumination with red light at cryogenic temperatures had a g value of 2.0026 and a linewidth of 8 gauss. These parameters are very similar to those for other chlorophyll cation radicals. The light-induced signal was observed only at high oxidation-reduction potentials ($E_m > +450$ mV) and under these conditions cytochrome b_{559} is known to be fully oxidized prior to illumination and therefore to be incapable of undergoing photooxidation. Subsequent optical studies showed an irreversible absorbance change in the 820-nm spectral region under conditions comparable to those used in the EPR studies, and it was concluded that the irreversibly formed optical component also arose from a chlorophyll cation radical (Mathis and Vermeglio, 1973). Recently, it has been possible to detect absorbance changes of this species in the red spectral region, and a component with an absorbance maximum at 676 nm was characterized by Visser (1975).

The EPR and optical studies described above provided evidence that the PSII oxidation is most likely of a species that is a chlorophyll molecule. Whether this component is a direct electron donor to P680 is, however, not yet clear. It has been shown by Bearden and Malkin (1973) and by Visser (1975) that an unknown component with an E_m of about +450 mV must be oxidized before the chlorophyll cation radical can be photoinduced at cryogenic temperatures. This component may be identical to one observed during titration of the PSII fluorescence yield at 77°K, where a quenching component with an E_m of about +475 mV was noted (Malkin et al., 1974).

Two models incorporating these additional secondary electron donors to P680 appear to be possible. In one, the alternate donors would interact with one reaction center:

$$\text{Chlorophyll}\,a \nearrow \begin{array}{l} \text{P680} \leftarrow \text{cytochrome } b_{559} \\ \uparrow \\ \text{X} + 450\,\text{mV} \end{array} \tag{5}$$

According to this model, the chemical oxidation (by ferricyanide) of any one donor would allow the next thermodynamically favorable component to function as a donor. A second model has been proposed in which a heterogeneity of the PSII reaction centers exists so that cytochrome b_{559} functions with some reaction centers and the other donors function with another group of reaction centers (Visser, 1975):

$$P680 \leftarrow \text{cytochrome } b_{559} \qquad \begin{array}{c} P680 \leftarrow X + 450 \text{ mV} \\ \uparrow \\ \text{chlorophyll} \end{array} \qquad (6)$$

In support of the latter model are observations by Visser (1975) that cytochrome b_{559} was photooxidized in only half of the PSII reaction centers. In addition, Mathis *et al.* (1974) and Visser (1975) reported that the rate of photooxidation of cytochrome b_{559} and the rate of reduction of C-550 or Q was different at low light intensities. It is not clear in the latter model if the heterogeneity of the reaction centers is considered to be an intrinsic property of the photosynthetic system or if this heterogeneity arises from some experimental artifact, such as the freezing procedure necessary for measurements at cryogenic temperatures. Heterogeneity in the PSI reaction center at cryogenic temperatures has been observed (Lozier and Butler, 1974), but the finding has not been adequately explained in terms of possible physiological significance.

B. ELECTRON DONORS AT PHYSIOLOGICAL TEMPERATURE

Although various electron donors for P680 at cryogenic temperature have been described, their role as intermediates in water oxidation under physiological conditions is not clear. Because the molecular nature of several of the components that function at cryogenic temperature is not known with certainty, it has been difficult to study their possible involvement under physiological conditions. We will not attempt in this section to describe the work on the kinetics of oxygen evolution, a subject that was recently reviewed (Radmer and Kok, 1975; Joliot and Kok, 1975). However, a brief discussion of electron donors to the PSII reaction center will be presented because of significant recent advances in our knowledge of this area and because of our previous discussion of the kinetics of $P680^+$ reduction at 300°K.

As previously described, the best-documented electron donor to P680 at cryogenic temperature is cytochrome b_{559}. The role of this cytochrome under physiological conditions has been controversial. A recent review describes the present status of various opinions regarding the function of the cytochrome (Cramer and Horton, 1975), and only a few key observations will be made at this point. Several groups have reported that little or no cytochrome b_{559} photoreaction occurs in untreated chloroplasts (Knaff and Arnon, 1969a; Hiller *et al.*, 1971; Cramer and Böhme, 1972), whereas other groups have reported cytochrome turnover (Ben-Hayyim and Avron, 1970). Cytochrome b_{559}

photoreactions have also been reported in oxygen-evolving membrane fragments from a blue-green alga (Knaff, 1973; Tsujimoto *et al.*, 1976). This diversity of results has led to a certain amount of confusion about the exact site of function of the carrier, and no consensus has been reached at this time. Much of the evidence appears to indicate that the cytochrome does not function as an intermediate in the oxidation of water, although it is known to be closely associated with the PSII reaction center. In a study of various treatments that inactivated oxygen evolution, Cox and Bendall (1972) found no correlation between the high-potential form of cytochrome b_{559} and oxygen evolution; significant amounts of oxygen were evolved (30% of control rates) even in the absence of the high-potential form of the cytochrome. Also consistent with the view that the high-potential form of cytochrome b_{559} is not an intermediate in the oxidation of water is the finding that during the greening process oxygen evolution appears well before the high-potential form of cytochrome b_{559} can be detected (Henningsen and Boardman, 1973). Some workers have placed the cytochrome on a side path around PSII (Boardman *et al.*, 1971). The possible physiological function of such a pathway is obscure.

The assignment of another chloroplast component, represented by EPR "Signal II" to the oxidizing side of PSII has been supported by recent studies in chloroplasts. Signal II refers to an EPR signal which is light-induced and has a g value of 2.0046, a linewidth of approximately 20 gauss, and a complex hyperfine structure (Kohl, 1972), as shown in Fig. 10. This signal is not found in photosynthetic bacteria or in algal mutants unable to evolve oxygen (Weaver, 1962; Weaver and Bishop, 1963). It is found in subchloroplast fragments enriched in PSII activity but is absent from PSI subchloroplast fragments (Kohl, 1972). These findings have led to the suggestion that Signal II is associated with the oxygen-evolving system of plant and algal photosynthesis.

Early evidence suggested that Signal II arose from plastosemiquinone or a closely related species. This conclusion was based primarily on the results of Kohl and Wood (1969), who showed that the removal of plastoquinone by organic solvent extraction resulted in a decrease of Signal II intensity and that readdition of plastoquinone led to an increase in light-induced Signal II. More important, reconstitution of extracted fragments with totally deuterated plastoquinone led to a narrowing of the characteristic Signal II spectrum. Although this evidence indicates that plastoquinone somehow is required for the formation of Signal II, direct kinetic evidence is lacking for the association of Signal II with the plastoquinone pool that functions between the two chloroplast light reactions. No antagonistic effect of red and far-red light on Signal II has been demonstrated. The extremely long lifetime of Signal II (about 1 hr

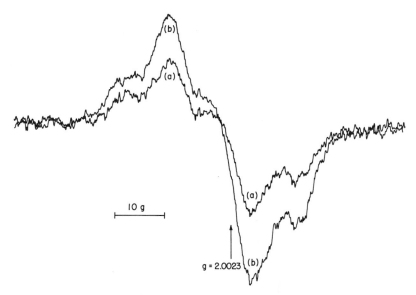

10 g

g = 2.0023

FIG. 10. EPR spectra at 300°K of dark-adapted spinach chloroplasts in the dark before (a) and immediately after (b) illumination. From Babcock and Sauer (1973a).

in untreated chloroplasts) led to speculations that the component responsible for Signal II is not involved in overall electron transfer in chloroplast photosynthesis.

More recent work has led to the conclusion that the site of function of Signal II is on the oxidizing side of PSII. Initial observations that led to a reconsideration of the site of function were made by Butler and co-workers (Okayama et al., 1971; Lozier and Butler, 1973) when it was shown that treatments that specifically affect the oxidizing intermediates involved in oxygen evolution also affect the decay of Signal II. Tris-treatment or NH_2OH treatment of chloroplasts, two procedures known to block specifically oxygen evolution, resulted in an acceleration of Signal II decay (Lozier and Butler, 1973; Warden and Bolton, 1974). These types of observations suggested that Signal II might be related to an intermediate functioning between P680 and water during oxygen evolution.

Although the initial studies that led to the proposal that Signal II functions on the oxidizing side of PSII were mainly with various modified chloroplast preparations, Babcock and Sauer (1973a,b,1975a,b) have reported an extensive series of studies on the behavior of Signal II in untreated chloroplasts in both continuous and flashing light. These experiments have provided strong evidence in favor of a role for the Signal II component as an electron donor to P680.

Babcock and Sauer (1973a) first showed that Signal II can be induced by PSII illumination in a reaction insensitive to DCMU. Far-red light was ineffective in producing Signal II. The photoinduced change also occurred with as high a quantum efficiency as that observed for the photooxidation of P700. From kinetic studies with flash-activation it was concluded that Signal II is in equilibrium with oxidizing equivalents generated by PSII. A general problem that still existed in these measurements was that the kinetic response of Signal II was too slow to be from a componet that is an intermediate in O_2 evolution. However, in subsequent studies, Babcock and Sauer (1975a,b) identified several different kinetic forms of Signal II, and some of these forms showed more rapid kinetics.

One form of Signal II showed flash-induced changes that are rapid enough to be from a component involved as an electron donor to the PSII reaction center (Blankenship *et al.*, 1975; Warden *et al.*, 1976). This Signal II species has been detected in O_2-evolving chloroplasts at physiological temperatures; it arises faster than 100 μsec and has a decay half-time of 700 μsec. The association of this rapid Signal II component with the donor end of PSII, in contrast to the acceptor end, was determined by studies of the effect of Tris treatment on the transient signal and by the effect of specific PSII inhibitors.

Blankenship *et al.* (1975) and Warden *et al.* (1976) have proposed that Signal II arises from the electron donor to P680, based on the above kinetic information. This assignment, however, must be considered tentative until the rise-time for the Signal II species has been measured and until discrepancies in the data reported on the kinetics of reduction of $P680^+$ at 300°K have been resolved. It also is not apparent what, if any, is the relationship of Signal II to plastoquinone, although Okayama (1974) recently reported findings that indicate a possible role of plasto-quinone on the oxidizing as well as on the reducing side of PSII.

V. Concluding Remarks

Considerable progress has been made in recent years toward an understanding of the nature of the reactants in the primary photochemical event of PSII. It now appears certain that the primary donor of this photosystem, P680, is a specialized chlorophyll *a* dimer that is oxidized in the light to the corresponding cation radical. A substantial amount of evidence has been accumulated to indicate that the primary electron acceptor is a specialized plastoquinone molecule that functions between the levels of the fully oxidized and semiquinone anion. The oxidation–reduction properties of the primary acceptor and the kinetics of its reoxidation by the secondary plastoquinone pool have been well characterized. However, several problems await solution; among these are the following:

1. Characterization of the oxidation–reduction properties of P680. It is known only that P680$^+$ must be a strong enough oxidant to result ultimately in the oxidation of water to oxygen (E_m greater than 0.82 V at pH 7.0). Investigations of this question have been hampered because all known oxidants electropositive enough to oxidize P680 destroy many chloroplast components.

2. Identification of physiological electron donors to P680$^+$. Discrepancies between reported decay times of P680$^+$ remain unresolved, and the relationship of the decay times to such intermediates as Signal II remains to be elucidated.

3. The role of other required cofactors, such as manganese and chloride ions (Cheniae, 1970; Radmer and Kok, 1975) on the oxidizing side of PSII. Recent studies on the involvement of manganese indicate that this will be an active area of investigation (Blankenship and Sauer, 1974; Wydrzynaski et al., 1975).

4. Chemical characterization of X-320 to establish unambiguously that the chloroplast component producing this spectral change has the oxidation–reduction properties expected for the primary electron acceptor, based on measurements of Q and C-550.

5. A consideration of the possibility that a transient species accepts the electron lost by P680 in the picosecond time range prior to transfer to plastoquinone. Such an intermediate, possibly involving reduced bacteriopheophytin (Dutton et al., 1975; Parson and Cogdell, 1975; Rockley et al., 1975: Fajer et al., 1975; Tiede et al., 1976), is now known to function in the 5–250 psec time range in purple photosynthetic bacteria, and the consideration should be investigated that such short-lived intermediates may occur in the photosynthetic reaction centers of higher plants.

It is anticipated that during the next several years some of these questions will be answered and our insight into the mechanism of the charge separation of PSII will be advanced. This knowledge should be a basis for the characterization of secondary electron-transfer reactions associated with PSII, including the reaction that is the key to the life of animals on our planet, oxygen evolution.

ACKNOWLEDGMENT

The authors' work cited in this paper was supported in part by grants from the National Science Foundation.

REFERENCES

Amesz, J. (1973). *Biochim. Biophys. Acta* **301**, 35–51.

Arnold, W., and Clayton, R. K. (1960). *Proc. Natl. Acad. Sci. U.S.A.* **46**, 769–776.

Ausländer, W., and Junge, W. (1974). *Biochim. Biophys. Acta* **357**, 285–298.

Babcock, G. T., and Sauer, K. (1973a). *Biochim. Biophys. Acta* **325**, 483–503.

Babcock, G. T., and Sauer, K. (1973b). *Biochim. Biophys. Acta* **325**, 504–519.

Babcock, G. T., and Sauer, K. (1975a). *Biochim. Biophys. Acta* **376**, 315–328.

Babcock, G. T., and Sauer, K. (1975b). *Biochim. Biophys. Acta* **376**, 329–344.

Bearden, A. J., and Malkin, R. (1973). *Biochim. Biophys. Acta* **325**, 266–274.

Bearden, A. J., and Malkin, R. (175). *Q. Rev. Biophys.* **7**, 131–177.

Ben-Hayyim, G., and Avron, M. (1970). *Eur. J. Biochem.* **14**, 205–213.

Bensasson, R., and Land, E. J. (1973). *Biochim. Biophys. Acta* **325**, 175–181.

Berkner, L. V., and Marshall, L. C. (1968). Brookhaven Lect. Ser. No. 64 (BNL-50114). Brookhaven Natl. Lab., Upton, New York.

Blankenship, R. E., and Sauer, K. (1974). *Biochim. Biophys. Acta* **357**, 252–266.

Blankenship, R. E., Babcock, G. T., Warden, J. T., and Sauer, K. (1975). *FEBS Lett.* **51**, 287–293.

Boardman, N. K. (1972). *Biochim. Biophys. Acta* **283**, 469–482.

Boardman, N. K., Anderson, J. M., and Hiller, R. G. (1971). *Biochim. Biophys. Acta* **234**, 126–136.

Borg, D. C., Fajer, J., Felton, R. H., and Dolphin, D. (1970). *Proc. Natl. Acad. Sci. U.S.A.* **67**, 813–820.

Butler, W. L. (1972a). *Proc. Natl. Acad. Sci. U.S.A.* **69**, 3420–3422.

Butler, W. L. (1972b). *Biophys. J.* **12**, 851–857.

Butler, W. L. (1973). *Acc. Chem. Res.* **6**, 177–184.

Butler, W. L., and Kitajima, M. (1974). *Proc. Int. Congr. Photosynth., 3rd, 1974* pp. 13–24.

Butler, W. L., and Kitajima, M. (1975). *Biochim. Biophys. Acta* **376**, 116–125.

Butler, W. L., and Okayama, S. (1971). *Biochim. Biophys. Acta* **245**, 237–239.

Butler, W. L., Visser, J. W. M., and Simons, H. L. (1973a). *Biochim. Biophys. Acta* **292**, 140–151.

Butler, W. L., Visser, J. W. M., and Simons, H. L. (1973b). *Biochim. Biophys. Acta* **325**, 539–545.

Chance, B., and Bonner, W. D. (1963). *In* "Photosynthetic Mechanisms of Green Plants" (B. Kok and A. Jagendorf, eds.), pp. 66–81. Natl. Acad. Sci.—Natl. Res. Counc., Washington, D. C.

Cheniae, G. M. (1970). *Annu. Rev. Plant Physiol.* **21**, 467–498.

Cox, R. P., and Bendall, D. S. (1972). *Biochim. Biophys. Acta* **283**, 124–135.

Cox, R. P., and Bendall, D. S. (1974). *Biochim. Biophys. Acta* **347**, 49–59.

Cramer, W. A., and Böhme, H. (1972). *Biochim. Biophys. Acta* **256**, 358–369.

Cramer, W. A., and Butler, W. L. (1969). *Biochim. Biophys. Acta* **172**, 503–510.

Cramer, W. A., and Horton, P. (1975). *Photochem. Photobiol.* **22**, 304–308.

Den Haan, G. A., Warden, J. T., and Duysens, L. N. M. (1973). *Biochim. Biophys. Acta* **325**, 120–125.

Den Haan, G. A., Duysens, L. N. M., and Egberts, D. J. N. (1974). *Biochim. Biophys. Acta* **368**, 409–421.

Döring, G. (1975). *Biochim. Biophys. Acta* **376**, 274–284.

Döring, G., and Witt, H. T. (1972). *Photosynth., Two Centuries Its Discovery Joseph Priestley, In Proc. Int. Congr. Photosynth. Res., 2nd, 1971* pp. 39–45.

Döring, G., Stiehl, H. H., and Witt, H. T. (1967). *Z. Naturforsch., Teil B* **22**, 639–644.

170   DAVID B. KNAFF AND RICHARD MALKIN

Döring, G., Renger, G., Vater, J., and Witt, H. T. (1969). *Z. Naturforsch., Teil B* **24**, 1139-1143.
Dutton, P. L. (1971). *Biochim. Biophys. Acta* **226**, 63-80.
Dutton, P. L., Leigh, J. S., and Wraight, C. A. (1973). *FEBS Lett.* **36**, 169-173.
Dutton, P. L., Kufmann, K. J., Chance, B. C., and Rentzepis, P. M. (1975). *FEBS Lett.* **60**, 275-280.
Duysens, L. N. M., and Sweers, H. E. (1963). In "Studies on Microalgae and Photosynthetic Bacteria" (S. Miyachi, ed.), pp. 353-372. Univ. of Tokyo Press, Tokyo.
Erixon, K., and Butler, W. L. (1971a). *Photochem. Photobiol.* **14**, 427-433.
Erixon, K., and Butler, W. L. (1971b). *Biochim. Biophys. Acta* **234**, 381-389.
Erixon, K., and Butler, W. L. (1971c). *Biochim. Biophys. Acta* **253**, 483-486.
Erixon, K., Lozier, R., and Butler, W. L. (1972). *Biochim. Biophys. Acta* **267**, 375-382.
Etienne, A. L. (1974). *Biochim. Biophys. Acta* **333**, 497-508.
Evans, M. C. W., Lord, A. V., and Reeves, S. G. (1974). *Biochem. J.* **138**, 177-183.
Fajer, J., Brune, D. C., Davis, M. S., Forman, A., and Spaulding, L. D. (1975). *Proc. Natl. Acad. Sci. U.S.A.* **72**, 4956-4960.
Floyd, R. A., Chance, B., and De Vault, D. (1971). *Biochim. Biophys. Acta* **226**, 103-112.
Gläser, M., Wolff, C., Buchwald, H.-E., and Witt, H. T. (1974). *FEBS Lett.* **42**, 81-85.
Gould, J. M., and Izawa, S. (1973). *Eur. J. Biochem.* **37**, 185-192.
Govindjee, ed. (1975). "Bioenergetics of Photosynthesis." Academic Press, New York.
Haveman, J., and Lavorel, J. (1975). *Biochim. Biophys. Acta* **408**, 269-283.
Haveman, J., and Mathis, P. (1976). *Biochim. Biophys. Acta* **440**, 346-355.
Haveman, J., Mathis, P., and Vermeglio, A. (1975). *FEBS Lett.* **58**, 259-261.
Henningsen, K. W., and Boardman, N. K. (1973). *Plant Physiol.* **51**, 1117-1126.
Hiller, R. G., Anderson, J. M., and Boardman, N. K. (1971). *Biochim. Biophys. Acta* **245**, 439-452.
Jackson, J. B., Cogdell, R. J., and Crofts, A. R. (1973). *Biochim. Biophys. Acta* **292**, 218-225.
Joliot, P., and Kok, B. (1975). In "Bioenergetics of Photosynthesis" (Govindjee, ed.), pp. 388-412. Academic Press, New York.
Katoh, S., and Kimimura, M. (1974). *Biochim. Biophys. Acta* **333**, 71-84.
Ke, B., Sahu, S., Shaw, E., and Beinert, H. (1974). *Biochim. Biophys. Acta* **347**, 36-48.
Ke, B., Hawkridge, F. M., and Sahu, S. (1976). *Proc. Natl. Acad. Sci. U.S.A.* **73**, 2211-2215.
Kitajima, M., and Butler, W. L. (1973). *Biochim. Biophys. Acta* **325**, 558-564.
Knaff, D. B. (1973). *Biochim. Biophys. Acta* **325**, 284-296.
Knaff, D. B. (1975a). *FEBS Lett.* **60**, 331-335.
Knaff, D. B. (1975b). *Biochim. Biophys. Acta* **376**, 583-587.
Knaff, D. B., and Arnon, D. I. (1969a). *Proc. Natl. Acad. Sci. U.S.A.* **63**, 956-962.
Knaff, D. B., and Arnon, D. I. (1969b). *Proc. Natl. Acad. Sci. U.S.A.* **63**, 963-969.
Knaff, D. B., and Arnon, D. I. (1969c). *Proc. Natl. Acad. Sci. U.S.A.* **64**, 715-722.
Knaff, D. B., and Arnon, D. I. (1971). *Biochim. Biophys. Acta* **226**, 400-408.
Knaff, D. B., and McSwain, B. D. (1971). *Biochim. Biophys. Acta* **245**, 105-108.
Knaff, D. B., and Malkin, R. (1976). *Arch. Biochem. Biophys.* **174**, 414-419.
Knaff, D. B., Malkin, R., Myron, J. C., and Stoller, M. (1977). *Biochim. Biophys. Acta* **459**, 402-411.
Kohl, D. H. (1972). In "Biological Applications of Electron Spin Resonance" (H. M. Swartz, J. R. Bolton, and D. C. Borg, eds.), pp. 213-264. Wiley, New York.
Kohl, D. H., and Wood, P. M. (1969). *Plant Physiol.* **44**, 1439-1445.
Kok, B., Malkin, S., Owens, O., and Forbush, B. (1963). *Brookhaven Symp. Biol.* **19**, 446-459.
Kok, B., Forbush, B., and McGloin, M. (1970). *Photochem. Photobiol.* **11**, 457-475.

Lozier, R. H., and Butler, W. L. (1973). *Photochem. Photobiol.* **17**, 133–137.
Lozier, R. H., and Butler, W. L. (1974). *Biochim. Biophys. Acta* **333**, 465–480.
Malkin, R., and Bearden, A. J. (1973). *Proc. Natl. Acad. Sci. U.S.A.* **70**, 294–297.
Malkin, R., and Bearden, A. J. (1975). *Biochim. Biophys. Acta* **396**, 250–259.
Malkin, R., and Knaff, D. B. (1973). *Biochim. Biophys. Acta* **325**, 336–340.
Malkin, R., Knaff, D. B., and McSwain, B. D. (1974). *FEBS Lett.* **47**, 140–142.
Mathis, P., and Vermeglio, A. (1974). *Biochim. Biophys. Acta* **368**, 130–134.
Mathis, P., and Vermeglio, A. (1975). *Biochim. Biophys. Acta* **369**, 371–381.
Mathis, P., Michel-Villaz, M., and Vermeglio, A. (1974). *Biochem. Biophys. Res. Commun.* **56**, 682–688.
Mathis, P., Haveman, J., and Yates, M. (1976). *Brookhaven Symp. Biol.* **28**.
Mauzerall, D. (1972). *Proc. Natl. Acad. Sci. U.S.A.* **69**, 1358–1362.
Murata, N., Itoh, S., and Okada, M. (1973). *Biochim. Biophys. Acta* **325**, 463–471.
Okada, M., Kitajima, M., and Butler, W. L. (1976). *Plant Cell Physiol.* **17**, 35–43.
Okayama, S. (1974). *Plant Cell Physiol.* **15**, 95–101.
Okayama, S., and Butler, W. L. (1972a). *Biochim. Biophys. Acta* **267**, 523–529.
Okayama, S., and Butler, W. L. (1972b). *Plant Physiol.* **49**, 769–774.
Okayama, S., Epel, B. L., Erixon, K., Lozier, R., and Butler, W. L. (1971). *Biochim. Biophys. Acta* **253**, 476–482.
Parson, W. W., and Cogdell, R. J. (1975). *Biochim. Biophys. Acta* **416**, 105–149.
Prince, R. C., and Dutton, P. L. (1976). *Arch. Biochem. Biophys.* **172**, 329–334.
Pulles, M. P. J., Kerkhof, P. L. M., and Amesz, J. (1974). *FEBS Lett.* **47**, 143–145.
Pulles, M. P. J., Van Gorkom, H. J., and Verschooer, G. A. (1976). *Biochim. Biophys. Acta* **440**, 98–106.
Radmer, R., and Kok, B. (1975). *Annu. Rev. Biochem.* **44**, 409–433.
Redfearn, E. R., and Friend, J. (1962). *Phytochemistry* **1**, 147–151.
Renger, G., and Wolff, C. (1976). *Biochim. Biophys. Acta* **423**, 610–614.
Rockley, M. G., Windsor, M. W., Cogdell, R. J., and Parson, W. W. (1975). *Proc. Natl. Acad. Sci. U.S.A.* **72**, 2251–2258.
Sauer, K. (1975). *In* "Bioenergetics of Photosynthesis" (Govindjee, ed.), pp. 116–181. Academic Press, New York.
Seki, H., Arai, S., Shida, T., and Imamura, M. (1973). *J. Am. Chem. Soc.* **95**, 3404–3405.
Stiehl, H. H., and Witt, H. T. (1968). *Z. Naturforsch., Teil B* **23**, 220–224.
Stiehl, H. H., and Witt, H. T. (1969). *Z. Naturforsch., Teil B* **24**, 1588–1598.
Tiede, D. M., Prince, R. C., Reed, G. H., and Dutton, P. L. (1976). *FEBS Lett.* **65**, 301–304.
Trebst, A., Harth, E., and Draber, W. (1970). *Z. Naturforsch., Teil B* **25**, 1157–1159.
Tsujimoto, H. Y., McSwain, B. D., Hiyama, T., and Arnon, D. I. (1976). *Biochim. Biophys. Acta* **423**, 303–312.
Van Gorkom, H. J. (1974). *Biochim. Biophys. Acta* **347**, 439–442.
Van Gorkom, H. J., and Donze, M. (1973). *Photochem. Photobiol.* **17**, 333–342.
Van Gorkom, H. J., Tamminga, J. J., and Haveman, J. (1974). *Biochim. Biophys. Acta* **347**, 417–438.
Van Gorkom, H. J., Pulles, M. P. J., and Wessels, J. S. C. (1975). *Biochim. Biophys. Acta* **408**, 331–339.
Van Gorkom, H. J., Pulles, M. P. J., Haveman, J., and Den Haan, G. A. (1976). *Biochim. Biophys. Acta* **423**, 217–226.
Vater, J., Renger, G., Stiehl, H. H., and Witt, H. T. (1967). *Naturwissenschaften* **55**, 220–221.
Vermeglio, A., and Mathis, P. (1973). *Biochim. Biophys. Acta* **314**, 57–65.
Visser, J. W. M. (1975). Ph.D. Thesis, University of Leiden.
Warden, J. T., Jr., and Bolton, J. R. (1974). *Photochem. Photobiol.* **20**, 245–250.

Weaver, E. C. (1962). *Arch. Biochem. Biophys.* **99,** 193–196.
Weaver, E. C., and Bishop, N. I. (1963). *Science* **140,** 1095–1097.
Wessels, J. S. C., Van Alphen-Van Wavern, O., and Voorn, G. (1973). *Biochim. Biophys. Acta* **292,** 741–752.
Witt, H. T. (1975). *In* "Bioenergetics of Photosynthesis" (Govindjee, ed.), pp. 493–554. Academic Press, New York.
Witt, H. T., Müller, A., and Rumberg, B. (1961). *Nature (London)* **192,** 967–969.
Witt, K. (1973). *FEBS Lett.* **38,** 116–118.
Wolff, C., Gläser, M., and Witt, H. T. (1974). *Proc. Int. Congr. Photosynth., 3rd, 1974* pp. 295–305.
Wydrzynski, T., Zumbulyadis, N., Schmidt, P. G., and Govindjee (1975). *Biochim. Biophys. Acta* **408,** 349–354.
Yamashita, T., and Butler, W. L. (1968). *Plant Physiol.* **43,** 1978–1986.
Yamashita, T., and Butler, W. L. (1969). *Plant Physiol.* **44,** 1342–1346.

Electron Transport and
Photophosphorylation

Photosynthetic Electron-Transport Chains of Plants and Bacteria and Their Role as Proton Pumps

A. R. CROFTS[1]
Department of Biochemistry
University of Bristol, Medical School
Bristol, England

P. M. WOOD
Department of Biochemistry
University of Cambridge
Cambridge, England

I. Introduction

Our understanding of photosynthesis has made remarkable progress over the past 10 years, especially in the areas of energy conversion and conservation, which form the subjects of the present article. In writing this review we have been aware of a very extensive review on bacterial photosynthesis that will appear shortly (Clayton and Sistrom, 1977), and we are grateful to Drs. Colin Wraight, Richard Cogdell, and Britton Chance for a preview of their contribution to this volume. We are also grateful to Drs. Bendall and Cramer for making available to us their reviews on electron and proton transfer in chloroplasts (Bendall, 1977),

---

[1] Present address: Department of Physiology and Biophysics, University of Illinois, Urbana, Illinois.

and on photosynthetic cytochromes (Cramer, 1977) before publication. These excellent reviews have covered much of the area to be discussed here; it has not been possible to avoid a certain amount of overlap, but we have endeavored to lay a different emphasis on those parts of the field that are common ground. The present work is therefore far from a complete survey, and we recognize that we have skimped over important areas. Where appropriate, we have referred to more extensive reviews published recently, which, together with those mentioned above, cover the ground more thoroughly (Joliot and Kok, 1975; Radmer and Kok, 1975; Witt, 1975; Cheniae, 1970; Avron, 1975; Trebst, 1974; Jagendorf, 1975; Lavorel, 1975).

II. Comparative Photosynthetic Electron Transport

A. The Scheme for Higher Plants

Current ideas of photosynthetic electron transport in higher plants have been covered in several recent reviews (Trebst, 1974; Avron, 1975; Bendall, 1977), with particularly thorough treatment of the primary photoreactions and the linkage of electron transport to photophosphorylation by energy conservation. Rather than cover this ground again, a brief description with a somewhat chemical bias will be given here. Space does not permit much discussion of points of controversy, and several of the references will be to specialized reviews dealing with a particular topic.

Figure 1 presents a general scheme for the electron transport system of higher plant chloroplasts. Two photochemical reactions are linked to a chain of electron or hydrogen carriers (the Z scheme) such that water is oxidized to molecular oxygen with concomitant reduction of $NADP^+$ to NADPH. The relation of this series of carriers to photophosphorylation will be considered elsewhere; here one must simply bear in mind that the carriers are considered to be arranged in a vectorial manner as demanded by the chemiosmotic hypothesis, so that oxidation of water and plastoquinol both result in proton release into the thylakoid interior, while both the primary photochemical acts cause electron transfer from the inner side of the thylakoid membranes to the outer side (Trebst, 1974). An analogous vectorial arrangement is presumed to hold for cyclic electron transport.

It is convenient to trace the carriers in the order followed by the reducing equivalents. Kok et al. (1970) deduced from oscillations in the yield of oxygen upon illumination with a series of flashes that four successive electron transfers by one reaction center are needed for the production of one molecule of oxygen from water. In their model these centers are transformed from state S_1 through S_2 to S_3 (higher oxidation states) by illumination with two flashes of light. On a third flash, S_3 is

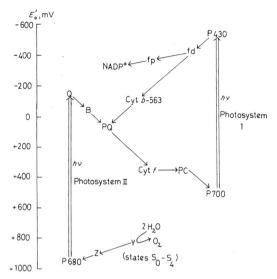

FIG. 1. Photosynthetic electron transport in higher plants. PQ, plastoquinone; PC, plastocyanin; fd, ferredoxin; fp, ferredoxin-NADP reductase; Cyt, cytochrome.

promoted to S_4, which releases oxygen and protons (Fowler and Kok, 1974a, but see below) and returns to state S_0, capable of promotion to S_1 by a fourth flash. Damping is introduced by "double hits" and "misses." The most stable state is S_1, and consequently the centers mainly return to S_1 during a dark period. [For reviews of this topic, to which justice cannot be done in a short space, see Joliot and Kok (1975) and Radmer and Kok (1975).] There is a lack of uniform nomenclature for the water-splitting complex, for which the letters E, M, Y, and Z are used by various authors; the symbols $S_0–S_4$ simply denote different oxidation states. The chemistry is obscure, but the process is known to require manganese (present at a stoichiometry of several atoms per reaction center), and also chloride ions (Cheniae, 1970). Recent work has shown that the stimulatory effect of bicarbonate ions (HCO_3^-) on photosystem II (PS II) is not due to interaction at the oxygen evolving site and is mainly explained by action on the reducing side of PS II (Stemler and Radmer, 1975; Jursinic et al., 1976). On the basis of experiments in which water-splitting activity is inhibited, but artificial PS II donors are still active, it seems that there is an additional manganese center Z between the S states and PS II which is less labile than the water-splitting site (Cheniae and Martin, 1970). [Some authors use the letters Y for Z as defined above, and Z for the water-splitting complex; this nomenclature seems somewhat perverse when used in conjunction with the widespread term ADRY agents for chemicals that *a*ccelerate the *d*eactivation *r*eactions of the water-splitting complex *Y* (Renger, 1972a).]

P680 (Bearden and Malkin, 1975) is the reaction center of PS II; the optical, and electron paramagnetic resonance (EPR) spectra due to the oxidized form (Malkin and Bearden, 1975; Van Gorkom *et al.*, 1975) are closely similar to those from oxidized P700, and therefore P680, like the better characterized P700 (see below), is probably a chlorophyll dimer, which is converted to the dimer cation radical on illumination.

Much recent work has been concerned with the reactions between the water-splitting complex (the S states) and the primary electron donor, P680. Space does not permit a full account of these investigations, but they are summarized below and in Fig. 2.

The reactions between the water-splitting complex and P680 depend on the oxidation state of the complex and are reflected in oscillations of period 4 in the behavior of phenomena associated with these reactions, on illumination by successive brief flashes from the dark-adapted state. Such oscillations have been studied in oxygen evolution (see reviews cited above), delayed fluorescence (see review by Lavorel, 1975), EPR signal IIvf, probably due to Z in an oxidized state (Warden *et al.*, 1976; Babcock *et al.*, 1976; Blankenship *et al.*, 1977), microsecond fluorescence yield rise associated with reduction of P680$^+$ (Delosme, 1972; Den Haan *et al.*, 1974), and release of internal H$^+$ (Fowler and Kok, 1974a). It may be concluded from these studies that the reaction between Z and P680 occurs with $t_{1/2} \leq 1$ μsec, that the reaction between the S states and Z occurs with $t_{1/2}$ values that vary, becoming slower with increasing oxidation level, between <100 μsec for donation from S$_0$ and S$_1$, to ~2 msec for donation from S$_3$ (Diner, 1974; Babcock *et al.*, 1976). The level of oxidation of Z also appears to vary with the S states, indicating some reversibility in the equilibrium between Z and the higher oxidation states. In addition to the secondary donor Z, other donors to this part of the electron transport chain are also indicated. One of these (Den Haan *et al.*, 1974) is apparent when the water-splitting reactions are inhibited by hydroxylamine treatment; another is identified as the reductant for the EPR signals IIf and IIs (Babcock and Sauer, 1975a–c); a further donor is apparent at subzero temperatures from the fluorescence yield rise (Butler *et al.*, 1973; Mathis and Vermeglio, 1975), and, finally,

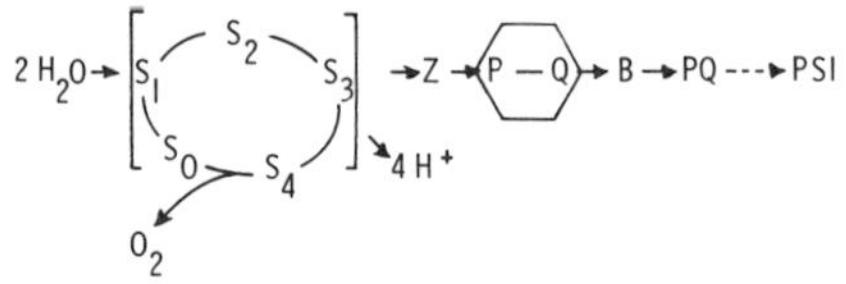

$$2\ H_2O \rightarrow \left[\begin{array}{c} S_2 \\ S_1 \quad \quad S_3 \\ S_0 \quad S_4 \end{array} \right] \rightarrow Z \rightarrow \langle P{-}Q \rangle \rightarrow B \rightarrow PQ \dashrightarrow PSI$$
$$\searrow 4\ H^+$$
$$O_2$$

FIG. 2. Photosystem II and the oxygen-evolving apparatus.

cytochrome b-559$_{HP}$ acts as a donor at liquid N_2 temperature (Okayama and Butler, 1972; Butler *et al.*, 1973). The relation between these donors is obscure; the last three all have potentials (E_m) in the region 350–400 mV, but cytochrome b-559 is certainly distinct from the other two. The donor D which mediates hydroxylamine reduction has a half-time of 35 μsec for reduction of P680$^+$.

The electron transfer to P680$^+$ can be studied directly from the absorbancy change only with difficulty because of the fluorescence changes that accompany flash excitation. Glaser *et al.* (1976) have recently extended the work of Witt and his colleagues (Döring *et al.*, 1967; see also Witt, 1975) and concluded that at least three kinetic components contribute to the reduction of P680$^+$, with half-times of ≤ 1 μsec, 35 μsec, and 200 μsec. It seems likely that these represent, respectively, the involvement of Z, D, and electron transfer through Z from the S states. The kinetics seen following a flash of 20 μsec duration probably reflected in part a second turnover of the photochemistry, suggesting that a rapid reaction on the acceptor side may also occur on the first flash after a dark period (Glaser *et al.*, 1976).

The relaxations of PS II and the water-splitting complex have also been much investigated, the former from delayed fluorescence studies, and the latter from work on the relaxation of the higher S states as measured by the flash yield of oxygen for successive flashes with varied intervals. The intrinsic back reaction of the photoreactants that gives rise to delayed fluorescence appears to have a half-time of $\sim$120 μsec

$$P \cdot Q \xrightarrow{\ h\nu\ } P^+ \cdot Q^- \xrightarrow[\ \ ]{\ \sim 120\ \mu sec\ } P \cdot Q$$
$$\searrow DF$$

(Haveman and Lavorel, 1975), as judged by the dominant component in the decay kinetics under conditions in which the donation of electrons to P$^+$ and the acceptance of electrons from Q$^-$ are both inhibited. The relaxations of the S states occur over a much longer time scale in uninhibited chloroplasts, but can be greatly accelerated by various compounds, including the ADRY reagents mentioned above (Joliot *et al.*, 1971; Renger, 1972a–d), or stabilized by NH_4Cl (Velthuys, 1975, 1976).

Electron transfer from PS II to the plastoquinone pool has been the subject of much recent work, discussed by Bendall (1977). The primary acceptor of PS II is the quencher of its fluorescence, designated Q for this reason by Duysens and Sweers (1963). [In this review the word "primary" means the first carrier apart from transient excited states or anions of chlorophyll (Rockley *et al.*, 1975).] An ultraviolet absorption with the kinetic properties expected for the PS II acceptor was named X-320 by Stiehl and Witt (1968), and attributed to reduction of a special bound form of plastoquinone to the semiquinone anion. It now seems likely that Q and X-320 are the same species, while a third candidate for primary acceptor of PS II, C550 (named from an optical absorption change), is likely to be an unidentified molecule acting as an indicator of the oxidation state of Q without being directly involved (Bendall, 1977). Several different estimates of the redox potential of Q have been made (Cramer and Butler, 1969; Knaff, 1975) showing mid potentials of ~0 at pH 7, varying by −60 mV per pH unit. Knaff (1975) concluded that Q showed a pK for the reduced form at 8.9 and E^0 for the couple Q/Q⁻ of −130 mV, and that this latter value was the likely operating potential of the couple. Ke *et al.* (1976) have recently deduced E_0' = −325 mV for Q at pH 7 by redox titrations of fluorescence yield; this is much lower than earlier estimates, and it is difficult to reconcile this value with the potential of ~1 V required for P680 to oxidize water at pH 5, and the energy available (see review by Crofts *et al.*, 1971).

For electron transfer from Q to the plastoquinone pool (referred to as A, for acceptor, in biophysical studies) the present evidence (Bouges-Bocquet, 1973; Velthuys and Amesz, 1974; Bendall, 1977) indicates an intermediate two-electron carrier B (also called R), which may be a plastoquinone molecule bound to a specific protein. The inhibitor 3-(3,4-dichlorophenyl)-1,1-dimethylurea (DCMU) acts between Q and B. No *b*-type cytochrome shows absorption changes appropriate for involvement at this point.

Plastoquinone (Amesz, 1973), a lipophilic quinone with an isoprenoid side chain, acts as an electron buffer between the two light reactions, and there is evidence that it translocates H atoms across the membrane as required for a Mitchellian loop (Trebst, 1974). There are typically about five active molecules of plastoquinone (capable of holding ten electrons) for each molecule of Q, and there is evidence that a given plastoquinone molecule can receive electrons from several PS II units (Siggel *et al.*, 1972). The redox potential *in vivo* has recently been measured (Okayama, 1976) as +80 mV at pH 7, with a slope of −60 mV per pH unit, changing to −30 mV per pH unit below pH 5 (attributed to protonation of the quinone form although quinones are normally unprotonated at ordinary phs).

The order of the components between plastoquinone and P700 has been a subject of considerable controversy in the past, but there is now a large body of evidence favoring the pathway shown in Fig. 1 (Siedow *et al.*, 1973a; Trebst, 1974; Wood and Bendall, 1975; Bendall, 1977). The first characterized component after plastoquinone is cytochrome *f* (Bendall *et al.*, 1971), a membrane bound *c*-type cytochrome named from *frons*, Latin for foliage. It has been extracted and purified from a range of higher plants, in all cases by methods involving organic solvents. It has an α-band at 554-555 nm, a γ-band for the reduced cytochrome at 421-422 nm, and a midpoint potential of about +360 mV. The cytochrome extracted from Cruciferae is monomeric in solution with a molecular weight of 32,000-33,000 (Takahashi and Asada, 1975; Matsuzaki *et al.*, 1975); the product from other plants is oligomeric (Forti *et al.*, 1965; Bendall *et al.*, 1971; Nelson and Racker, 1972), but measurement of the molecular weight by sodium dodecyl sulfate (SDS) gel electrophoresis has given 34,000 for spinach (Nelson and Racker, 1972). The extracted cytochrome shows a very high rate of electron transfer to plastocyanin (Wood, 1974), but no activity in a plastoquinol-1-plastocyanin reductase assay (Wood and Bendall, 1976); it seems likely that other uncharacterized components are needed for passage of electrons from reduced plastoquinone (a two H-atom carrier) to cytochrome *f* (a one-electron carrier).

Plastocyanin is an acidic copper protein, blue when oxidized. It is very loosely adsorbed to the inner face of the thylakoids (or free in the intrathylakoid space), and all treatments which break the thylakoids result in loss of much of the plastocyanin (Trebst, 1974). It has been sequenced from several species, and in all cases the result corresponds to a molecular weight of about 10,500 with one copper atom per mole (Scawen *et al.*, 1975), so reports of a molecular weight of about 22,000 are likely to be explained by dimerization (Boulter *et al.*, 1977). The coordination of the copper atom is distorted, the three nearest ligands being a cysteine sulfur and two imidazole rings from histidines (Solomon *et al.*, 1976). When purified the redox potential is about +370 mV; the implications of a lower value *in vivo* are discussed by Bendall (1977).

There is good evidence that P700, the reaction center of PS I, is a chlorophyll dimer in a special environment which is bleached by light to form the radical cation (Bearden and Malkin, 1975). It has not yet been obtained free from other chlorophyll. Shipman *et al.* (1976) present a model for its structure, based on theoretical considerations and the properties of a chlorophyll-ethanol adduct with an absorption peak at 700 nm. A recent measurement (Knaff and Malkin, 1973) of the redox potential gives a value of +520 mV, higher than a much-quoted earlier value obtained from chloroplasts treated with acetone.

The primary acceptor of PS I (Bearden and Malkin, 1975) has been identified by two separate techniques: optical kinetic studies led to the discovery of the pigment P430, while low-temperature EPR indicated a bound form of an iron-sulfur protein, "bound ferredoxin." There is now good evidence that these species are the same (Bearden and Malkin,

1975). The redox potential is about -550 mV. See the article by Ke, in this volume, for a more extended discussion.

A number of species have been proposed to act between P430 and the soluble ferredoxin, for instance ferredoxin-reducing substance, oxygen-reducing substance, cytochrome-reducing substance, S_{L-ETH} (from lyophilized, ether-treated chloroplasts), 310 factor, D2 factor, phosphodoxin, and protein factor. These compounds have been reviewed (Siedow et al., 1973b; Bishop, 1971), but while they may in some circumstances mediate electron transfer between the primary acceptor and ferredoxin, none is generally accepted as being necessary.

The next well-defined component is the conventional soluble ferredoxin (Palmer, 1975), a 2 Fe–2 S protein, redox potential -420 mV, molecular weight 10,500. NADP$^+$ reduction by reduced ferredoxin is mediated by a FAD-containing flavoprotein, ferredoxin–NADP reductase, EC 1.6.7.1 (oxidoreductase) and 1.6.1.1 (transhydrogenase), which is bound to the chloroplast lamellae but can be removed by repeated washing in hypotonic medium (Schneeman and Krogmann, 1975). Two recent papers on this enzyme, including references to older work, are Schneeman and Krogmann (1975) and Fredericks and Gehl (1976).

Cytochrome b-563 (or b_6) has a redox potential of about -100 mV. Many studies have suggested that it functions in a cyclic pathway around PS I; see Heber et al. (1976) for a recent paper with discussion of earlier work. The cyclic pathway is mainly known by evidence for cyclic photophosphorylation (Simonis and Urbach, 1973), for which recent work confirms the involvement of both ferredoxin and plastoquinone, and also shows that the rate is critically dependent on the general redox poising of the carriers (Arnon and Chain, 1975; Kaiser and Urbach, 1976). In agreement with the involvement of ferredoxin, cytochrome b-563 photoreduction seems to have been observed only in chloroplast preparations that include ferredoxin (Cramer and Butler, 1967; Murata and Fork, 1971; Böhme, 1977), despite the frequent inclusion of direct arrows from P430 to cytochrome b-563 in schemes like Fig. 1. Cyclic photophosphorylation has also been recently demonstrated in Neottia nidus-avis (Menke and Schmid, 1976), a mycotrophic orchid which lacks PS II (cf. certain algal cells, below).

In addition to the components already mentioned, there are a number of others that must at present be regarded as "homeless." For instance, bound iron–sulfur proteins include one (Evans et al., 1974) or two (Ke et al., 1973) low-potential centers, $E_0' = -500$ to -600 mV, in addition to P430, and a high-potential center, $g = 1.90$, at $E_0' = +290$ mV (Malkin and Aparicio, 1975). The remaining b-type cytochromes also fall into this category.

Cytochrome b-559 has been the subject of a recent review by Cramer and Horton (1975). In its best characterized form it is tightly bound to PS II, and is photooxidized by PS II at 77°K, but not under physiological conditions at room temperature. In freshly prepared chloroplasts it is mostly present in a high-potential state (b-559$_{HP}$), $E_0' = +360$ mV, but aging or a variety of treatments (most of which disrupt membrane structure) reduce this by as much as 300 mV (to give b-559$_{LP}$). Its role is unknown.

The existence of a further type of cytochrome b-559 (another b-559$_{LP}$), that remains bound to cytochromes b-563 and f on detergent treatment, is controversial; see Cramer and Horton (1975) and Heber et al. (1976) for two points of view. For unfractionated chloroplasts, the interpretation depends on whether any of the b-559 associated with PS II is in the low-potential state (Bendall et al., 1971; Henningsen and Boardman, 1973; Cramer and Horton, 1975). In fractionated preparations, it is the shapes of the spectra at 77°K that are difficult to rationalize. Thus Anderson and Boardman (1973) found a peak at 557 nm in a ferrooxalate minus quinone spectrum at 77°K of PS I particles, while Stuart and Wasserman (1973) found that electrophoretically pure cytochrome b-563 gave a split α-band (557 and 561 nm) at 77°K. It has also been suggested that the α-band of cytochrome b-563 can shift to shorter wavelengths than 563 nm (Nelson and Neumann, 1972), and the redox potential of cytochrome b-563 may be variable and hence possibly nonuniform in a chloroplast preparation (Böhme and Cramer, 1973).

The intracellular location of yet another b-type cytochrome with α-band at 559–560 nm, cytochrome b_3 (Bendall et al., 1971) or b-560 (Matsuzaki and Kamimura, 1972), has not been ascertained; unlike the others it is a soluble protein, and its redox potential has been measured as +130 mV (Matsuzaki and Kamimura, 1972).

B. DIFFERENCES IN THE ALGAE

The various classes of algae (there are at present about 14) are distinguished primarily by their secondary pigments (and blue-greens by being prokaryotic), but they all contain chlorophyll a, and once the light energy has reached chlorophyll a, most of the chain is essentially identical to that of higher plants. This extreme conservation of electron transport with respect to evolution has permitted results obtained with algae to be used in several cases to unravel the pathway in higher plants: cells of Chlorella were used in many of the experiments that led to the S-state model (Joliot et al., 1969), the red alga Porphyridium has been used to demonstrate a lack of direct participation of b-type cytochromes in noncyclic electron flow (Amesz et al., 1972b), and results from mutants of Chlamydomonas and Scenedesmus have been of general importance

(Bishop, 1973). As another aspect of this similarity, which extends to the prokaryotic blue-green algae (cyanobacteria), cytochrome b-559$_{HP}$ may have no known function in higher plants, but it has been found in a wide range of algae, e.g., *Euglena* (Ikegami *et al.*, 1968), *Chlamydomonas* (Epel *et al.*, 1972), *Anabaena* (Fujita, 1974), and *Nostoc* (Aparicio *et al.*, 1974), the last two being blue-greens, and several algal mutants are known for which defective noncyclic electron transport was later found to be associated with altered characteristics of b-type cytochromes (Levine, 1971; Epel *et al.*, 1972). However, there are differences, and a tendency to neglect them is felt to justify the space allotted here.

The importance of cyclic electron flow in algae shows wide variation, an extreme case being cells that only possess PS I and cyclic photophosphorylation. This state of affairs is found in certain green algae (Droop, 1974) which cannot fix CO_2 photosynthetically or assimilate organic carbon in the dark (e.g., *Chlamydobotrys* strains), and also in heterocysts of blue-green algae for which the rationale is that oxygen evolution would prevent nitrogen fixation (Fogg, 1974).

Whereas nitrogenase has only been found in blue-green algae, the enzyme hydrogenase is found in members of all the major groups, and, according to Kessler (1974), is present in about half the species examined (see also the chapter by Bishop and Jones). It has not been found in higher plants. Hydrogenase can reversibly catalyze evolution or uptake of hydrogen gas ($2RH \rightleftharpoons H_2 + 2R$), given a suitable donor or acceptor. It is only active at very low partial pressures of oxygen (i.e., anaerobic or microaerobic conditions). It permits hydrogen to be used as a source of electrons for reducing $NADP^+$, especially when ATP is being formed by photophosphorylation, and also leads to hydrogen evolution when illumination follows a period of dark anaerobic incubation. These properties are discussed by Kessler (1974); their relation to noncyclic electron transport is unclear, as is the importance of hydrogenase *in vivo*. Hydrogenase has been purified from the purple bacterium *Chromatium* and shown to be an iron–sulfur protein (Gitlitz and Krasna, 1975); the purified enzyme was active with artificial donors or acceptors, but not with ferredoxin or NADP.

Many species of blue-green algae can grow in anaerobic habitats, in which free sulfide is liable to occur as a product of sulfate-reducing bacteria, such as *Desulphovibrio*. [The nonenzymic reaction of HS^- with oxygen generally prevents coexistence of oxygen and sulfide (Cline and Richards, 1969).] Until recently the ability of any blue-green alga to photooxidize hydrogen sulfide to sulfur in the manner typified by the purple and green sulfur photosynthetic bacteria has been controversial (Kessler, 1974). Thus, Stewart and Pearson (1970) found that photosynthesis by *Anabaena flos-aquae* in the presence of hydrogen sulfide led to sulfide oxidation, but DCMU was inhibitory, which makes it difficult to decide whether H_2S was donating to the oxidizing side of PS II, or sulfur was formed by reduction of oxygen liberated by normal water splitting. Recently Cohen *et al.* (1975a) have conclusively shown that *Oscillatoria limnetica* exhibits both oxygenic and anoxygenic photosynthesis; in the absence of sulfide, oxygen is evolved and DCMU inhibits, but in the presence of sulfide they were able to observe the formation of sulfur granules (Cohen *et al.*, 1975b), DCMU was no longer inhibitory and 700 nm (PS I) illumination was adequate. They worked at pH 7, close to the pK for H_2S/HS^-, so either species could be the donor, the uncharged H_2S being more permeant. The steps leading to entry of electrons from H_2S into the noncyclic system are unknown; ambient redox potentials in the presence of sulfide are typically -100 to -300 mV (Baas Becking *et al.*, 1960), with the sulfur/sulfide couple forming a redox buffer, so entry of electrons at any point after PS II is feasible. The most probable mechanism is perhaps a 2-electron coupling of sulfide and plastoquinone, with protons released on plastoquinol oxidation in the normal way. This type of electron transport clearly links blue-green algae with green bacteria, considered below.

In certain algae ferredoxin is partially replaced by flavodoxin (also called phytoflavin). Flavodoxin has been found in the blue-greens *Anabaena cylindrica* (Entsch and Smillie, 1972) and various species of *Synechococcus* (Entsch and Smillie, 1972; Norris *et al.*, 1972) including "*Anacystis nidulans*" (Stanier *et al.*, 1971), in the green alga *Chlorella fusca* (Zumft and Spiller, 1971) and the red alga *Chondrus crispus* (Husain *et al.*, 1976). Flavodoxins are reviewed by Mayhew and Ludwig (1975); the redox component is FMN, which acts as a 1-electron carrier operating between the semiquinone and fully reduced states, with a redox potential of about -450 mV (similar to ferredoxin). Like ferredoxin the flavodoxins are acidic proteins, but the molecular weights (at 19,000–22,000) are considerably higher. Outside the algae such proteins are found only in bacteria (Mayhew and Ludwig, 1975), where the ferredoxin–flavodoxin duality is surprisingly widespread (e.g., *Azotobacter, Desulphovibrio, Escherichia coli, Pseudomonas*, and *Rhodospirillum*). They can substitute for almost all reactions of ferredoxin (of which there are a large number in bacteria), and synthesis is often stimulated by iron deficiency, as was found in several of the algal studies (Zumft and Spiller, 1971; Entsch and Smillie, 1972; Norris *et al.*, 1972). In general it seems that ferredoxin is preferred in a rich medium either because of slightly superior properties or because of the lower biosynthetic input needed for its synthesis, but since ferredoxin constitutes a significant part of the cell's requirement for iron, flavodoxin is a useful standby for growth when iron supplies are limiting.

Whereas plastocyanin has been isolated from so many higher plants and at such a constant stoichiometry to other photosynthetic components as to make it reasonable to regard it as universal, this is not the case for algae. Plastocyanin was first isolated from *Chlorella* (Katoh, 1960; Kelly and Ambler, 1974), and is also well known in the other common unicellular green algae, *Chlamydomonas* (Gorman and Levine, 1966a) and *Scenedesmus* (Powls *et al.*, 1969). The only other reports of its isolation from eukaryotic algae are for the green seaweeds *Ulva* (as a footnote in Katoh *et al.*, 1961; no details are given) and *Enteromorpha* (Boulter *et al.*, 1977). In blue-greens, plastocyanin has been isolated and sequenced from *Anaebaena variabilis* (Lightbody and Krogmann, 1967; Aitken, 1975) and *Plectonema boryanum* (syn. *Phormidium luridum*; Biggins, 1967; Aitken, 1976). A. Aitken (personal communication) has recently obtained evidence for its presence in many other blue-green algae. It is difficult to assess to what extent the distribution of plastocyanin in algae is restricted, or whether people simply have not looked very hard. There have been several reports claiming the presence of plastocyanin on the basis of less direct evidence, e.g., for *Euglena gracilis* (Brown *et al.*, 1965), *Porphyridium aerugineum* (Visser *et al.*, 1974) and "*Anacystis nidulans*" (Visser *et al.*, 1974), but in the case of *Euglena* (Wildner and Hauska, 1974a) and *Anacystis* (Aitken, 1975, 1976) subsequent work indicates at least that plastocyanin is not normally present in significant amounts. Another alga for which a thorough study has failed to reveal the presence of plastocyanin is the diatom *Bumilleropsis* (Kunert and Böger, 1975).

The most striking difference between higher plant and algal electron transport is the seemingly universal presence in algae of a soluble c-type cytochrome (often called cytochrome f) which has not been found in higher plants (i.e., bryophytes and upward). It resembles higher plant cytochrome f in redox potential, a frequently asymmetrical α-band, and isoelectric point. However, in all other respects it differs; its α-band maximum is typically at 552–553 nm, its γ-band maximum at 415–417 nm, and its molecular weight is invariably about 11,000. [As with plastocyanin, a few reports of a molecular weight of about 22,000 are probably explained by dimerization, since all the chains sequenced have the low molecular weight (Ambler and Bartsch, 1975).] Its kinetic properties are also quite distinct from higher plant cytochrome f, with only the soluble algal cytochrome being an efficient donor to P700 (Wood and Bendall, 1975). In reconstitution experiments the soluble cytochrome behaves in very much the same way as plastocyanin, and there is good evidence that it acts *in vivo* as donor to P700 (Murano and Fujita, 1967; Amesz *et al.*, 1972a; Wildner and Hauska, 1974a; Kunert and Böger, 1975). From what follows, it will be clear that to call the soluble cytochrome "cytochrome f" suggests false analogies and can cause confusion. Following the usage of Dickerson and Timkovich (1975), it is therefore best designated by its α-band (as c-552, c-553 etc), to emphasize its sequence homology and mitochondrial cytochrome c, while distinguishing it from the mitochondrial cytochrome present in the same cell [c-550 in general, c-558 in *Euglena* (Dickerson and Timkovich, 1975)]. For convenience in this review, it will be referred to as cytochrome c-552. Its distribution is clearly of importance in consideration of its function and the following list should give some idea of its widespread occurrence [sequencing studies are indicated (s)]: blue-greens, *Anabaena variabilis* [Lightbody and Krogmann, 1967; Aitken, 1976 (s)], *Plectonema boryanum* [also called *Phormidium luridum*, Biggins, 1967; Aitken, 1976 (s)], *Spirulina maxima* [Ambler and Bartsch, 1975 (s)], *Synechococcus* spp. including "*Anacystis nidulans*" [Holton and Myers, 1967; Crespi *et al.*, 1972; Aitken, 1976 (s)], *Tolypothrix tenuis* (Katoh, 1959); red algae, 17 marine species including *Porphyra tenera* (s) [Sugimura *et al.*, 1968; Ambler and Bartsch, 1975 (s)], *Porphyridium cruentum* (Nishimura, 1968); Chrysophyceae, *Monochrysis lutheri* [Laycock and Craigie, 1971; Ambler and Bartsch, 1975 (s)]; Xanthophyceae, *Vaucheria* sp. (Sugimura *et al.*, 1968); brown algae, 5 marine species including *Alaria esculenta* (s) [Sugimura *et al.*, 1968; Ambler and Bartsch, 1975 (s)]; diatoms, *Navicula pelliculosa* (Yamanaka *et al.*, 1967) and *Bumilleropsis filiformis* (Kunert and Böger, 1975); *Euglena gracilis* [Peroni *et al.*, 1964; Ambler and Bartsch, 1975 (s)]; green algae, 10 marine species (Sugimura *et al.*, 1968), *Chlamydomonas reinhardtii* (Gorman and Levine, 1966b), *Chlorella* spp. (Honda *et al.*, 1961; Grimme and Boardman, 1974), *Kirchneriella obesa* (Nalbandyan, 1972), and *Scenedesmus* spp. (Powls *et al.*, 1969; Kunert *et al.*, 1976). The author is not aware of any alga in which a search for the soluble cytochrome has been unsuccessful, although less than 1 molecule per 10^5 chlorophylls in log phase *Chlamydomonas* was reported by Gorman and Levine (1966b).

The fact that the soluble cytochrome is the only well known algal cytochrome with a redox potential resembling higher plant cytochrome f, plus a similar asymmetry commonly fond in their α-bands, has given rise to a common belief that they are analogous. This would be decidedly surprising, given that higher plants have presumably evolved from organisms resembling green algae, since somewhere on the way an extrinsic cytochrome of molecular weight 11,000 would have to change into an intrinsic one of molecular weight 32,000.

In 1966 Gorman and Levine (1966b) reported that *Chlamydomonas reinhardtii* contained a high-potential membrane-bound c-type cytochrome, but in an accompanying paper (1966c) described a mutant (*ac*-

206) which lacked both soluble and membrane bound c-type cytochromes while showing no other deficiency. Later similar findings were reported for *Scenedesmus* (Powls *et al.*, 1969), for which mutant 50 lacked both the soluble and membrane-bound cytochromes. These results have led to a general assumption that the membrane-bound cytochrome is merely a form of the soluble one. Similar membrane-bound cytochromes have also been reported from *Chlorella* (Grimme and Boardman, 1974) and for the blue-green alga *Plectonema boryanum* (Biggins, 1967).

Recently the author has shown (Wood, 1976) that chloroplast membranes from *Euglena* have properties very similar to higher plant chloroplasts in a plastoquinol-1 reductase assay, which in higher plants has been shown to proceed via cytochrome f (Wood and Bendall, 1976). Moreover, a bound c-type cytochrome was discovered, as well as evidence against an earlier scheme in which b-type cytochromes were on the main noncyclic pathway between plastoquinone and PS I; on the contrary, the b-type cytochromes behaved exactly as in higher plants. A low ratio of bound:soluble cytochrome probably explained why the bound cytochrome had not been noticed before, although its cytochrome:chlorophyll ratio was not much less than for cytochrome f in some higher plants. It was also possible to extract and partially purify the membrane-bound cytochrome from *Chlamydomonas, Euglena*, and "*Anacystis nidulans*," by an adaptation of a method for obtaining cytochrome f from higher plants (Wood, 1977). The spectrum of the product from *Chlamydomonas* or *Euglena* was virtually identical to that of parsley cytochrome f, with an α-band maximum near 554 nm, a very asymmetrical β-band, and a γ-band maximum at 421 nm. The cytochrome from *Anacystis* had α- and γ-bands both shifted to slightly longer wavelengths. The redox potential of the cytochrome from *Chlamydomonas* was determined as +350 mV, and its molecular weight as 31,000 in SDS. The cytochrome from *Euglena* showed very different kinetic properties from the soluble *Euglena* cytochrome c-552 and was unaffected by *Euglena* cytochrome c-552 antiserum (Wood, 1977). It seems clear that this membrane-bound cytochrome is the algal equivalent of cytochrome f, and as such the name cytochrome f should be reserved for it. An explanation for the *Chlamydomonas* and *Scenedesmus* mutants lacking both cytochromes is then required; a plausible one is that a chloroplast enzyme is needed to attach protoheme to apocytochromes c, and it is this (uncharacterized) enzyme that is defective in these mutants. The need for such an enzyme has been inferred for animal mitochondria (Kadenbach, 1971) and recently for *Euglena* chloroplasts (Wildner, 1976).

So for *Euglena* there is now clear evidence (Wildner and Hauska,

1974a; Wood, 1976, 1977) for the order: plastoquinone–cytochrome f–cytochrome c-552–P700. This is very similar to the mitochondrial sequence, ubiquinone–cytochrome c_1–cytochrome c–cytochrome oxidase; in both cases the membrane-bound cytochrome mediates electron transport between a lipophilic quinone and an extrinsic cytochrome of about the same redox potential as the bound cytochrome. In both cases b-type cytochromes and other components are also associated with quinol oxidation in a way that is not well understood; thus for the plastoquinol-1 reductase reaction of chloroplasts (Wood and Bendall, 1976), although no redox changes of b-type cytochromes could be observed, it was not possible to prepare an active fraction lacking b-type cytochromes. Sequence homology has been shown between the soluble algal cytochrome and mitochondrial cytochrome c (Pettigrew, 1974; Ambler and Bartsch, 1975). The heme peptide of cytochrome c_1 has a molecular weight in SDS of 31,000 (Ross and Schatz, 1976) or 32,000 (Katan et $al.$, 1976) when purified from yeast, and 30,600 from bovine heart (Trumpower and Katki, 1975), very similar to the values cited above for cytochrome f. The similar function and similar molecular weight point strongly to cytochrome f being homologous with cytochrome c_1. Furthermore, given that a bound c-type cytochrome has now been detected in $Euglena$, three genera of green algae and two of blue-greens, it seems likely that cytochrome f will be found to be universal throughout organisms containing plastoquinone: that is, all plants and algae.

If one accepts that cytochrome c-552 in $Euglena$ has a role equivalent to that of plastocyanin in higher plants then what can one say about the roles of these two proteins in algae that contain both? In fact, there seem to be no significant kinetic differences between plastocyanin from higher plants and from $Chlamydomonas$, or between cytochrome c-552 from $Chlamydomonas$ and from $Euglena$ (P. M. Wood, unpublished). This is borne out by the experiments of Hauska et $al.$ (1971) on protection of cyclic photophosphorylation by plastocyanin or cytochrome added before sonication; they used spinach chloroplasts and found similar results with spinach or $Scenedesmus$ plastocyanin, and with $Scenedesmus$ or $Euglena$ cytochrome c-552. One is led to the inescapable conclusion that in green algae that contain both, the two proteins have the same role. [There is nothing new in this idea (Lightbody and Krogman, 1967), although in the past it has suffered through indiscriminate use of the term cytochrome f.] The great similarity of these two proteins in $Chlamydomonas$ (Gorman and Levine, 1966b) and $Scenedesmus$ (Powls et $al.$, 1969) is borne out by their elution from DEAE-cellulose by a salt gradient one just after the other, in addition to similar redox potentials and molecular weights. [The fact that $Chlamydomonas$ mutant ac-208, which lacked plastocyanin, had photosynthetic electron transport largely inhibited (Gorman and Levine, 1966c) is not incompatible with this view, since the mutant cultures contained no more soluble cytochrome than the wild type in log phase, about 1 molecule per 10^5 chlorophylls.] It has recently been shown (P. M. Wood, unpublished) that when C. *reinhardtii* is grown under conditions in which the soluble cytochrome is barely detectable, a shortage of available copper leads to cessation of plastocyanin synthesis (the main copper requirement) and a concurrent large increase in soluble cytochrome content. This provides the best possible proof that they constitute an interchangeable pair, analogous to ferredoxin and flavodoxin. Kinetic measurements showed that the rate of electron exchange between the two proteins is quite slow, and therefore unlikely to be important.

In blue-green algae in which plastocyanin has been characterized, both it and the soluble cytochrome are basic (Lightbody and Krogmann, 1967; Biggins, 1967; Aitken, 1976). This is somewhat surprising, since in eukaryotes both are always acidic [as is the soluble cytochrome in some other blue-greens such as *Synechococcus* species (Holton and Myers, 1967; Aitken, 1976)], but sequence studies show homology with the corresponding proteins in eukaryotes (Aitken, 1975, 1976; Ambler and Bartsch, 1975). Two groups have shown (Lightbody and Krogmann, 1967; Tsuji and Fujita, 1972) that either protein from *Anabaena* will catalyze electron transport reactions of *Anabaena* PS I preparations, given a suitable donor and acceptor.

The advantages of having an interchangeable iron and copper protein pair derive not only from the well known problem plants face through deficiency of available iron, but also from the question of copper availability. Copper forms stronger complexes with organic chelating agents than any other biologically important metal [bearing in mind that for iron at pH 7 the relevant equilibria involve $Fe(OH)_2^+$ rather than Fe^{3+} (Sillén and Martell, 1964)], and in complexes such as Cu·EDTA the metal is essentially unavailable for algal uptake (Spencer, 1957; Manahan and Smith, 1973). The second situation in which copper is unavailable is in the presence of H_2S, since the solubility products of copper sulfides are spectacularly low: for cuprous sulfide $K_s = 10^{-48}$ and for cupric sulfide $K_s = 10^{-35}$, far lower than solubility products for FeS ($K_s = 10^{-17}$) and MnS ($K_s = 10^{-12}$) (Sillén and Martell, 1964). Plastocyanin has not been detected under any growth conditions in a second *Chlamydomonas* species, C. *mundana* 'Boron', originally isolated from an anoxic pond with free H_2S (P. M. Wood, unpublished). A wide range of algae, not just blue-greens, can grow in habitats where anaerobiosis is liable to occur for at least part of the time (Eppley and MaciasR, 1963; Fenchel and Riedl, 1970; Elliott and Bamforth, 1975), with consequent formation of sulfides and loss of copper from solution. Thus copper uptake is likely to be a problem under conditions of strong organic chelation or anaerobiosis, conditions where iron deficiency is unlikely since in the first case the chelators will keep a pool of ferric iron in solution, while in the second iron is reduced to the more soluble ferrous state.

In primordial seas, copper would have been locked up in extremely insoluble copper or mixed sulfides (e.g., chalcopyrite) (Österberg, 1974), and copper proteins could only have been synthesized once the sea and air became more oxidized through photosynthetic evolution of oxygen by blue-green algae. Thus the first blue-green algae could not have used plastocyanin, which must have developed later, and the pathway in *Euglena* or other algae lacking plastocyanin is the primitive one. (Plastocyanin has not been found in green or purple photosynthetic bacteria.) In many algae the two proteins coexist with the advantages described above, while in land plants, with ancestors lacking a deep root system and unlikely to be subject to anaerobiosis, but perhaps troubled by iron deficiency, plastocyanin has entirely taken over. The available scanty evidence indicates that plastocyanin is superior to the soluble cytochrome in reconstituted systems, for *Anabaena* (Lightbody and Krogmann, 1967; Tsuji and Fujita, 1972), spinach (Hauska *et al.*, 1971) and even *Euglena* (Wildner and Hauska, 1974a,b).

A copper–iron protein pair with interesting parallels is azurin and cytochrome *c*-551 in *Pseudomonas aeruginosa* or *P. fluorescens*, which are denitrifying bacteria. These proteins show similar kinetics with *Pseudomonas* nitrite reductase (assayed as cytochrome oxidase; Horio *et al.*, 1961; Wharton *et al.*, 1973), and *Pseudomonas* cytochrome *c* peroxidase (Soininen and Ellfolk, 1972). Sequence homology has been shown between azurin and plastocyanin (Ryden and Lundgren, 1976) and between cytochrome *c*-551 and the soluble algal cytochrome (Ambler and Bartsch, 1975; Dickerson and Timkovich, 1975).

In the prokaryotic blue-green algae there are no discrete respiratory organelles, ubiquinone has not been detected, and the soluble cytochrome *c* described above is the only cytochrome known to show sequence homology with mitochondrial cytochrome *c*. Respiratory activity is generally low, but oxygen uptake in the dark and ATP synthesis linked to dark oxidation of pyridine nucleotides have both been demonstrated (Carr, 1973). It seems likely that the section of the chain from plastoquinone to cytochrome *c*-552 (and plastocyanin?) is common to photosynthesis and respiration (cf. the Rhodospirillaceae, below), but further experimentation is needed.

C. PHOTOSYNTHETIC BACTERIA

The current taxonomy of the photosynthetic bacteria (Pfennig and Trüper, 1974) is that they belong to one order, Rhodospirillales, which is divided into two suborders. The first suborder, Rhodospirillineae (purple bacteria) contains the families Rhodospirillaceae and Chromatiaceae (formerly Athiorhodaceae and Thiorhodaceae). The second suborder, Chlorobiineae (green bacteria), contains the families Chlorobiaceae and the newly named Chloroflexaceae (Trüper, 1976). Lipid analyses show that the green bacteria *Chlorobium* and *Chloroflexus* are more closely related to the blue-green algae than are the purple bacteria (Kenyon and Gray, 1974), and it is therefore convenient to consider them first.

1. Green Bacteria

The green bacterium which has been most studied is *Chlorobium limicola* f. *thiosulfatophilum*. ["*Chloropseudomonas ethylicum*" has been shown to be a syntrophic mixture of *Chlorobium limicola* and a colorless sulfate-reducing bacterium (Gray *et al.*, 1973).] It is an autotrophic anaerobe, in which photosynthesis is associated with cigar-shaped "*Chlorobium* vesicles." Its photosynthetic electron transport (Fig. 3) resembles the noncyclic and cyclic pathways of PS I of plants and algae (see Fig. 1), but with the components on the oxidizing side of the photoreaction (as far back as the quinone pool) shifted to lower redox potentials. There is good evidence (Evans, 1969; Knaff and Buchanan, 1975) for noncyclic electron transport from Na_2S (i.e., H_2S or HS^- at neutral pH) to $NADP^+$, sensitive to antimycin A but unaffected by the uncouplers gramicidin D and carbonyl cyanide *m*-chlorophenylhydrazone. The components will now be considered in the order adopted for higher plants.

Two cytochromes not shown on Fig. 3, cytochrome *c*-551 ($E_0' = +135$

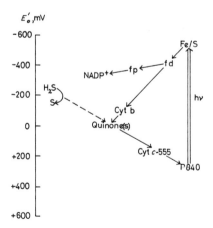

FIG. 3. Photosynthetic electron transport in *Chlorobium*. Note that other S compounds can act as donor in place of sulfide, and in all cases the means of entry of electrons into the conventional chain is uncertain. For abbreviations see Fig. 1.

mV; Meyer *et al.*, 1968) and flavocytochrome c-553 ($E_0' = +100$ mV) have been suggested as points of entry for electrons from thiosulfate and sulfide, respectively, by Kusai and Yamanaka (1973b). They worked with purified cytochromes and their scheme, which does not include quinones or b-type cytochromes, is almost certainly too simplistic, though the manner of linkage of substrates with the quinone pool is not known.

The quinones are unusual, with approximately equal amounts of the naphthoquinones menaquinone-7 ($E_0' = -80$ mV) and "*Chlorobium* quinone" ($E_0' = +40$ mV), differing from menaquinone-7 in the omission of the first methylene in the isoprenoid chain, which leaves the double bond in the first isoprenoid unit conjugated (Redfearn and Powls, 1968; Powls and Redfearn, 1969).

There are two molecules of cytochrome c-555 close to each reaction center, which are photooxidized with differing kinetics (Prince and Olson, 1976). This cytochrome has a very asymmetrical α-band, a redox potential of $+165$ mV (Prince and Olson, 1976) or $+220$ mV (Knaff *et al.*, 1973), and is bound slightly more firmly than the soluble algal cytochrome. It has been sequenced by Van Beeumen and Ambler (1973), and the molecular weight is about 10,000. (In view of the data presented above for algal cytochrome f, one wonders whether the two cytochromes per reaction center are one and the same protein.)

Light is harvested by bacteriochlorophyll c (*Chlorobium* chlorophyll 660), and the energy is transferred to bacteriochlorophyll a. The reaction center, P840 (presumed to be a special form of bacteriochlorophyll a), has a redox potential of $+250$ mV (Prince and Olson, 1976) or $+330$ mV (Knaff *et al.*, 1973). The primary acceptor is a bound iron–sulfur center, $E_0' = -550$ mV (Knaff and Malkin, 1976; Jennings and Evans, 1977), from which electrons for reduction of $NAD(P)^+$ pass first to ferredoxin (Evans, 1969) and then to a flavoprotein reductase (Kusai and Yamanaka, 1973a) as in plants and algae (Knaff and Buchanan, 1975).

A b-type cytochrome, b-562 or b-564, $E_0' = -90$ mV (Knaff and Buchanan, 1975) is implicated in cyclic electron flow (like cytochrome b-563 of plants and algae) and shows reversible light induced oxidation (Fowler, 1974) or reduction (Knaff and Buchanan, 1975), affected by antimycin A. Further iron–sulfur centers include one of "Rieske" type ($g = 1.90$), $E_0' = +160$ mV (Knaff and Malkin, 1976); cf. the $+290$ mV center in chloroplasts.

The type species of the Chloroflexaceae is *Chloroflexus aurantiacus*, first described in 1974 (Pierson and Castenholz, 1974). It differs from other members of the Chlorobiineae in its gliding motility and filamentous growth. Furthermore, whereas the ordinary green bacteria are strictly anaerobic autotrophs, *Chloroflexus* is nutritionally a very versatile organism, capable of aerobic growth in the dark (Madigan *et al.*, 1974).

Its photosynthetic apparatus seems to be similar to that of *Chlorobium*, with "*Chlorobium* vesicles," bacteriochlorophylls *a* and *c*, and elemental sulfur as the major product of photosynthetic sulfide oxidation (Madigan and Brock, 1975). The electron transport components have not yet been described.

2. *Purple Bacteria*

As stated above, these can be divided into two types. The Chromatiaceae are obligate phototrophic anaerobes. Sulfide is the usual donor of reducing equivalents, though other sulfur compounds can be used, and some species can utilize instead hydrogen or organic donors (e.g., pyruvate). The Rhodospirillaceae do not oxidize sulfur compounds, but instead make use of a variety of organic compounds and under anaerobic conditions tend to dismute rather than oxidize. They are not all restricted to anaerobic habitats, and several species will grow aerobically in the dark.

The primary photochemistry has been the subject of much research in recent years, reviewed by Parson and Cogdell (1975). They summarize the evidence (and discuss a dissenting view) for the primary acceptor being at a much higher potential than in the organisms described hitherto, making it impossible for these bacteria to reduce ferredoxin directly, at least at a significant rate under physiological conditions (Prince and Dutton, 1976b). In accordance with this, it is now generally believed that reduction of $NAD(P)^+$ by purple bacteria occurs solely by energy-linked reversed electron flow (Keister and Yike, 1967; Jackson and Crofts, 1968; Gest, 1972), unless a very reducing substrate such as hydrogen is present (cf. algal hydrogenase, above). The photosynthetic electron transport system (Fig. 4) is then almost entirely a cyclic one; the extent to which it is used to transfer electrons from redox couples of a moderately high potential to ones of lower potential is not known. The vectorial arrangement of the carriers, and their capacity for proton-pumping, is considered elsewhere (see Crofts *et al.*, 1974a,b, 1975a,b, and below).

The nature of the reaction center and primary acceptor is discussed by Parson and Cogdell (1975). Detergent fractionation has been more successful for certain Rhodospirillaceae than for any other photosynthetic organisms, and has given a particle containing four molecules of bacteriochlorophyll *a* and two of bacteriopheophytin (Straley *et al.*, 1973). The bleached form of the reaction center, $P870^+$, corresponds to a bacteriochlorophyll dimer cation. Redox titrations indicate a midpoint potential of +450 to +490 mV for a variety of purple bacteria (Parson and Cogdell, 1975). There is now a large body of evidence (Okamura *et al.*, 1975; Parson and Cogdell, 1975; Romjin and Amesz, 1976) that the

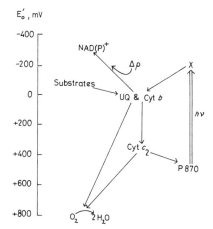

FIG. 4. Some electron-transport pathways in *Rhodopseudomonas capsulata*. Based in part on La Monica and Marrs (1976). UQ, ubiquinone; Δp, protonic potential difference; Cyt, cytochrome.

primary acceptor (called X) is a ubiquinone molecule associated with an iron atom (Fe^{2+}), and reduced to the semiquinone by a flash of light (cf. PS II of plants). The redox potential has been somewhat controversial; in a recent determination Prince and Dutton (1976b) found a kinetically operational potential of -160 mV for *Chromatium*, -180 mV for *Rhodopseudomonas sphaeroides* and -200 mV for *Rhodospirillum rubrum*. [The reader is referred to Rockley *et al.* (1975) and to the chapter by Ke, in this volume, for discussion of transient species linking P870 and the primary acceptor.]

In most Rhodospirillaceae the donor to $P870^+$ is cytochrome c_2. This has a redox potential usually between $+300$ and $+380$ mV, an α-band at 549–552 nm, and a γ-band at 416–418 nm (Bartsch, 1977). Its sequence shows distinct similarities with mitochondrial cytochrome c (Ambler *et al.*, 1976). It is more or less loosely bound to the membrane, depending on the species, and is located either on the outer face of the cytoplasmic membrane or in the periplasmic space (Prince *et al.*, 1975b; Hochman *et al.*, 1975). In some species there is evidence for two molecules per reaction center (Dutton *et al.*, 1975).

Cytochrome c_2 has not been found in *Rhodopseudomonas gelatinosa* or in any of the Chromatiaceae; in these organisms (and also in *Rhodopseudomonas viridis*) two membrane-bound cytochromes are closely associated with photosynthesis, cytochromes c-556 and c-552, with redox potentials of about $+330$ and -10 mV, respectively (Bartsch, 1977). Neither has been successfully purified. At room temperature both of these cytochromes can be photooxidized directly by P870, with very similar kinetics in the case of *Chromatium* (Case *et al.*, 1970). Surprisingly, photooxidation of the low potential member of the pair, cytochrome c-552, is still possible at 77°K or 4°K (cf. cytochrome b-559$_{HP}$ in higher plants); this has been observed for all these bacteria (Kihara and Chance, 1969; Dutton, 1971). Low temperature photooxidation of poorly characterized cytochromes has also been detected in some other purple bacteria (Kihara and Chance, 1969).

The quinone in purple bacteria is ubiquinone, with the number of isoprenoid units varying in different species from 7 to 10 (Carr and Exell, 1965). Ubiquinone is generally regarded as the secondary acceptor.

The Rhodospirillaceae often contain several b-type cytochromes (Dutton and Jackson, 1972; Saunders and Jones, 1975). As an example of their involvement in electron transport, in *Rhodopseudomonas sphaeroides* the next component after ubiquinol is reported to be a b-type cytochrome with α-band at 560 nm, $E_0' = +50$ mV (Dutton and Jackson, 1972; Prince and Dutton, 1975), and Prince and Dutton (1975) have shown that in the uncoupled state reoxidation of this cytochrome occurs at the same rate as reduction of cytochrome c_2, but is inhibited by antimycin A. Cytochromes of b-type in purple bacteria have also recently been implicated in "Q-cycles" (Crofts *et al.*, 1975a), and for this reason cytochrome b and ubiquinone have simply been grouped together in Fig. 4. (The cyclic electron transport schemes for other photosynthetic organisms are probably subject to similar reservations, but have not been studied so intensively in recent years.) Cytochromes of b-type are less well known for the Chromatiaceae, but Knaff and Buchanan (1975) have recently shown that *Chromatium* does contain one, with $E_0' = 5$ mV and photoreduction enhanced by antimycin A. A component "Z," not detected spectroscopically, is frequently included in cyclic schemes between b-type cytochrome(s) and cytochrome c_2 (Crofts *et al.*, 1974a); it would be tempting to equate this with the "Rieske" iron–sulfur center, $E_0' = +285$ mV, found in both *Rhodopseudomonas sphaeroides* and *Chromatium* (Prince *et al.*, 1975a), but the observed kinetics of the latter are too slow.

There is a plethora of c-type cytochromes in most purple bacteria (Dickerson and Timkovich, 1975), and only a small proportion have any assigned role. Presumably in the Chromatiaceae most are connected with the complex redox chemistry of sulfur (Dickerson and Timkovich, 1975), and in the Rhodospirillaceae they are needed for nutritional versatility. In aerobically grown Rhodospirillaceae the respiratory chains are decidedly complex (see Fig. 4), with several branching points (King and Drews, 1975; Zannoni *et al.*, 1976; La Monica and Marrs, 1976). Ubiquinone is common to photosynthesis and respiration, and this appears to be true also of cytochrome c_2, and therefore for the carrier between them. Marrs (1976) has recently reported a mutant strain of *Rhodopseudomonas capsulata* in which electron flow from cytochrome b_{50} to cytochrome c_2 is absent in both photosynthetic and respiratory pathways. Presumably a common component (Z, see above) is lost. Van Grondelle *et al.* (1976) have recently made a study of changes in cytochrome and reaction center on flash excitation of whole cells of *Rhodospirillum rubrum*. They suggest that a second type of photochemical reaction center may be present at low concentration.

Throughout this survey, attention has frequently been drawn to analogies between the central regions of photosynthetic and mitochondrial electron transport. Apart from the presence of isoprenoid quinones, all the organisms described here share with complex III of mitochondria (Rieske, 1976), the existence of b-type cytochrome(s) with redox potentials around 0 mV, and a "Rieske" g = 1.90 iron–sulfur center with redox potential 160–290 mV. All except higher plants and purple bacteria lacking cytochrome c_2 contain a c-type cytochrome homologous with mitochondrial cytochrome c. If all these components do indeed have a dual role in photosynthesis and respiration in Rhodospirillaceae and blue-green algae, and one accepts that all photosynthetic organisms have a common ancestry, then these analogies between the dark and light-driven pathways are inevitable.

III. Protolytic Sites in Photosynthetic Electron-Transport Chains and Their Role in Proton Pumping

In reviewing the components of electron transport chains above, we have been concerned mainly with a comparison of pathways for photosynthetic electron flow. These have been well characterized in relatively few systems. In the following section, we will be concerned mainly with the evidence that suggests that the arrangement of photosynthetic electron transport chains in these systems determines that they act as proton pumps, so that the trapped photon energy is conserved in part as a proton gradient which can be used to drive secondary energy conversion processes, such as ATP synthesis, ion transport, motility.

A. PROTONS AS CHEMICAL REACTANTS

In view of the ubiquitous involvement of protons in the reactions of electron transport, energy transfer, and phosphorylation, it seems worthwhile to discuss briefly some pertinent aspects of the physical chemistry of such reactions. Protons react with other chemical species through two classes of reaction—acid dissociation and oxidation or reduction of H carriers. The reduced components of H-carrying couples are often weak acids; indeed such reactions may be formally treated as two partial reactions, electron donation followed by H^+ association, so that the characterization of pH-depenent effects, and of reactions leading to binding or release of H^+ in biochemical systems, must come to terms with this initial degree of complexity. The physical chemistry of these simple proton equilibria is well understood (Clark, 1960; Dutton and Wilson, 1974), and appropriate experiments can be designed which make it possible to resolve contributions from acid dissociation and redox reactions. However, with biochemical molecules and in supramolecular structures, additional complexities may be introduced through mechanisms which are less well understood. Several general categories of complexity can be defined:

1. *"Cooperative" phenomena* in proteins or protein complexes. The example most fully characterized is the Bohr effect in hemoglobin. The binding of a proton at one site in the molecule affects, and is affected by, the binding of other ligands at separate sites, or by the redox state of the molecule, or by the state of photoactivated groups. Chance *et al.* (1970) pointed out the possibility of this sort of mechanism for rapid proton changes in photosynthetic bacteria and coined the name "membrane Bohr effect." In general, the proton binding site in such systems will be a dissociable group showing well defined physicochemical properties (pK, etc.). However, when the proton binding is associated exclusively with either the oxidized or reduced form of a redox protein, the behavior observed may be that of H carrier (H^+-binding on reduction), or its inverse (H^+-binding on oxidation), and both types of behavior have been characterized (Pettigrew *et al.*, 1976). In the case of these redox-linked changes, the formal equilibria are similar to those of dissociable simple molecules, with multiple pKs. It is doubtful, therefore, whether any useful purpose is served by pursuing the mechanistic distinction implied above, except insofar as evidence from approaches other than those based on redox effects demand it. The work of Petty and Dutton (1976b) on the pH-dependent redox properties of the *b* cytochromes of *Rhodopseudomonas sphaeroides* shows an example of this sort of study. As these authors pointed out, the question of physiological interest is whether the H^+ changes involved occur within the time scale of the redox reactions, a question of importance with respect to the second category of mechanisms.

2. *Failure of equilibration of protons* between the reaction site and the reference medium. Within this category two separate causes of a failure of equilibration may be defined:

a. Intervention of a proton insulating phase between the reference aqueous phase and an aqueous phase with which the proton reaction site is in equilibrium. Typically, the coupling membrane separates the internal aqueous phase from the external phase containing a measuring electrode, pH indicator dye, or redox-mediating couple.

A number of experimental techniques have been developed to cope with this sort of problem. Several studies have exploited the possibility of introducing indicators into the insulated phase, the most promising of these being neutral red (Ausländer and Junge, 1975). The more usual expedient adopted is to make observations in the presence or absence of a concentration of protonophore sufficient to allow complete and rapid equilibration of proton activity across the insulating phase. Any discrepancy between results with and without protonophore suggests a failure of equilibration. However, in making measurements of this sort great care must be taken to ensure that any changes in buffering capacity consequent upon allowing buffering groups in the previously "masked" phase to come into play are properly controlled.

It should perhaps be emphasized that difficulties due to nonequilibration of protons between aqueous phases will be experienced in making any measurements in which protons are involved, including those in which H-carrying redox mediators are used.

b. The reaction site itself is insulated from both aqueous phases, so that measurement of the pH activity of the aqueous phases gives no indication of the H^+ activity at the reaction site. This is usually evident as a kinetic problem, and Prince and Dutton (1976b), and Knaff (1975) have drawn attention to the consequence of such a situation for the primary acceptors in bacterial photosynthesis, and in PS II, respectively. In these cases, the reduction of a quinone in the reaction center is not accompanied on a kinetic time scale by the uptake of an equivalent of proton, and spectrometric evidence (either optical, or EPR, or both) shows that an anionic semiquinone is the species formed. Nevertheless, titration of photochemical reactions using redox mediators indicates that in chromatophores (or in chloroplasts), the reaction site equilibrates with protons of the medium. The discrepancy between these observations almost certainly reflects the different time scale of measurements for the kinetic and equilibrium phenomena involved.

Other instances of failure of equilibration on a kinetic time scale, almost certainly exist, but have not been so well characterized (see below).

The area of redox potentiometry and the role of protonic equilibria has been excellently reviewed by Dutton and Wilson (1974).

B. THE PROTOLYTIC REACTIONS IN GREEN PLANT PHOTOSYNTHESIS

It seems fairly established (see reviews by Bendall, 1977; Witt, 1971, 1975; Junge, 1975; Trebst, 1974) that the protons released on oxidation of H_2O and of plastodihydroquinone appear in the inner aqueous phase of the thylakoid, and that the protons involved in the reduction of plastoquinone and of system I acceptors are lost from the external medium. In none of these cases can it be claimed that the mechanism of the reaction involved is known, even at the level of the identity of reaction partners. Nevertheless, the general "architecture" of the non-cyclic electron transport pathway is clear, and the role of protons in these reactions can be convincingly explained in terms of classical Mitchellian proton pumps operating by virtue of an anisotropic distribution of redox components arranged to act as alternating H-carrying and electrogenic arms of two H^+ pumping loops. The H-carrying arms are the water/oxygen couple and the plastoquinone/plastohydroquinone couple, with a balancing stoichiometry of external H^+ involved in the reactions on the acceptor side of PS I. The electrogenic arms of the loops are the photochemical reactions with their immediate donors and acceptors.

In the following sections we will review the evidence for this point of view, and examine in greater detail the data available on the sites at which protons are thought to interact.

1. Arrangement of the Electron Transport Chain

  a. The Oxygen-Evolving Apparatus. Parts of the electron transport chain on the oxygen side of PS II (Girault and Galmiche, 1974; Giaquinta and Dilley, 1975; Lien and Racker, 1971; Cramer and Horton, 1975; Horton and Cramer, 1974; Izawa and Ort, 1974; Selman et al., 1973; Giaquinta et al., 1974; see review by Arntzen and Briantais, 1975) appear to be available to reagents on either side of the thylakoid membrane. The protolytic reaction sites are in equilibrium with the aqueous phase inside the thylakoid (see below for detailed discussion). Manganese released by relatively mild treaments appears to be discharged into the inner aqueous phase (Blankenship and Sauer, 1974). However, in treated chloroplasts added Mn^{2+} can act as an effective electron donor to PS II, as can ferrocyanide (Izawa and Ort, 1974). These are both highly charged ions; in the case of the ferri–ferro cyanide couple, it is known that neither component is able to cross liposome membranes (Deamer et al., 1972), and in the case of Mn, the ion released on Tris treatment of chloroplasts does not become available to the external phase for hours (Blankenship and Sauer, 1974). Cytochrome b_{559}, which acts as a donor to PS II at low temperature is itself readily and rapidly oxidized by ferricyanide (Horton and Cramer, 1974; Cramer et al., 1974; Ikegami and Katoh, 1973); the water-splitting reactions and cytochrome b-559 are modified by trypsin, and by a variety of other reagents which are unable to cross the membrane (see references above). We may conclude that the oxygen-evolving apparatus is bulky, and that it spans the membrane, but that the sites of proton release are within the thylakoid. Babcock and Sauer (1975c) have proposed a topological model which accounts for many of these observations. Since little more is known of the detailed reaction mechanism it would be premature to speculate further.

    b. The Photosystems. Witt and his colleagues (see reviews by Witt, 1971, 1975) following Mitchell's suggestion (Mitchell, 1968) showed that both PS II and PS I act to transfer charge across the thylakoid membrane. The charge transfer, measured by the 515 nm change, was so rapid as to preclude the involvement of dark "thermal" reactions, and Junge and Witt (1968) concluded that the photochemical reactions themselves were so arranged in the membrane that the oxidized donor remained on the inner side while the charge of the electron on the acceptor was "felt" by the external aqueous phase, and was available for reaction with protons from that phase. These arguments are now generally accepted and have been extensively reviewed elsewhere (Walker and Crofts, 1970; Witt, 1971, 1975; Trebst, 1974; Jagendorf, 1975); we will not pursue them further here. The evidence from the 515 nm change has been supported by more direct measurements of the electrical field using an electrode technique (Fowler and Kok, 1974b; Witt and Zickler, 1973), and indirectly in the case of PS II from effects of ionophores or of artificial gradients on delayed fluorescence (for references, see Crofts *et al.*, 1971; Joliot and Joliot, 1974; Ellenson and Sauer, 1976). Although the role of the photosystems in transferring charge across the thylakoid membrane is not in dispute, the precise arrangement of the reaction components has recently been questioned. Two complicating features have emerged. First, a second spectral change of the carotenoids has been identified, again by Witt's group, associated with the role of the carotenoids in dumping excess absorbed photon energy—the so-called valve effect (Wolff and Witt, 1969). The spectrum, and more especially, the intensity saturation curves of this effect are sufficiently distinct from the electrochromic change to allow for their separation, by appropriate experimentation (Mathis, 1966; Mathis and Galmiche, 1967; Cox and Delosme, 1976; Joliot *et al.*, 1977). However, overlapping of changes due to the valve effect and those due to electrochromic phenomena may have complicated earlier kinetic measurements. Second, the work from Witt's group involved signal averaging following repeated flashes, so that no interpretation could be drawn from the work about the transitions occurring on excitation of systems from the dark-adapted state. More recently Joliot and his colleagues (Joliot and Delosme, 1974; Joliot *et al.*, 1977) developed techniques for studying both these aspects of the problem. Their results, which suggest that the original simple model may require modification, may be summarized as follows:

1. The electrochromic change on excitation of dark adapted material showed at least three phases: (a) a rapid phase within the apparatus response time ($\sim$3 μsec); (b) a phase of half time $\sim$25 μsec; (c) a phase of half-time $\sim$1–50 msec.

2. The contribution from PS II could be resolved by experiment and appropriate manipulation and shown to consist of part of phase a only.

3. The contribution from PS I-linked reactions contained all three phases. Phase c was seen in *Chlorella* only in dark-adapted material and disappeared after a few flashes; it may be due to electron flow in pathways off the main chain which are filled or depleted rapidly on illumination.

4. PS I itself contributed two phases—a part of phase a, and phase b. Joliot *et al.* (1977) suggest that these biphasic kinetics may reflect the fact that the oxidized primary donor (P700$^+$) is buried within the membrane, and that electrons from the secondary donor (plastocyanin) must traverse a part of the overall electrogenic pathway to reduce the P700$^+$. The rapid phase would then reflect the oxidation of P700; and phase b, its reduction by plastocyanin. This situation recalls the interpretation previously suggested for a similar biphasic kinetics observed in the carotenoid change of *Rhodopseudomonas sphaeroides* by Jackson and Dutton (1973). However, Junge and Ausländer (1975) were unable to find any kinetic component in the electrochromic change measured at 524 nm, which reflected reduction of P700$^+$ as measured from the change at 705 nm recorded under similar conditions, and preferred a model in which the photoactive chlorophyll lay close to the inner surface of the membrane.

5. In hydroxylamine-treated chloroplasts (in which the O_2-evolving apparatus is disconnected from the photosystem) supplemented with an electron donor, the electrochromic change associated with PS II was much reduced. Joliot and Joliot (1976) have suggested that either (a) the hydroxylamine treatment leads to a major disruption of the arrangement of the components in the membrane, or (b) the arrangement of components in the PS II reaction center is similar to that for PS I, with the primary donor (P680) buried in the membrane toward the *outer* side, and the major contribution to the electrogenic effect arising from electron transfer from Z, the secondary donor to P680$^+$. This electron transfer would have to be faster than 1 μsec (or faster than 20 nsec) (Wolff, *et al.*, 1969; Witt, 1971), as was already suggested from the rapid changes in fluorescence yield on reduction of P680$^+$, measured by Delosme (1972), and by Duysens and collaborators (1974; den Haan *et al.*, 1974). Such an arrangement of the photosystem is in excellent agreement with

the model previously proposed by Babcock and Sauer (1975c) to explain the paradoxical ability of large charged groups to donate electrons to PS II, as discussed above.

c. *The Electron Transfer Chain between the Photosystems*. The picture of the arrangement of the photosystems derived from the approaches discussed above is complemented by work on the intermediate electron transport chain. Aspects of this are considered elsewhere in the article and have been reviewed extensively by others (see reviews cited above). In particular, Bendall (1977) discusses in detail the control of electron transport by pH and ΔpH, and the roles of cytochrome f and plastocyanin in mediating electron flow between PQH_2 and P700. Trebst (1974) has previously covered much of this ground, and deals more fully in his elegant work with Hauska with the proton-pumping loops introduced through use of artificial redox mediators (Hauska *et al.*, 1974, 1975).

2. The Involvement of Protons in the Reactions on the Oxygen Side of PS II

The oxidation of water by electron transfer to PS II leads to the release of $4H^+$ per O_2. There has been strong evidence from direct measurements of pH changes (Schwartz, 1968; Fowler and Kok, 1974a; Graber and Witt, 1975, 1976; Ausländer and Junge, 1974; Junge and Ausländer, 1974), and from the effects of internal pH on electron transport from water (Siggel, 1975; Bamberger *et al.*, 1973; Wraight *et al.*, 1972) and on delayed fluorescence (Wraight and Crofts, 1971), that the protons released on oxidation of water are in equilibrium with the inner aqueous phase. They appear rapidly outside the thylakoid only in the presence of excess amine, or ionophores such as gramicidin, nigericin, or dianemycin, which catalyze the rapid equilibration of H^+-concentration gradients without inhibiting the water-splitting reactions (Renger, 1972a–d; Etienne, 1974). We will discuss the involvement of protons in the reactions of the water-splitting apparatus by reference to the model proposed by Kok, Forbush, and McGloin (1970) and Fowler and Kok (1974a).

$$S_4^{4+} + 2H_2O \rightarrow S_0 + 4H^+ + O_2 \qquad (1)$$

a. *The Reactants in the Transition of S_4 to S_0*. The mechanism for O_2

evolution is usually shown (see scheme above) as involving the four-stage accumulation of charge to give a complex able to react directly with water so that a concerted reaction occurs leading to the liberation of oxygen and the regeneration of the "uncharged" state as shown above. While this is obviously a convenient and simple representation of a reaction generally acknowledged to be considerably more complex, it may also be misleading in an important sense, since the assumed reaction of *water* with the S_4 state of the system has no basis in experimental fact. It could equally well be the case that H_2O reacts with the system in its lower oxidation states, and is oxidized in consecutive one-electron steps as suggested in the model system of Renger (1972b,c). If this were the case, the release of protons might be expected to accompany the single-electron oxidation steps, and to be reflected in effects of pH on various other PS II parameters. Any discrepancy between these effects and the directly measured release of protons might then reflect cryptic proton production at sites not readily in equilibrium with the aqueous phase.

Sinclair and Arnason (1974) and Arnason and Sinclair (1976) have attempted to measure directly the involvement of water in these reactions, by substitution of ^2H_2O for H_2O under conditions in which inhibition would have been expected if reactions involved in the splitting of OH bonds in the system had been rate limiting. They observed no marked isotope effect on the transitions of any of the S states and concluded that the reactions of water involving the splitting of OH (or O^2H) bonds were not rate-determining at any step in the cycle. Arnason and Sinclair (1976) discussed this lack of a deuterium isotope effect in terms of some activated charge complex, the build-up of which would represent the rate-determining step in the transition $S_3 \rightarrow S_0$, and the reaction of which with H_2O would be rapid. They also pointed out that their results are consistent with the hypothesis of Metzner (1975) that water is not the direct source of photosynthetic oxygen, but that some activated form of water, possibly involving CO_2 is involved.

b. Thermodynamic and Kinetic Considerations. The E_m for the primary acceptor Q seems to be ~ -130 mV (but see Ke *et al.*, 1976), the maximal work expected from 1 quantum of light after photochemical conversion is about 1.2–1.3 V so that we may expect the maximal potential of the P^+/P couple to be between 0.9 and 1.2 V depending on how much energy is lost for kinetic reasons, and on work terms other than redox work, which need to be satisfied (see Crofts *et al.*, 1971; Ross *et al.*, 1976). The mid-potential of the water/O_2 couple in equilibrium with air is 0.806 at pH 7, varying by -60 mV/pH unit at 30°C. We would expect that the water/oxygen couple would therefore be in equilibrium with the P^+/P couple at some pH value between $\sim$5 and $\sim$3, but that at pH values above ~5 the redox gradient between the two couples would be sufficient to ensure that the concentrations of the reactants and products offered no kinetic limitation so long as the mean potential is considered. However, experimental evidence suggests that maximal rates of power conversion (measured as rates of phosphorylation) occur at external pH values ~ 8.5 in conjunction with a pH gradient of >3.5 units. Under these conditions, water oxidation would be occurring in equilibrium with the internal phase at pH < 5.

Rumberg and Siggel (1969), Siggel (1975), Rottenberg *et al.* (1972), and Bamberger *et al.* (1973) have measured the dependence of electron transport on pH and shown that the uncoupled rate becomes inhibited at values of pH below 5. This inhibition appears to be controlled by protonation of a group within the thylakoid, of pK ~ 5.7, associated with the electron-transport chain between PQ and P700. Electron transfer to PQ appeared to be limited by a group with a pK ~ 4.7 in equilibrium with internal proton. Bendall (1977) interpreted the results of Bamberger *et al.* (1973) as showing that, in addition, electron transport may be controlled by an external site associated with PS II, becoming inhibited as the external pH rises above pH 8.2. The rate of oxygen evolution also becomes inhibited at low pH, and this is associated with an inhibited rate of flow to P680$^+$ (see below). However, it is not clear whether the inhibition of the rate of electron flow is due to thermodynamic (rate constants unaffected by pH, concentrations of reactants limiting) or kinetic (rate constants changed by pH, concentrations of reactants not limiting) constraints, or a mixture of these. Kok *et al.* (1975; Radmer and Kok, 1975) have estimated the relative equilibrium constants of the transitions of the S states with respect to the oxidized photosystem, by measuring forward and reverse rate constants. Bendall (1977) has used these values to compute a potential of 1.1–1.3 V for the oxidized reaction center. However, these measurements have yet to be made over a sufficiently wide range of pH to provide information about the role of protons in these transitions. Wraight (1972) measured the effect of pH on the yield of oxygen from each of a series of flashes, and compared these to the steady-state value. Below pH 5.5 a marked deficit of yield was apparent, which Wraight interpreted as being due to a failure of oxidation of state S_3. However, oscillations of period 4 were apparent in the yield down to pH 5, although with a modified relative yield per flash. Diner and Joliot (1976) showed a marked decline for the turnover time for the photosystem as measured by the transition $S_1^* \rightarrow S_2$ below pH 6.45, but concluded that this reflected an inhibition of the reoxidation of Q^- at lower values of pH.

In Eq. (1) above, four protons are released per transition of state S_4^{4+} to S_0. If we regard the equation as representing a conventional chemical reaction, we may express the thermodynamic balance by an appropriate equilibrium constant.

$$K_{eq} = \frac{[S_0] \cdot [H^+]^4 \cdot [O_2]}{[S_4^{4+}] \cdot [H_2O]^2}$$

Recognizing that the activities of O_2 and water in equilibrium with the water-splitting reactions will not vary greatly under physiological conditions, we may write an expression,

$$[S_4^{4+}]/[S_0] = K'[H^+]^4, \quad \text{where} \quad K' = 1/K_{eq} \cdot [O_2]/[H_2O]^2$$

relating the relative concentrations of states S_4 and S_0 to the concentration of protons. It is known that oxygen evolution occurs over a wide range of values of pH (from 8.5 to 5 or less) for the phase in equilibrium with these reactions. The equilibrium concentration ratio of S_4^{4+}/S_0 as expressed above would change by $>10^{14}$ over this pH range. We may assume that in practice the reactions occur some way away from equilibrium at neutral pH since kinetic evidence suggests that the production of oxygen occurs in a spontaneous, rapid reaction. Nevertheless, the range of values of the ratio S_4^{4+}/S_0 expected for kinetically competent oxygen evolution over the pH range observed, assuming the reaction described by Eq. (1), seems unrealistic, especially in view of the kinetic evidence for equal populations of states S_0–S_3 in the steady state. This may be seen more clearly if we include the reaction leading to formation of S_4^{4+} from S_3^{3+}, and represent the overall change as a redox half-cell:

$$2H_2O + S_3^{3+} \rightarrow S_0 + 4H^+ + O_2 + e^-$$

$$E_h = E^0 - 4(2.303RT/F)pH + RT/F \ln \frac{[S_0]}{[S_3^{3+}]} \cdot \frac{[O_2]}{[H_2O]^2}$$

We can see that the operating mid-potential for the oxidation of S_3^{3+} would have to vary by -240 mV/pH unit. Assuming that in the steady state $[S_0] = [S_3]$, and taking the range of pH over which oxygen evolution occurs, this change of E_m would be 960 mV on going from pH 8.5 to 4.5; all the extra work involved in oxidizing water at lower pH would be loaded onto a single electron transition and would involve the necessity for an oxidizing agent with a potential at least this much more positive than the mean midpoint for the O_2/H_2O couple at the highest operating pH. Since this would require more energy than is available, it seems unlikely unless very special mechanisms are invoked.

Kinetic arguments may also be given to show that a reaction mechanism which is fourth order in protons is unrealistic, although these may be annulled by proposing a multistage mechanism of proton release; a further difficulty is posed by the energetics of accumulation of four positive charges in a local environment. It may, therefore, be worth enquiring into what modifications of this reaction mechanism might be thermodynamically and kinetically more appealing, and whether the experimental evidence supports the very dramatic controlling effect predicted for protons on the $S_4 \rightarrow S_0$ transition by the Fowler–Kok mechanism.

   c. *Direct Measurement of H^+ Changes Associated with PS II.* The most detailed studies of these changes have been those of Fowler and Kok (1974a, 1976a; Fowler, 1977). They observed that in the presence of excess methylamine or gramicidin the release of protons on excitation with successive flashes of short duration showed a periodicity of 4, similar to that observed for oxygen evolution. They suggested that the protons were released in synchrony with the oxygen, so that the reaction of the appropriate stage of the cycle of S states involved could be described by Eq. (1) above.

   In making this suggestion, Fowler and Kok (1974a) pointed out some discrepancies between their results and those anticipated on the basis of the equation above. In particular, the second flash after dark adaptation (leading to the transition $S_2 \rightarrow S_3$ for the greater proportion of the states) sometimes showed a greater yield of protons than of oxygen, and the yields on other transitions did not match precisely.

   The estimation of protons released inside the thylakoids due to reactions on the donor side of PS II is complicated by simultaneous uptake of protons in reactions on the acceptor side. In their first paper, Fowler and Kok (1974a) were able to show an unambiguous 4-stage oscillation by using high concentrations of methylamine as an uncoupler, since this appeared to inhibit proton uptake associated with the acceptor reactions. However, no explanation for this effect of methylamine was suggested, and the observation of Velthuys (1976) that at high concentrations methylamine reacts with the higher S states, suggests that this reagent should be used with caution in studying system II reactions. Recently, in papers dealing with flash stoichiometries of protons and electrons, Fowler and Kok (1976) and Fowler (1977) published data that highlight these difficulties. In the presence of other uncouplers, the release of protons did not appear to be in synchrony with the release of oxygen. A subsequent paper discussing these points has yet to appear. Saphon and Crofts (1977) have extended the work of Fowler and Kok (1974a, 1976) and Junge and colleagues (Ausländer and Junge, 1974, 1975), by careful use of pH indicator techniques to observe pH changes, induced by a series of flashes, both outside and inside the thylakoids of dark-adapted chloroplasts. Saphon's results confirm those of the more recent papers of Fowler and Kok (1976; Fowler, 1977) in a qualitative fashion. More detailed analysis shows that on illumination of dark adapted chloroplasts, protons are released on all transitions of the S states, with the possible exception of the transition $S_1 \rightarrow S_2$, rather than solely or predominantly on the transition $S_3 \rightarrow S_0$ as suggested by the earlier work of Fowler and Kok (1974a).

d. Effects of pH on Delayed Fluorescence. The intensity of delayed fluorescence associated with the back reaction of PS II under a wide variety of conditions appears to be dependent on the pH of phases in equilibrium with reactions on both the donor and acceptor side (Wraight and Crofts, 1971; Crofts *et al.*, 1971; Kraan, 1971; Wraight *et al.*, 1972; Lavorel, 1975). In general, we may consider that the intensity of delayed fluorescence will depend upon the concentration of the reaction center in the state $P^+ \cdot Q^-$

$$L \propto [P^+ \cdot Q^-] = \phi_L \cdot k \cdot [P^+ \cdot Q^-]$$

where ϕ_L is the luminescence yield, and k is a pseudo first-order rate constant for recombination of $P^+ \cdot Q^-$.

This apparently simple relationship is complicated by the fact that all three terms on the right are to some extent independently variable (see Crofts *et al.*, 1971; Wraight, 1972; also review by Lavorel, 1975). The major term known to affect k is the electrical gradient across the reaction center (Fleischman, 1971; Wraight and Crofts, 1971; Joliot and Joliot, 1974), while the concentration of P^+ and Q^- will depend on reactions with donor and acceptor pools. Since these latter involve protons, the stability of P^+ and Q^- will depend on the activity of protons in the phases with which the secondary reactions are in equilibrium:

$$\tfrac{1}{2}H_2O + P^+ \rightleftharpoons \tfrac{1}{4}O_2 + P + H_{in}^+ \tag{2}$$

$$Q^- + H_{out}^+ + \tfrac{1}{2}PQ \rightleftharpoons Q + \tfrac{1}{2}PQH_2 \tag{3}$$

If we are to consider these reactions in greater detail, we must include in Eq. (2) a set of intermediate reactions that will differ for each of the 4 oxidation states of the water-splitting apparatus. We may anticipate that effects of pH change will be seen only on those reactions in which a proton is involved. Similar considerations apply to reactions on the acceptor side [Eq. (3)].

Since the early work of Mayne (1967, 1968) a number of studies have attempted to clarify the separate effects of pH on donor and acceptor reactions, and of the pH gradient $(-\Delta pH)$ when this was allowed to develop. Several different approaches have been used (see below).

i. *Effects of pH on delayed fluorescence in the millisecond to second time range*. Wraight and Crofts (1971) observed that, in chloroplasts inhibited with DCMU, the delayed fluorescence from PS II was stimulated when a proton gradient was produced through the operation of PS I. They suggested that the stimulation of delayed fluorescence reflected the involvement of internal protons in the equilibria of the water-splitting reactions. Joliot and Joliot (1974) have extended this work by a thorough study of the effect on delayed fluorescence of the electrical field generated through PS I. They concluded that the field was delocalized at least over a membrane area sufficient to allow it to be "felt" by PS II. The report by Fowler and Kok (1974a) of a proton release from the water-splitting mechanism which was synchronous with that of O_2, rendered the observation of Wraight and Crofts (1971) somewhat ambiguous. P. Joliot (personal communication) has pointed out that since chloroplasts inhibited with DCMU are able to undergo only a single transition of S state, no stimulation of delayed fluorescence would be expected from the Fowler–Kok hypothesis on illumination of dark-adapted chloroplasts under the conditions of Wraight and Crofts (1971), since protons would not be involved in the predominant transitions ($S_1 \rightarrow S_2$, $S_0 \rightarrow S_1$). J. M. Bowes and A. R. Crofts (unpublished observations) have recently made a study of the dependence of these effects on flash preillumination. They found that the production of a proton gradient through activation of PS I led to stimulation of delayed fluorescence in chloroplasts inhibited by DCMU added a few seconds before illumination, either after prolonged dark adaptation (20 minutes), or after illumination by 1–3 short saturating flashes following dark adaptation. The stimulated delayed fluorescence was considerably greater after flash preillumination, showing a marked periodicity of four when the number of flashes was varied, with maxima after 2 or 6, and minima after 0, 4, or 8 preilluminating flashes. These observations suggest that the internal protons are in equilibrium with all transitions of the S states, with the possible exception of that of S_1 to S_2.

ii. *Effects of pH on delayed fluorescence and fluorescence yield in the microsecond range*. The rapid reactions of secondary donor and acceptor pools with the photochemical reactants of system II, as expressed in the rapid phases of delayed fluorescence decay and fluorescence yield rise following a flash, are not greatly affected by change in pH over the pH 5–8 range (Haveman and Lavorel, 1975; Van Gorkom *et al.*, 1976; den Haan, 1976). A diminution observed in the fast phase of the fluorescence yield rise (Van Gorkom *et al.*, 1976; Pulles *et al.*, 1976a), and a slowing of faster components in the decay of delayed fluorescence intensity (Haveman and Lavorel, 1975), are consistent with the displacement of the equilibrium concentrations of reactants in the secondary donor pools as a result of the change in concentration of protons, and with the effects of these on the relative rates of the forward and backward reactions. The effects of pH changes in this range do not appear to be restricted to a single reaction of the S states, since damped oscillations with a period of 4 are apparent throughout, and at each end of the range, at least in fluorescence yield changes (Van Gorkom et al., 1976; S. Itoh and A. R. Crofts, unpublished results). This observation is in contrast to the strong controlling effect of pH expected if protons were released solely in the transition $S_4 \rightarrow S_0$.

Below pH 4.5 and above pH 8.5 the reactions of the donor pools appear to be irreversibly disconnected from the photosystems (Van Gorkom et al., 1976; Pulles et al., 1976a).

  e. ADRY Reagents and Effects of Ammonium on the Water-Splitting Reactions. We do not intend here to review at length the involvement of these reagents in the reactions of the S states, except to consider their relation to the involvement of protons. Renger and colleagues (1973; Renger, 1972a–d) and Lemasson and Etienne (1975) have shown that ADRY reagents accelerate the decay of the states S_3 and S_2 to S_1. If S_3 and S_2 are more oxidized states, then the decay involves successive electron (or hydrogen) transfers from a donor. This latter is thought ultimately to be a component(s) of the acceptor pool, but the pathway for the transfer is not known (Renger, 1973; Renger et al., 1973). The accelerating effect of the ADRY reagents is in contrast to a stabilizing effect of NH_4Cl on these reactions (Velthuys, 1975, 1976; Delrieu, 1976).

ADRY reagents are in general uncoupling agents of the "weak acid-lipid soluble anion" type (Renger, 1972a, 1975), whose protonophoric mechanism was predicted by Mitchell (1961). Other uncoupling agents or conditions (gramicidin, nigericin, low concentrations of detergents or amines) do not have an ADRY effect. This has led Renger (1972b) to suggest that the mechanism of the ADRY effect is unrelated to the uncoupling effect. Indeed, Renger (1972b) showed that the accelerating effect was not due to the ability of the reagents to destroy the proton gradient, collapse the electrical gradient, induce irreversible structural changes, or act as electron donors. However, for several reagents (Renger, 1975) the ADRY effect occurred over the same concentration range as the uncoupling effect. The important feature of molecular structure for the ADRY reagents was the presence of a group, either $-NH$ or $-OH$, able to undergo acidic dissociation; the ADRY effect increased with increasing concentration of the anion. Renger (1972b, 1973) and Renger et al. (1973) suggest that the ADRY reagents act as mobile catalysts accelerating electron transfer from the acceptor pools to the oxidized states of the water-splitting system, possibly through a mechanism involving nucleophilic reaction of the anion. It is worth pointing out that the mechanism of uncoupling by ADRY reagents differs from that of the non-ADRY reagents, in particular because of the involvement of a lipid-soluble anionic species as charge carrier (Renger, 1972b, 1975). It seems possible that their effect on the relaxation of the S states may be associated with this property, and that it is not coincidence that these reagents are also uncoupling agents. In this view ADRY reagents would act to equilibrate the H^+ activity between a cryptic site and the aqueous phases, and this would be the basis of their mechanism of action. Their specificity would then arise from the nature of the cryptic site. They could be envisaged as acting by one of the following mechanisms: (1) to allow H^+ to move away from a site at which its high activity stabilizes reactions in the higher states. In this context, the stabilizing effect of NH_4Cl on the S states might be envisaged as arising from the fact that NH_3, which appears to be the active species (Velthuys, 1975, 1976), would be able to penetrate to a site of high H^+ activity and react there to give NH_4^+, a nonpenetrant, thereby stabilizing the charge. In this view it would be the electrical component of the H^+ activity which stabilizes the higher S states. (2) To allow H^+ to a site at which it was required for reactions in the reverse direction. This would occur on electron donation to a H-carrier with limited access to the aqueous phase. (3) Following the suggestion of Kok (Kok *et al.*, 1975) to allow electrons to return to the S states from the acceptor pool via a component C (assumed to be cytochrome b–559) the potential of which would be more reducing in the presence of uncoupling agents of the ADRY type. The effect of the reagent in changing the potential of the component might then reflect the accessibility of H^+ to the redox site.

f. The Protolytic Reactions on the Oxygen Side—Conclusions. It seems very likely from the above, both from theoretical considerations and from the observed effects of pH, that protons are involved in the oxygen-evolving reactions in more than one state of the mechanism. From the evidence summarized above, and from our own results, it appears most likely that protons are released at all the transitions of the S states, except possibly that of S_1 to S_2. This would suggest a mechanism in which water reacts with the apparatus in the lower S states, and in which oxidized equivalents are accumulated in a complex, where bound or activated water associates with Mn at the active site, and undergoes transitions through the intermediate oxidation states of the water/oxygen couple. Mechanisms of this sort have been previously considered by Cheniae (1970; Cheniae and Martin, 1970), Renger (1972b,c), and Wydrzyrski *et al.* (1975). Since the role of manganese is at present unknown, we show a general mechanism of this sort in Fig. 5.

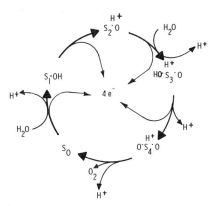

FIG. 5. Possible mechanism showing the involvement of protons in the transitions of the states of the oxygen-evolving apparatus.

The well-known requirement for Cl^- in the reaction might reflect the need to compensate the positive charge of the bound proton in states $S_2 \rightarrow S_4$.

3. The Involvement of H^+ on the Acceptor Side of PS II

Recent work by Bouges-Bocquet (1973) and Velthuys and Amesz (1974; Velthuys, 1976) has shown that the flow of electrons from the primary acceptor Q to the plastoquinone pool occurs by a two-stage mechanism. Little work has been done on the involvement of protons in this reaction, though it is clear that the full reduction of the quinone to dihydroquinone must involve the uptake of 2 equivalents of H^+. Spectrophotometric and other evidence suggests that both the primary acceptor Q and the secondary acceptor B are plastoquinone molecules in a special environment, each in a fixed stoichiometry of 1 per reaction center, and that each undergoes reduction to the plastosemiquinone anion (Stiehl and Witt, 1968; Witt, 1973; Van Gorkom, 1974; Pulles *et al.*, 1976b). After dark adaptation, the state QB^- arising from an odd number of flashes appears to be stable over a time scale of seconds (Velthuys and Amesz, 1974; Pulles *et al.*, 1976b). A fuller discussion of possible equilibria between the components of the Q/B/A system will be found in the review by Bendall (1977).

Witt and Junge and their colleagues have made an extensive study of the kinetics of flash-induced H^+ uptake associated with the acceptor pools (Witt, 1971, 1975; Junge, 1975) and concluded that (i) 1 equivalent of H^+ was taken up for each electron flowing to the pool; (ii) in unmodified chloroplasts, the kinetics of H^+ uptake were slower by two orders of magnitude than the rate of electron flow to the plastoquinone pool; (iii) in chloroplasts treated by sand grinding, digitonin, or by addition of low concentrations of uncouplers, the half-time for H^+ uptake increased, but was still somewhat slower than the appearance of dihydroplastoquinone. The change in initial rate induced by uncouplers was less obvious than the change in half-time, and a more detailed kinetic analysis is required before the half-times quoted can be related to the plastoquinone reduction time. Junge and Ausländer (1974, 1975; Ausländer et al., 1974) interpret their results as showing that some barrier exists to the rapid equilibrium of H^+ between its site of uptake in the reduction of plastoquinone, and the measuring indicator dye in the external bulk aqueous phase. Sand grinding was thought to partially remove the barrier, and uncoupling agents or detergent to facilitate the diffusion of protons across it. The precise nature of this diffusion barrier was not apparent; it seems unlikely that the selection against protons occurs on the basis of size, since much larger molecules (acceptors and inhibitors, such as DCMU and DBMIB) appear to react rapidly. Further-more, the reactions of PS I acceptors such as ferricyanide $[Fe(CN)^{3-}]$ and benzyl- and methylviologen (BV^{2+} and MV^{2+}), which are physically larger and more highly charged, are not impeded. It seems likely that the barrier is to some degree electrically insulating and that the reaction site for H^+ with plastoquinone is not in rapid equilibrium with the phase to which the PS I acceptor has access. Nevertheless, there are strong arguments for believing that the barrier is distinct from the chloroplast membrane, and the most plausible suggestion seems to be that it is local to the PS II acceptor site. The recent results of Renger et al. (1976) and Renger (1976) on the effect of trypsin on reactions on the acceptor side strongly suggest that the insulating barrier is proteinaceous.

The experiments of Junge and colleagues involved the use of repeti-tive flashing and averaging techniques, and so throw no light on the involvement of protons in the two-stage mechanism of electron transfer, since the associated oscillation of 2 damps out after a number of flashes.

More recently, Fowler and Kok (1976; Fowler, 1977) have applied their electrode technique to the study of these reactions, and their results are to some extent in contrast to those of Ausländer and Junge (1974, 1975). Fowler and Kok (1976) confirmed the kinetic observations of Ausländer and Junge (1974, 1975) under comparable conditions, but showed a stoichiometry of H^+ uptake per flash which was greater than that measured by the latter group. They discuss their results in terms of a stoichiometry of at least three protons per electron traversing the noncyclic pathway, and suggest a mechanism in which the protons are transferred across the membrane on pumps less directly linked to electron flow than in the Mitchellian mechanism. One difficulty arising from use of the electrode method is that of calibrating the observed H^+ changes. An "internal" calibration was used by Fowler and Kok (1976) based on the stoichiometry of H^+ release in the oxygen-evolving reactions, as measured with ferricyanide as an electron acceptor, either in chloroplasts in which rapid proton equilibrium between the aqueous phases was ensured by the inclusion of an uncoupling agent, or from the steady-state rate of proton release. Under these conditions, buffering groups within the thylakoids would have been available for titration by protons in both phases. However, in the absence of uncouplers (the condition used for measuring the external stoichiometry of H^+ changes), the buffering groups within the thylakoids would have equilibrated much more slowly with protons in the external phase. This could have led to an overestimate of the number of protons taken up rapidly from the external phase.

The electrode method (Schwartz, 1968) as developed by Fowler and Kok (1974a, 1976) has nevertheless several advantages over the indicator methods used by others. The changes are unambiguously due to protons, they are independent of any measuring light, so that experiments can be done more readily with dark-adapted chloroplasts, and the range of pH available for test is greater than with any single indicator.

Although not specifically designed to test this point, the results reported by Fowler and Kok (1976) show that even in dark-adapted chloroplasts, the *uptake* of H^+ from the external medium showed no strong oscillation of 2, suggesting that electron transfers from Q^- to B, and from QB^{2-} to PQ are both linked to the uptake of protons. In contrast, a strong biphasic oscillation was observed in the appearance of H^+ within the chloroplasts; this is well accounted for in terms of the involvement of a two-stage mechanism in the transfer of "H" to the plastoquinone pool, and subsequent release of $2H^+$ on oxidation of plastodihydroquinone by electron acceptors. Fowler (1977) has more recently extended these observations and discussed his results in similar terms, but prefers a model in which additional protons are transported on a pumping mechanism, to account for the apparent excess stoichiometry. These results have been confirmed by S. Saphon (unpublished observation) using indicator techniques, but because of the problems of calibration discussed above, no unambiguous stoichiometry could be demonstrated.

Fowler and Kok (1974a) also showed that DCMU inhibited the uptake of protons associated with PS II activation and suggested that Q^- is not protonated on a physiological time scale (but see Diner and Joliot, 1976).

An alternative approach to the study of protons in the reactions between the photosystems has been suggested by Haehnel (1976), who studied the kinetics of redox changes associated with Q (X-320) and P700. By kinetic analysis, the rates of electron flow to and from the plastoquinone pool were estimated, and the effect on these rates of the ambient pH were measured. Haehnel found that decreasing pH accelerated the rate of reduction of the plastoquinone of the acceptor pool as deduced from the lag in P700 reduction, but that the effect of pH was much less than expected from a simple kinetic model involving H^+ in a rate-determining step. The rate of oxidation of the reduced primary acceptor was not accelerated by decreased pH (in fact, an increase in half-time was observed), suggesting that no proton in equilibrium with the external aqueous phase was involved in this reaction. Haehnel (1976) suggests that "at least one additional reaction step, which is not accelerated by increasing proton concentration, precedes the rate limiting step," and discusses several mechanisms to account for the anomalous behavior with respect to pH. He also points out the kinetic and spectral discrepancy between the values observed, and those expected for protonation of the primary acceptor. Since Haehnel (1976) used an averaging technique, his results throw no new light on the involvement of protons in the two-stage mechanism. However, Bouges-Bocquet (1974) found that the rates of transition Q^-B to QB^- and Q^-B^- to QB^{2-} were unaffected by pH over the range 5.2–8, suggesting that these reactions do not involve any proton—they are pure electron transfers.

The Protolytic Site on the Acceptor Side—Conclusions. It will be clear from the above that the site of reaction of external H^+ with the components of the acceptor pools for PS II has not yet been established. Most interpretations assume a chemiosmotic model in which the bulk plastoquinone operates as the H-carrier in the neutral arm of a Mitchellian loop (Mitchell, 1966, 1968). The sceptical will be inclined to point out that the evidence for such a detailed molecular mechanism is at present weak. However, we may conclude that:

1. Protons are taken up stoichiometrically with electrons on reduction of the acceptor pools, probably with a stoichiometry of $1H^+/e$ (but see Fowler and Kok, 1976; Fowler, 1977).

2. The reduction of plastoquinone to the dihydroquinone involves 2 protons per molecule ($1H^+/e$).

3. There is a major discrepancy between the rate at which plastoquinone is reduced, and the rate at which H^+ leaves the external aqueous phase, suggesting that the protons involved in the reduction do not come immediately from that phase.

4. There is a discrepancy between the observed uptake of 1 equivalent of H^+/flash by dark-adapted chloroplasts (with *no* strong oscillation of period 2; Fowler, 1977) and the observation that the electron from odd numbered flashes is stabilized in the semiquinone anion, B^- (Pulles *et al.*, 1976b). Since the rate of H^+ uptake does not appear to vary markedly with flash number under these conditions (Fowler and Kok,

1974a, 1976; S. Saphon, unpublished observations), we might conclude that the proton taken up from the external medium does not react immediately at the redox site.

While the discrepancy under (3) above can be rationalized as suggested by Ausländer and Junge (1974) by postulating a cryptic phase from which the protons react rapidly with the redox site, but which equilibrates relatively slowly with the external aqueous phase, the discrepancy under (4) above cannot be so explained. Clearly, further work is needed to clarify the situation. One possibility, suggested by a recent report by Wollman and Thorcz (1976) is that the proportion of B in the semiquinone form after dark adaptation is closer to 50% than the 20–30% suggested by others. However, if this were the case, no oscillation in the phenomena associated with these reactions would be observable. It is also necessary to take account of the involvement of HCO_3^- in the electron transport in this region, as demonstrated by the recent work of Govindjee and colleagues (Jursinic et al., 1976; Govindjee et al., 1976). Electron flow from $Q \cdot B^{2-}$ to bulk plastoquinone was almost completely inhibited in chloroplasts depleted of HCO_3^-. The mechanism of the HCO_3^- effect could clearly have an important bearing on the equilibration of H^+ with the reactants in this part of the chain.

4. A Model for the Proton Pumps of Green Plant Photosynthesis

A model for the topology of the components of the noncyclic electron transport chain is shown in Fig. 6. The components have been arranged as two proton pumps in conventional Mitchellian loops, and the model serves as a summary of the work reviewed above. No doubt future work will show up the inadequacies of this scheme. It nevertheless represents a consensus of opinion which reflects the dominant role of a chemiosmotic viewpoint in current thinking, and it seems likely that its inadequacies lie mainly in the direction of ignorance about details of mecha-

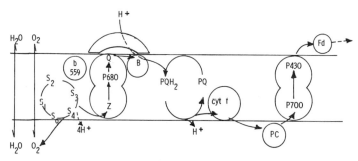

FIG. 6. The electron transport chain of green plant photosynthesis arranged as two proton pumps (the protons shown are not meant to indicate any particular stoichiometry).

nism. The area of greatest ignorance seems to be in our understanding of the biochemistry of oxygen evolution. This is despite a great deal of study by the most elegant of biophysical methods, which has described the kinetic and thermodynamic framework within which the biochemistry must operate, but without clothing this skeleton in the flesh and blood of reactants and intermediates.

C. THE PROTON PUMPS OF BACTERIAL PHOTOSYNTHESIS

The major review of bacterial photosynthesis edited by Clayton and Sistrom (1977) contains several articles of obvious relevance to the present topic; and in particular a very comprehensive article by Wraight *et al.* (1977) already referred to above. We will restrict ourselves here to a consideration of the Rhodospirillaceae (formerly Athiorhodaceae), since electron transport in this group has been most fully studied.

1. Topology of the Electron Transport Chain

The species studied in greatest detail are *Rhodopseudomonas sphaeroides* and *R. capsulata*, which appear to be very similar in their photosynthetic apparatus. The electron transport chain appears to be arranged anisotropically in the membrane so as to act as two proton pumps in series when electrons traverse the cyclic pathway (Crofts, 1974; Wraight *et al.*, 1977). The photochemical reaction center spans the membrane, with the reaction sites for cytochrome c_2, and for the secondary acceptor and the proton associated with its reduction, on opposite sides of the membrane (Jackson and Crofts, 1971; Prince *et al.*, 1975b; Dutton *et al.*, 1975). In chromatophores, cytochrome c_2 and its reaction site are inside, while the site of proton uptake is outside; in whole cells the cytochrome c_2 is outside the cell membrane in the periplasmic space. This arrangement of the photochemical apparatus is now generally accepted, and the evidence for it has been reviewed at length elsewhere (Parsons and Cogdell, 1975; Crofts, 1974; Wraight *et al.*, 1977). The arrangement of the "dark" electron transport chain by which electrons return from the primary acceptor to oxidized cytochrome c_2 is much less certain, and there is some doubt as to the redox carriers contributing to the chain. However, it is clear that electron flow through this section of the chain is also linked to the pumping of protons, and through the proton gradient to phosphorylation.

In the following sections we will review the evidence available for the constitution and arrangement of these sections of the electron transport chain, and consider the sites at which protons react.

Components of the Cyclic Electron Transport Chain. The photochemical reactions in *R. sphaeroides* and *R. capsulata* leave P870 oxidized and the iron–quinone complex of the primary acceptor reduced. In both species the E_m for the P870+/P870 couple is ~450–470 mV, and the E_m, for the X/X$_{red}$ couple (where X is probably ubiquinone closely associated with iron [Fe–UQ]) in equilibrium with protons from the aqueous phase is ~−60 mV at pH 7. However, no proton uptake occurs over the time range of reduction of the couple (Chance et al., 1970), so it would be expected that the initial reaction would produce X$^-$, and this is supported by EPR spectroscopic, and spectrophotometric evidence (Slooten, 1972; Clayton and Straley, 1970; Wraight et al., 1975; Dutton et al., 1973). The E_m for this couple depends on the pK for the dissociation of the protonated form

$$XH \rightleftharpoons X^- + H^+$$

Prince and Dutton (1976b) have estimated pKs for the primary acceptors of a number of different species by observing the effect of pH on the redox titration curve of the photochemistry. The method assumes that over the time period of a redox experiment, the primary acceptor equilibrates with the aqueous protons, so that the pK can be estimated from the break in the dependence of E_m on pH. Above the pK the E_m does not vary with pH, and therefore corresponds to the value of E_m for the X/X$^-$ couple. Using this approach Prince and Dutton (1976b) showed that although the values of E_m measured below the pK values varied widely for the primary acceptors of a number of species, the values of E_m for the couple X/X$^-$ (the values above the pKs) varied very much less; the variation in E_m below the pK values reflected variations in pK. They pointed out that the "operating" E_m was in all cases likely to be that of the X/X$^-$ couple, the value for this being ~−160 mV. From these results we may conclude that the maximal redox span encompassing the components of the cyclic pathway could be from −200 mV to +500 mV. Of the redox components identified in R. sphaeroides or R. capsulata the vast majority fall within this range (Dutton and Jackson, 1972; Crofts, 1974; Dutton and Wilson, 1974), so that identification of the reactants in the chain can be achieved only by combining redox measurements with kinetic measurements (Jackson and Dutton, 1973; Evans and Crofts, 1974b). The components identified by EPR spectroscopy are for the most part detectable only at low temperature (Dutton et al., 1973; Dutton and Wilson, 1974; Prince and Dutton, 1976a), and the kinetic resolution of the signals is a technically forbidding task. For this reason, most of our information about the components of the cyclic pathways has come from spectrophotometric measurements. These have involved either (a) direct measurement of the components (P870, the cytochromes, X$^-$), (b) measurement of electrogenic events by the carotenoid change, or (c) measurement of H$^+$ changes by use of appropriate indicator dyes.

2. Photochemical Reaction Centers

Several excellent, recent reviews have covered the biophysics and biochemistry of bacterial reaction centers (Parsons and Cogdell, 1975; Sauer, 1975); we will restrict ourselves here to points pertinent to the role of reaction centers as proton pumps, and avoid discussion of those transient events and intermediates (Rockley *et al.*, 1975; Kaufman *et al.*, 1975; Tiede *et al.*, 1976; Shuvalov *et al.*, 1976; Prince *et al.*, 1976) which precede the stabilization of the energy conversion process through oxidation of P870 and reduction of the iron–ubiquinone complex.

Most of the work reported has been done with reaction centers extracted from a carotenoid-deficient mutant of *R. sphaeroides* (R26), using LDAO (Clayton and Wang, 1971). Reaction centers from carotenoid-containing strains extracted by similar procedures are similar in all essentials, apart from the additional presence of 1 molecule of carotenoid per reaction center (Jolchine and Reiss-Husson, 1974, 1975). Reaction centers isolated from *R. capsulata* and *R. rubrum* by similar procedures also appear to be comparable in essential features with the R26 reaction centers (Prince and Crofts, 1973; Noel *et al.*, 1972). Most workers have used detergent-solubilized preparations. In these, the reactions with added *c*-type cytochromes (either mammalian or native) are second order, and strongly dependent on ionic strength (Prince *et al.*, 1974). In more complex preparations (Ke *et al.*, 1970) or those in which the reaction centers have been incorporated into liposomes (Dutton *et al.*, 1976), the reaction may be of a lower order due to binding of cytochrome *c*.

Reactions on the acceptor side are more complicated, and depend both on the state of the complex and on the method of preparation. In preparations purified by $(NH_4)_2SO_4$ precipitation, following a single flash from the dark-adapted state the charge is stabilized in a species absorbing at 430 nm, thought to be the anionic semiquinone of endogenous ubiquinone (Slooten, 1972; Wraight *et al.*, 1975). No H^+ uptake accompanied this change. A second flash leads to the uptake of 1 H^+ per reaction center. In the presence of added artificial acceptor (1,4-naphthoquinone), subsequent flashes lead to the uptake of 1 H^+ per reaction center as the acceptor is reduced. However, in these flashes after the first, no consistent pattern of behavior for the species absorbing at 430 nm is observed. In reaction centers that have never been exposed to $(NH_4)_2SO_4$, but purified by column chromatography, a more consistent pattern is observed. The first flash leads to formation of the anionic semiquinone, identified by the absorption at 430 nm and by EPR spectroscopy (Wraight, 1977; Vermeglio, 1977). The second flash removes the semiquinone and leads to the uptake of a proton (Wraight, 1977); in the presence of added acceptor, an oscillation with a period of

2 can be observed for many flashes, odd-numbered flashes producing, and even-numbered flashes removing, the semiquinone signal. This behavior is interpreted as reflecting a mechanism for charge accumulation on the acceptor side, between the one electron/molecule reaction of the primary acceptor (the tightly bound Fe·UQ), a second bound ubiquinone, and the two hydrogen/molecule reaction of loosely bound ubiquinone or added acceptor.

   a. The Reaction Center in Situ—Structural Aspects. Penna *et al.* (1974) and Vermeglio and Clayton (1976) have recently made a detailed study of the orientation of chromophores in the reaction center *in situ*, using and refining the techniques of linear dichroism introduced to the field by Breton and colleagues (Breton and Roux, 1971; Breton *et al.*, 1973; Breton, 1974). Vermeglio and Clayton (1976) found linear dichroic ratios of 0.5 at 600 nm, 1.6 at 860 nm and 1.5 at 1250 nm for the light-induced changes and suggested that these showed that the bacteriochlorophyll dimer, which acts as the primary donor, was nearly perpendicular to the plane of the membrane. Vermeglio and Clayton (1976) also showed that the polarization was of opposite sign for the changes at 790 nm and 810 nm. It had previously been supposed that the light-induced change at these wavelengths was due to a blue shift of the 803 nm band, but the opposite polarization was incompatible with this suggestion. They therefore proposed that the absorption decrease at 810 nm and 870 nm was due to bleaching of two bands of the bacteriochlorophyll dimer, while the 790 nm increase corresponded to the appearance of a monomeric chlorophyll band, a property of the singly oxidized dimer $(BChl)_2{}^+$.

   Feher (1971) and Holmes (1976) had previously observed the spectral complexity of the 800 nm band on analysis of the spectral components of purified reaction centers by derivative analysis and curve fitting. Their conclusions were necessarily less definitive than those of Vermeglio and Clayton (1976), since the data on the differential linear dichroism of the bands around 800 nm proved crucial. Holmes (1976) also showed that the room-temperature bands around 600 nm could be resolved into two components at 590 nm and 599.5 nm, only the higher wavelength band undergoing a partial bleaching on oxidation.

   Although the orientation of the reaction center in the membrane is now well characterized, little is known of its structure. D. Saddler and H. Celis (personal communication) have studied the shape of the reaction center in hexane solution using low-angle X-ray scattering. For these experiments the reaction centers were exchanged from a micellar suspension into hexane using the technique of Montal (1974, 1976). In hexane solution, the protein showed an absorption spectrum in the visible and near infrared similar to that in detergent solution, and still showed light-induced spectral changes, indicating that the structure was not seriously disrupted. The X-ray scattering data showed a radius of gyration for the molecule of ~30 Å, which is greater than the value calculated from the density and relative molecular mass, assuming a sphere. Data from low-angle neutron scattering (D. Saddler and H. Celis, personal communication) showed a similar radius. These results suggest that the reaction center protein does not take up the most compact shape in a lipid environment, and it is possible that this also holds for the reaction center in the membrane.

   b. *Reconstitution of the Proton-Pumping Activity of Reaction Centers.* The suggestion that the reaction center *in vivo* is so oriented as to form the electrogenic arm of a proton pump carries the implication that such an activity should be shown by reaction centers appropriately incorporated into artificial lipid membranes. Skulachev and his colleagues (Skulachev, 1972; Barsky *et al.*, 1976) have developed techniques for measuring the incorporation of electrogenic pumps into lipid membranes with high sensitivity. Using reaction centers prepared from *Rhodospirillum rubrum* they have demonstrated that when a suspension of liposomes containing bound reaction centers was allowed to equilibrate with a planar lipid membrane separating two aqueous phases, illumination caused the transfer of charge between the two phases. The mechanism of this charge transfer was not simple, and the polarity and location of the reaction centers with respect to the membrane were not well defined. Nevertheless, the experiments provide strong evidence for an orientation of the reaction centers in this artificial system similar to that observed *in vivo*. A similar electrogenic arrangement could be inferred from the light-induced extrusion of phenyldicarbaundecaborane anion from proteoliposomes containing reaction centers.

   Celis (see Crofts *et al.*, 1977) has recently developed a reconstituted system, which is better defined with respect to orientation and reaction mechanism. Reaction centers from *Rhodopseudomonas sphaeroides* (either from wild-type, Ga or R26 mutant cells) were incorporated into liposomes by a cholate dialysis method (Kagawa and Racker, 1971). Reactivity with reduced cytochrome *c* following a flash, or with ferricyanide in the dark, showed that 40–60% of the protein was oriented so as to present the site of reaction with P870 to the outside. In continuous light, the reactivity with added acceptors could be tested by observing the extent of oxidation of externally added cytochrome *c*. Only when reaction centers had access to both cytochrome *c* and acceptor would turnover lead to an oxidation of the cytochrome in excess of that in the absence of added acceptor. By this test, the majority of the reaction centers able to oxidize cytochrome *c* were able to reduce 1,4-naphthoquinone, but were unable to reduce 1,4-naphthoquinone 2-sulfonate unless the liposome membranes were disrupted by detergent. These experiments show clearly that the majority of the reaction centers were oriented *across* the membrane as *in vivo*, but without any clear selection for a particular direction; a similar conclusion had been reached previously by Dutton *et al.* (1976). However, by appropriate choice of reaction conditions (externally added cytochrome c and 1,4-naphthoquinone) a polarity of reaction could be imposed on the system such that electrons were pumped from outside to inside the liposomes. Under these conditions an extensive extrusion of H^+ occurred on illumination (up to 200 H^+/P870) indicating that the 1,4-naphthoquinone reduced inside was able to carry hydrogen across the membrane, and release H^+ on oxidation by cytochrome c outside.

Appropriate controls with bathophenanthroline, 1,4-naphthoquinone 2-sulfonate, uncoupling agents and ionophores showed that the reconstituted system was acting as an H^+ pump, with the photochemical reaction as the electrogenic arm and the 1,4-naphthoquinone as the H-carrying arm of a classical Mitchellian loop. When ubiquinone-10 was incorporated into the reaction center liposomes, 1,4-naphthoquinone was no longer necessary, but externally added $PMS-SO_3^-$ had to be added to catalyze the electron transfer from the ubiquinone to cytochrome c or directly to the reaction center. These experiments (H. Celis and A. R. Crofts, unpublished observations) show that membrane-bound ubiquinone-10 is also able to act as a secondary acceptor, and to function in an H-carrying capacity. The results imply that the reacting species of the ubiquinone/ubiquinol couple are sufficiently mobile in the membrane phase to react with reagents on both sides of the membrane, as required in the various postulated *in vivo* mechanism (Mitchell, 1966, 1968, 1976).

3. Reactions of the Acceptor Pools in Chromatophores

We have discussed above the evidence which suggests that in R-26 reaction centers purified without exposure to $(NH_4)_2SO_4$, a charge accumulation mechanism with a capacity for two electrons operates on the acceptor side (Wraight, 1977; Vermeglio, 1977). Such a mechanism would be essentially similar to that envisaged on the acceptor side of PS II (see above for references). In the latter case, the charge accumulation has been observed in the intact apparatus, and the acceptor pool of plastoquinone has been identified. In the case of chromatophores no such charge accumulation has been identified, possibly because appropriate experiments with dark-adapted material have not been done (but see Fowler, 1976).

The secondary acceptor in *Rhodopseudomonas sphaeroides* and *R. capsulata* is generally supposed to be ubiquinone (see Wraight *et al.*, 1977, for detailed discussion), but this has not been firmly established except by inference from the presence of ubiquinone in purified reaction centers or its use in reaction center liposomes where its role as a secondary acceptor is unambiguous. The electron transfer from primary to secondary acceptor has been studied in these species mainly from the H^+ uptake accompanying it. Earlier work (Cogdell et al., 1972; Cogdell and Crofts, 1974) had shown that the rapid H^+ uptake on flash illumination showed a stoichiometry of ~ 1 H^+ per reaction center, that the reaction was inhibited by ortho– or bathophenanthroline, that the rate of the reaction (~ 200 μsec half-time) was more rapid than that of electron flow to cytochrome b, the only other acceptor identified, and that redox titration of the effect showed a dependence on pH of the midpoint of -60 mV/pH unit. All these characteristics were compatible with the suggestion (Chance et al., 1970; Cogdell et al., 1972) that H^+ uptake reflected the reduction of an H-carrying secondary acceptor by electrons from the primary acceptor, and the most likely candidate for this role was ubiquinone. However, these results did not throw any light on the mechanism by which a one-electron donor interacted with a H carrier that would be expected to accept 2 H per molecule if the reaction went to full reduction. Fowler (1976), in a brief report on work using an electrode technique, has suggested that protons are released to the interior of chromatophores from *Rhodospirillum rubrum* in pairs, and this would be expected from a charge accumulation mechanism. On the other hand, many workers (but without using a dark-adaptation procedure) have observed uptake of H^+ on an initial flash, followed by reduction of cytochrome b, indicating that the electron has progressed beyond the quinone acceptor pool. In his work on dark-adapted materials, Fowler (1976) also found that no oscillation occurred on uptake of protons. Clearly the situation is one that requires further experimentation. Either a charge accumulation mechanism operates in chromatophores, or not. In the former case, both ubiquinone and a b-type cytochrome may participate in the charge accumulation. In the latter case, the mechanism observed in isolated reaction centers may simply reflect the fact that the only secondary acceptor present required 2 hydrogens per molecule, and that reaction with the primary acceptor in the semiquinone state is restricted by kinetic or thermodynamic factors (Bendall, 1977). Wraight et al. (1977) discuss the problem more fully.

4. Electron Transport from the Secondary Acceptor to Cytochrome c_2

Spectrophotometric measurements, both on a kinetic basis following flash activation, and from steady-state spectra of changes induced by continuous light, show that b-type cytochrome(s) are reduced in the light, while c-type cytochromes are oxidized. By performing such measurements under conditions of controlled redox potential (Dutton and Jackson, 1972; Dutton and Wilson, 1974; Evans and Crofts, 1974a,b), it has been possible to identify cytochrome b_{50}, and cytochrome c_{290} in *Rhodopseudomonas sphaeroides* (or cytochrome c_{340} in *R. capsulata*) as species kinetically competent for inclusion in the main cyclic pathway.

Both in kinetic experiments and in measurements of steady-state levels, the changes due to cytochrome *b* are smaller than those due to cytochrome *c*. In the presence of antimycin A both changes are enhanced, and in particular the cytochrome *b* changes become comparable with those of cytochrome *c*. In kinetic experiments antimycin can be seen to inhibit cytochrome *b* reoxidation and cytochrome *c* re-reduction, making it possible to measure the true rates of reduction of cytochrome *b* and oxidation of cytochrome *c* (assuming that these are unaffected by antimycin). Antimycin appears therefore to inhibit electron transport from cytochrome *b* to cytochrome *c*. By appropriate subtraction of kinetic traces in the presence and absence of antimycin (Prince and Dutton, 1975), and assuming again that antimycin does not affect electron transport to cytochrome *b* and from cytochrome *c*, the rate of electron transport between the cytochromes can be computed (Prince and Dutton, 1975; Crofts *et al.*, 1975a).

Several interesting features deserve note:

1. The maximal rates of reduction and oxidation of cytochrome *b* are comparable (though the former is more rapid), whereas the rate of cytochrome *c* oxidation is much faster than its reduction. This may account for the small changes in cytochrome *b* observed in kinetic or steady-state measurements.

2. The rate of reduction of cytochrome *b* appears to be faster than the rate of re-reduction of cytochrome *c*, and this is clearly the case in the presence of antimycin.

3. The rates of oxidation of cytochrome *b* and reduction of cytochrome *c* are similar. The work of Crofts *et al.* (1975a) was performed with chromatophores from *R. capsulata* which were coupled, and it was therefore possible to compare the rates of electrogenic events (as measured by the carotenoid change) with the electron transport rates. Although these rates were measured with coupled preparations, they approximate to uncoupled rates, since a single flash does not build up a sufficient proton potential to control electron flow, and addition of uncouplers or ionophores only marginally changed the rates. Prince and Dutton (1975) made their measurements using *R. sphaeroides* chromatophores under uncoupled conditions, and concluded, from the similarity of rates of electron flow from cytochrome *b* and to cytochrome *c*, that any kinetic impediment was shared in common by the two reactions under these conditions. Experiments with *R. sphaeroides* but without

uncoupler were difficult to interpret, since absorbance changes due to cytochrome b_{150} interfered with measurements of the changes due to cytochrome b_{50} (Dutton and Jackson, 1972; Jackson and Dutton, 1973).

Although addition of uncoupling agent made little difference to the kinetics observed following a single flash, the reverse was true when the kinetics of the cytochromes were measured under conditions in which a proton gradient had developed as a result of multiple turnovers of the electron transport chain (Jackson and Dutton, 1973; Dutton *et al.*, 1975; Petty and Dutton, 1976a; Crofts *et al.*, 1974a, 1975a). Under these conditions electron flow from cytochrome b and to cytochrome c became increasingly inhibited by successive flashes as the chromato-phore proton gradient built up, and uncouplers or ionophores like valinomycin, markedly stimulated these reactions. The stimulation by valinomycin was particularly interesting. In the absence of valinomycin, the membrane potential as indicated by the carotenoid change developed to a maximal value after relatively few turnovers (after 4–6 flashes), and inhibition of electron flow could be observed to develop in parallel. If valinomycin was present, neither the membrane potential nor the inhibition of electron flow developed, indicating that, for the first few flashes, control of electron flow was exerted by the membrane potential, as would be expected from the relatively low capacity of the membrane for charge as compared with the buffering capacity of the aqueous phases (see Mitchell, 1968, for a discussion of these capacity parame-ters). However, if the chromatophores were illuminated for several seconds after addition of valinomycin (Crofts *et al.*, 1974a, 1975a), no stimulation of electron flow was observed unless nigericin was added as well. Nigericin alone had no effect, but FCCP had an effect similar to that of valinomycin and nigericin. The latter combination, or FCCP alone, allows electrochemical equilibration of protons (Jackson *et al.*, 1969), so we may assume that electron flow from cytochrome b to cytochrome c is controlled by back pressure of the full electrochemical proton gradient when this is allowed to develop. These results show clearly that electrons traversing the chain between the b and c cyto-chromes are necessarily involved in driving a fully electrogenic proton pump. Before considering the mechanism of this pump we must discuss the measurement of electrogenic events as indicated by carotenoid spectral charges.

5. The Carotenoid Change as an Indicator of Membrane Potential

The empirical evidence relating the change in amplitude of the carotenoid difference spectrum (or the extent of the carotenoid change as measured at fixed wavelength) to the change in membrane potential, has been discussed at length elsewhere (Jackson and Crofts, 1969;

Holmes and Crofts, 1977a,b; Case and Parsons, 1973; Wraight *et al.*, 1977) as have the similar changes (the 515 nm change) in chloroplasts and algae (Witt, 1971, 1975). The absorbance changes in bacteria were shown by Vredenberg and Amesz (1967) to reflect shifts to the red of the spectrum of both carotenoid and chlorophyll bulk pigments. In both chloroplasts and chromatophores, the amplitude of the change appeared to be proportional to the membrane potential (Witt, 1971; Witt and Zickler, 1973, 1974; Jackson and Crofts, 1969). Attempts to provide a theoretical justification for this relationship have met with some difficulty. The most promising framework has been that provided by Witt and colleagues (Schmidt *et al.*, 1971, 1972; Reich *et al.*, 1976; Wraight *et al.*, 1977), who pointed out that the linear proportionality between the changes could be accounted for by an electrochromic mechanism (Liptay, 1969), if the absorbance changes were assumed to correspond to red shifts of the pigments induced by the electrical field. A linear relation between shift and field would be expected only for molecules with a strong permanent dipole. Typical photosynthetic pigments have no such dipole, and would have been expected to show a shift proportional to the square of the field. The discrepancy between the apparently linear shifts observed, and those expected, could be explained if they were represented as corresponding to points on a quadratic curve relating degree of spectral shift to applied field, as would be expected for molecules with no permanent dipole, and a major portion of the field was contributed by fixed (and constant) local charge effects, while only a small proportion came from the external, variable, and delocalized membrane potential. Such a hypothesis predicts quite rigorous limits for the values expected for the spectral change, and these are within the range of experimental test. It had been pointed out by several workers (Amesz *et al.*, 1973; Okada *et al.*, 170) that the spectral changes observed in wild-type cells of *R. sphaeroides* or chromatophores prepared from them, were different from those expected from a linear shift to the red of the bulk pigment. However, it could be argued that the contribution from several different chemical species of carotenoid present in wild-type cells would make unambiguous interpretation of these changes difficult. Crofts *et al.* (1974a), Holmes (1976), and Holmes and Crofts (1977a,b) have extended these observations using a strain of *R. sphaeroides* (mutant G1C) containing only one chemical species of carotenoid—neurosporene. Their findings essentially confirmed the interpretation of previous workers and showed that the maximal carotenoid change appeared to correspond to a shift of 10% of the bulk spectrum to the red by about 7 nm. Several noteworthy features of the change were that (a) an increase in amplitude of the change was not accompanied by a corresponding increase in degree of red shift; (b) the shifts induced by applied potentials opposite in sign to those produced on illumination, were to the blue, but showed the same apparent isosbestic point as the red shift, and their spectra were thus mirror images of the red-shifted spectra; (c) although there appeared to be a negligible departure from a true isosbestic point in the spectra of changes of differing amplitude measured at room temperature, a small shift in isosbestic point was observed when the amplitude of the change was increased stepwise by successive flashes of light (Crofts et al., 1974a).

Holmes (1976) and Holmes and Crofts (1977a,b) discussed these measurements in terms of several different possible mechanisms for the carotenoid change. They showed that the changes observed were not compatible with any simple electrochromic model in which the bulk pigment was involved, but that they appeared rather to suggest a "two population" model, as previously proposed by Amesz et al. (1973), in which the distribution between the populations was linearly related to the membrane potential. However, they were unable to suggest any molecular basis for such a relationship, and discussed variations of electrochromic models that might accommodate the data. De Grooth and Amesz (1976) measured carotenoid changes in wild-type R. sphaeroides at subfreezing temperatures, and were able to demonstrate a convincing shift to the red with increasing amplitude of change. However, the shift was of a small degree, as observed in the changes induced by successive flashes (Crofts et al., 1974a; Holmes and Crofts, 1977a). De Grooth and Amesz (1976) suggested that the change might correspond to a small shift to the red of a larger proportion of the pigment than the 10% previously suggested, but that the pigment involved had its absorption maximum several nanometers to the red side of the maximum of the overall spectrum. This suggestion would allow the observed changes to be accommodated in the electrochromic framework proposed by Schmidt et al. (1972). In discussing such a possibility, Holmes (1976) and Holmes and Crofts (1977a,b) pointed out that their detailed analysis showed no indication of multiple components contributing to the overall spectrum. However, such an analysis could not provide an exclusive test since the curve-fitting and fourth-derivative procedures used showed only the minimal complement of curves fitting the spectrum, and a close fit could be readily obtained by increasing this. In view of these considerations, the electrochromic mechanism seems still to be the most satisfactory explanation of the relation between the carotenoid change and membrane potential. However, as Chance (1977) points out, the postulate of contributions to the electrochromic field from both fixed charge and the delocalized potential begs the question as to which of these is undergoing a change when a change in the spectrum is observed.

The answer can be found only in experiment; in this respect, account must be taken of the empirically excellent correspondence between the membrane potential and the spectral change, found by a wide variety of different approaches (see above and review by Wraight *et al.*, 1977, for references). In the following discussion we will make the simplifying assumption that the light-induced change in the carotenoid spectrum is due predominantly to the electrical potential difference felt across the membrane between the opposing aqueous phases. Wraight *et al.* (1977) have a very complete discussion of the mechanism of the carotenoid change.

6. Electrogenic Spans of the Cyclic Electron-Transport Chain

The kinetics of the carotenoid change induced by a single turnover flash show at least three phases, a rapid phase ($t_{1/2} < 1$ μsec), associated with the photochemical reactions (Jackson and Crofts, 1971), an antimycin-sensitive slow phase, the half-time and extent of which vary with ambient redox potential (Jackson and Crofts, 1971; Jackson and Dutton, 1973), and an intermediate phase ($t_{1/2} < 30$ μsec) associated with the reduction of $P870^+$ by ferrocytochrome c_2 (Jackson and Dutton, 1973). The two faster phases can be interpreted as reflecting the transfer of charge across the membrane in the photochemical reaction (accounting for 60–70% of the displacement), and in the secondary donor reaction (30–40% of the change). Dutton *et al.* (1975) have suggested that the reaction center protein must be embedded in the membrane in such a way that the charge on $P870^+$ is not available to the aqueous phase until after electron donation from cytochrome c_2. This would imply that the faster phase of the change reflects a field that is more localized than that expected from a transmembrane potential. Case and Parson (1973) have made a study of the relation between the carotenoid shift and electron transfer from different cytochromes to the reaction center in *Chromatium* and have also concluded that these must contain contributions other than those from delocalized fields.

The slow phase of the carotenoid change appears to be associated with electron flow through the antimycin-sensitive site. The kinetics of the change are very similar to the kinetics of electron transfer from cytochrome b to c, and Crofts *et al.* (1972; Crofts, 1974) have interpreted this as indicating that this span of the electron transport chain must include an electrogenic process.

7. The Effect of Redox Potential on Electron Flow between the b and c Cytochromes

Work from several groups (Jackson and Dutton, 1973; Evans and Crofts, 1974b; Crofts *et al.*, 1975a; Prince and Dutton, 1976c) has shown that the rate of electron flow to cytochrome c, and the rate and extent of the slow phase of the carotenoid change in both $R.$ $sphaeroides$ and $R.$ $capsulata$ are markedly stimulated when the ambient redox potential is brought below ~200 mV. Maximal rates of electron flow (and the maximal extent of the slow phase of the carotenoid change) occurred at ambient potentials of ~90 mV at pH 7. The increased electron flow (observed either as a faster rate or cytochrome c re-reduction, or a greater extent of the slow phase of the carotenoid change after a set time) titrated in as the redox potential was lowered, with an E_m (apparent) of ~150 mV at pH 7 (Crofts et $al.$, 1975a; Prince and Dutton, 1976c), but this varied by -60 mV/pH over a wide range of pH, and showed a slope of $n = 2$, suggesting the involvement of a hydrogen carrying redox component of the quinone/quinol type. It has been suggested that the rate of reactions in this region of the chain may depend on the redox poise of the ubiquinone/ubiquinol couple, and we discuss mechanisms incorporating this suggestion in greater detail below.

Evans and Crofts (1974a,b) and Crofts et $al.$ (1975a) also observed that electron flow both to and from cytochrome b was accelerated over the same redox span in $R.$ $capsulata$. This curious behavior is not consistent with a simple linear electron transport chain of discrete stable redox couples. Either some mechanism exists by which redox poise affects the rate constant of the reactions, in addition to affecting the concentrations of reactants, or the chain must be so arranged as to allow interaction between different sections, through branched pathways, cyclic pathways, or shared intermediates. To explain this paradox, several models have been suggested, none of them entirely satisfactory. We will discuss the evidence reviewed above in terms of two current schemes shown in Fig. 7. Both these envisage the involvement of ubiquinone in two different types of reactions corresponding to those of the quinone/semiquinone and semiquinone/quinol couples. The first of these is a "Q-cycle" mechanism proposed by Prince and Dutton (1976c) and based on a principle suggested by Mitchell (1975, 1976) and modified by Crofts et $al.$ (1975a); the second is a linear mechanism proposed by Crofts et $al.$ (1975a).

$a.$ Q-$Cycle$ $Model.$ Mitchell's Q-cycle (1975, 1976) was originally proposed to overcome difficulties in the interpretation of mitochondrial experiments. The most obvious of these was the failure to detect any hydrogen carrier in the electron-transport chain between cytochromes b and c, as expected in a classical H^+-pumping loop. The second was the observation that under certain conditions, addition of oxidant reacting at cytochrome c led to the $reduction$ of cytochrome b. Several workers had suggested mechanisms to account for this latter phenomenon based on dual pathways and the involvement of ubisemiquinone (see reviews by

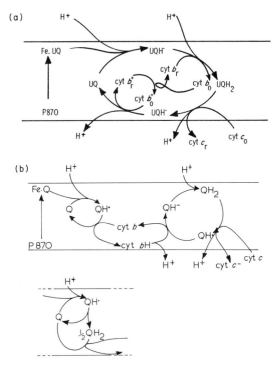

FIG. 7. The cyclic electron transport system of *Rhodopseudomonas sphaeroides* or *R. capsulata* shown as two proton pumps. (a) Quinone-cycle scheme (the asterisked cytochrome would be the low-potential cytochrome *b* of $E_m \sim -100$ mV. (b) Linear scheme. Inset shows a possible disproportionation reaction.

Rieske, 1976; Wikström, 1973), and Mitchell's scheme represented an elegant rationalization of these earlier models; it has been adapted by other workers to fit the circumstances of other situations (Garland *et al.*, 1975; Mitchell, 1976). In the context of the cyclic electron transport of *R. sphaeroides* and *R. capsulata* a Q-cycle provides a neat, integrated mechanism for the entire pathway outside the photochemical reactions. However, such a mechanism appears to be contradicted by experiment. Within the cycle, reduction of *b*-type cytochrome could occur only at the expense of ubisemiquinone produced by reduction of cytochrome *c*. Consequently, the reduction of cytochrome *c* should be observed always to be faster than the reduction of cytochrome *b*, and this is not the case. In the presence of antimycin, the reduction of cytochrome *b* is complete in a few milliseconds, while cytochrome *c* is reduced over a period of hundreds of milliseconds. In the absence of antimycin, time-resolved spectra (Prince and Dutton, 1975; Crofts *et al.*, 1977) suggest that cytochrome b goes reduced more rapidly than cytochrome c, but in this case, since the changes are small, the interpretation of data is confused by overlapping contributions from cytochromes b and c, and from the reaction center change at the wavelengths of interest. The resolution of these contributions requires more detailed study. To get over the difficulty presented by the kinetics in the presence of antimycin, Prince and Dutton (1976) suggested that in the presence of the inhibitor, the electron flow is redirected so that cytochrome b becomes reduced by semiquinone from the other side of the cycle. With this modification the cycle accounts well for most of the phenomena discussed above. In particular, the anomalous redox behavior is explained by the need for both quinone and quinol forms of ubiquonone at adequate concentrations for the cycle to operate. Nevertheless, it has to be assumed that the rapid reduction of the b cytochrome apparent in the absence of antimycin is artifactual or else that the semiquinone involved in the reduction of the b cytochrome is stable in the dark, and that the rapid reduction is a consequence of displacement of the equilibrium concentrations on removal of quinone by its reduction by the primary acceptor (see Crofts *et al.*, 1975a). However, these qualifications and that relating to the effect of antimycin, seem at present to be *ad hoc* interpolations, and their justification by further experimental data must precede acceptance of the cycle as an adequate mechanism.

b. Linear Model. In the linear electron transport scheme of Fig. 7b the ubiquinone is again envisaged as forming two separate reaction couples, but these are no longer involved in a cycle. This return to a simple sequence removes the necessity for *ad hoc* assumptions about the role of the b cytochromes and antimycin. The anomalous kinetic behavior as the redox potential is lowered may be rationalized in terms of changes in concentration of ubiquinone and ubiquinol, and the effects of these on the reactions of the semiquinone couples, and on the dismutation reaction (Crofts *et al.*, 1975a; Mitchell, 1976). However, the likelihood of such a mechanism depends heavily on assumptions about the stability of the semiquinone. The close coupling between the oxidation of cytochrome b and the reduction of cytochrome c can be explained if the semiquinone, produced as a result of the oxidation of quinol by cytochrome c, is stable only for a limited time in the dark. If this were the case, the oxidation of cytochrome b would be dependent on the production of the semiquinone, and if this latter were rate limiting, the two rates would appear to be closely matched. The mechanism may appear to encounter difficulties in explaining how the pathway could operate at high pH, where experiment suggests that the potential of the cytochrome b_{50} couple (independent of pH above the pK at 7.5) would be higher than that of the quinone/quinol couple (decreasing by 60 mV/pH unit increase). However, it should be noted that the reaction couple for oxidation of cytochrome b_{50} would be the semiquinone/quinol couple, the potential of which would be expected to be considerably higher than that of the parent couple. A further refinement of this model might be the inclusion of cytochrome b_{-90} between the quinone/semiquinone couple and cytochrome b_{50} as suggested in the model of Prince and Dutton (1976). In discussing the two models above, we have drawn attention to the crucial question of the stability of the semiquinone species. It is clear that mixed or intermediate models could be devised in which dismutation reactions were of importance, and also that the mobility of the different quinone species could determine the reaction pathway. Resolution of these difficulties clearly calls for more data, rather than further speculation. Mitchell (1976) has an extensive discussion of the adequacy of two models similar to those discussed here, as they apply to mitochondria.

8. The Protolytic Reactions of the Cyclic Electron-Transport Chain in Rhodopseudomonas sphaeroides and R. capsulata

We have discussed some aspects of these reactions above, in particular the involvement of H^+ in the reactions between primary and secondary acceptor. In summary, earlier work had shown that the kinetics and redox characteristics of the rapid H^+ uptake following excitation with a single turnover flash were compatible with the involvement of the H^+ in the reduction of a secondary acceptor, probably ubiquinone (see Cogdell et al., 1972; Crofts, 1974; also see above for details). This work had also shown that a second phase of H^+ uptake was present and was especially obvious when valinomycin was included in the reaction mixture. The second phase was equal in magnitude to the first, of variable rate, but slower than the first (fast) phase, and was sensitive to antimycin. On the basis of these observations, Cogdell et al. (1972) suggested that the second phase occurred as a result of H^+ uptake on reduction of an H-carrying redox component close to the antimycin-sensitive site in the pathway between cytochromes b and c. However, these workers recognized a paradox arising from the interpretation of these results and of the carotenoid change. As discussed above, the kinetics of the carotenoid change had shown that at a suitable ambient redox potential, the slow phase of the carotenoid change was similar in magnitude to the fast phase, suggesting that a second transfer of charge across the full extent of the chromatophore membrane had occurred. Nevertheless, a second phase of H^+ uptake of equivalent extent was not always observed in the absence of valinomycin, as would have been expected if the electrogenic arm indicated by the carotenoid slow phase had led to electron transfer to an H carrier as predicted for a H^+-pumping loop.

More recently, Petty and Dutton (1976a,b) have carried out an extensive study of the H^+ changes associated with the cyclic electron-transport chain of *R. sphaeroides*, and have extended the earlier work summarized above. In brief, they have shown that (a) the stoichiometry of H^+ uptake in the presence of antimycin, and at ambient redox potentials >200 mV, was close to 1 H^+/e delivered from the photoreactions; (b) over a wider potential range, but in the absence of antimycin and in the presence of valinomycin, the stoichiometry was close to $2H^+$/ e delivered; (c) with no antimycin or valinomycin, the stoichiometry was intermediate between that of (a) and (b) above; (d) the stoichiometry of the antimycin insensitive H^+ uptake dropped at higher pH, with an apparent pK of 8.4.

Petty and Dutton (1976b) also studied the role of protons in the reactions of cytochrome b_{50}. They concluded that cytochrome b_{50} acted as an H carrier at pH values below a pK of ~7.5. The H^+ taken up on reduction of the secondary acceptor was released inside the chromatophores only on oxidation of cytochrome b_{50}, unless the pH was higher than the pK of 7.5. At these higher values of pH, the H^+ was released inside even if cytochrome b_{50} oxidation was blocked by antimycin, presumably on oxidation of reduced ubiquinone by oxidized cytochrome b_{50} present before excitation. No H^+ uptake or release was seen if the pH was much higher than 8.4, the pK found for the rapid H^+ uptake.

These results are consistent with the linear scheme discussed above but are difficult to reconcile with the Q-cycle scheme for the dark electron-transport pathway. More recent work by K. M. Petty (unpublished observations made in collaboration with Dr. P. L. Dutton and Dr. J. B. Jackson) has shown a degree of complexity for the H^+ changes that is difficult to reconcile with either of the schemes discussed here. In the presence of antimycin, the E_m for redox titration of the rapid H^+ uptake (~85 mV at pH 7) was more positive by ~80 mV than in the absence of inhibitor and showed a pK of about 7.6. The antimycin-sensitive portion of the change showed a pK of about 7.4, and a value for E_m of -10 mV at pH 7, varying by -60 mV/pH unit below the pK. This E_m would identify the H^+ uptake as being associated with reduction of a component equipotential with the primary acceptor, rather than with components of E_m between 50 and 250 mV as expected for the antimycin site. This paradoxical behavior obviously requires further investigation.

9. The H^+ pumps of Rhodopseudomonas sphaeroides and
   R. capsulata—Conclusions

The role of the photochemical reaction in the generation of membrane potential is now well established, and it seems clear that the electrogenic reaction is part of the mechanism of an H^+ pump. It seems likely that a second electrogenic reaction is linked to electron flow through the part of the dark electron-transport chain between cytochromes b and c. Since the two electrogenic spans are part of a cyclic chain, a single electron completing the cycle must recross the membrane in neutral reactions occurring in sequence between the electrogenic stages. The protons of the protolytic reactions are likely to be associated with these neutral steps. The two schemes of Fig. 7 suggest possible mechanisms by which these events may occur. In these schemes, the common elements (the role of the photochemical reactions, the involvement of ubiquinone, and of cytochromes b_{50} and c_{290} or c_{340}, a second electrogenic site) may be taken as established. The major areas of ignorance are concerned with the protolytic reactions, the precise involvement of ubiquinone, the detailed kinetics of the b-type cytochromes, and the mechanism and site of action of antimycin. We have ignored the involvement of other components, in particular nonheme iron groups, since no experimental evidence as to their kinetic role is available.

ABBREVIATIONS

ADRY reagents: reagents which *a*ccelerate the *d*eactivation *r*eactions of the water-splitting complex, *Y*
DCMU: 3-(3,4-dichlorophenyl)-1,1-dimethylurea
SDS: dodecyl sulfate, sodium salt
BV^{2+}: benzyl viologen, 1,1′-dibenzyl-4,4′-bipyridinium dichloride
MV^{2+}: methyl viologen, 1,1′-dimethyl-4,4′-bipyridinium dichloride
LDAO: lauryl dimethylamine oxide
PMS: *N*-methyl phenazinium methylsulfate
$PMS-SO_3^-$: *N*-methyl phenazinium-3-sulfonate methylsulfate
FCCP: carbonyl cyanide *p*-trifluoromethoxy phenylhydrazone

ACKNOWLEDGMENTS

We should like to thank Dr. B. Chance for helpful correspondence, Dr. D. S. Bendall for valuable criticism of the manuscript, Ms. K. Petty and Dr. S. Saphon for helpful discussion, Drs. Saphon, Celis, Bendall, Wraight, Holmes, Chance, Petty, and Cramer for access to manuscripts and material before publication, and Ms. J. M. Bowes for help in preparation of the manuscript.

REFERENCES

Aitken, A. (1975). *Biochem. J.* **149**, 675–683.
Aitken, A. (1976). *Nature (London)* **263**, 793–796.
Ambler, R. P., and Bartsch, R. G. (1975). *Nature (London)* **253**, 285–288.
Ambler, R. P., Meyer, T. E., and Kamen, M. D. (1976). *Proc. Natl. Acad. Sci. U.S.A.* **73**, 472–475.
Amesz, J. (1973). *Biochim. Biophys. Acta* **301**, 35–51.
Amesz, J., Visser, J. W. M., van den Engh, G. J., and Pulles, M. P. J. (1972a). *Physiol. Veg.* **10**, 319–328.

Amesz, J., Visser, J. W. M., van den Engh, G. J., and Dirks, M. P. (1972b). *Biochim. Biophys. Acta* **256**, 370–380.
Amesz, J., 'tMannetje, A. H., and De Grooth, B. G. (1973). *Abstr. Symp. Prokaryotic Photosynth. Organisms, 1973* pp. 34–35.
Anderson, J. M., and Boardman, N. K. (1973). *FEBS Lett.* **32**, 157–160.
Aparicio, P. J., Ando, K., and Arnon, D. I. (1974). *Biochim. Biophys. Acta* **357**, 246–251.
Arnason, T., and Sinclair, J. (1976). *Biochim. Biophys. Acta* **449**, 581–586.
Arnon, D. I., and Chain, R. K. (1975). *Proc. Natl. Acad. Sci. U.S.A.* **72**, 4961–4965.
Arntzen, C. T., and Briantais, J.-M. (1975). In "Bioenergetics of Photosynthesis" (Govindjee, ed.), pp. 51–113. Academic Press, New York.
Ausländer, W., and Junge, W. (1974). *Biochim. Biophys. Acta* **357**, 285–298.
Ausländer, W., and Junge, W. (1975). *FEBS Lett.* **59**, 310–315.
Ausländer, W., Heathcote, P., and Junge, W. (1974). *FEBS Lett.* **47**, 229–235.
Avron, M. (1975). In "Bioenergetics of Photosynthesis" (Govindjee, ed.), pp. 373–386. Academic Press, New York.
Baas Becking, L. G. M., Kaplan, I. R., and Moore, D. (1960). *J. Geol.* **68**, 243–284.
Babcock, G. T., and Sauer, K. (1975a). *Biochim. Biophys. Acta* **376**, 315–328.
Babcock, G. T., and Sauer, K. (1975b). *Biochim. Biophys. Acta* **376**, 329–344.
Babcock, G. T., and Sauer, K. (1975c). *Biochim. Biophys. Acta* **396**, 48–62.
Babcock, G. T., Blankenship, R. E., and Sauer, K. (1976). *FEBS Lett.* **61**, 286–289.
Bamberger, E., Rottenberg, H., and Avron, M. (1973). *Eur. J. Biochem.* **34**, 557–563.
Barsky, E. L., Dancshazy, Z., Drachev, L. A., Il'ina, M. D., Jasaitis, A. A., Kondrashin, A. A., Samuilov, V. D., and Skulachev, V. P. (1976). *J. Biol. Chem.* **251**, 7066–7071.
Bartsch, R. G. (1977). In "The Photosynthetic Bacteria" (R. K. Clayton and W. R. Sistrom, eds.). Plenum, New York (in press).
Bearden, A. J., and Malkin, R. (1975). *Q. Rev. Biophys.* **7**, 131–177.
Bendall, D. S. (1977). *Biochem. Ser. One* **13**, 41–78.
Bendall, D. S., Davenport, H. E., and Hill, R. (1971). In "Methods in Enzymology" (A. San Pietro, ed.), Vol. 23A, pp. 327–344. Academic Press, New York.
Biggins, J. (1967). *Plant Physiol.* **42**, 1447–1456.
Bishop, N. I. (1971). *Annu. Rev. Biochem.* **40**, 197–226.
Bishop, N. I. (1973). *Photophysiology* **8**, 65–96.
Blankenship, R. E., and Sauer, K. (1974). *Biochim. Biophys. Acta* **357**, 252–266.
Blankenship, R. E., McGuire, A., and Sauer, K. (1977). *Biochim. Biophys. Acta* **459**, 617–619.
Böhme, H. (1977). *Eur. J. Biochem.* **72**, 283–289.
Böhme, H., and Cramer, W. A. (1973). *Biochim. Biophys. Acta* **325**, 275–283.
Bouges-Bocquet, B. (1973). *Biochim. Biophys. Acta* **314**, 250–256.
Bouges-Bocquet, B. (1974). *Proc. Int. Congr. Photosynth. Res., 3rd, 1974* Vol. 1, pp. 579–588.
Boulter, D., Haslett, B. G., Peacock, D., Ramshaw, J. A. M., and Scawen, M. D. (1977). *Biochem., Ser. One* **13**, 1–40.
Breton, J., and Roux, E. (1971). *Biochem. Biophys. Res. Commun.* **45**, 557–563.
Breton, J. (1974). *Biochem. Biophys. Res. Commun.* **59**, 1011–1017.
Breton, J., Michel-Villaz, M., and Paillotin, G. (1973). *Biochim. Biophys. Acta* **314**, 42–56.
Brown, J., Bril, C., and Urbach, W. (1965). *Plant Physiol.* **40**, 1086–1090.
Butler, W. L., Visser, J. M. W., and Simons, H. L. (1973). *Biochim. Biophys. Acta* **292**, 140–151.
Carr, N. G. (1973). In "The Biology of Blue-green Algae" (N. G. Carr and B. A. Whitton, eds.), pp. 39–65. Blackwell, Oxford.
Carr, N. G., and Exell, G. (1965). *Biochem. J.* **96**, 688–692.
Case, G. D., and Parson, W. W. (1973). *Biochim. Biophys. Acta* **325**, 441–453.

Case, G. D., Parson, W. W., and Thornber, J. P. (1970). *Biochim. Biophys. Acta* **223**, 122–128.

Chance, B. (1977). *Annu. Rev. Biochem.* (in press).

Chance, B., Crofts, A. R., Nishimura, M., and Price, B. (1970). *Eur. J. Biochem.* **10**, 226–237.

Cheniae, G. M. (1970). *Annu. Rev. Plant Physiol.* **21**, 467–498.

Cheniae, G. M., and Martin, I. F. (1970). *Biochim. Biophys. Acta* **197**, 219–239.

Clark, W. M. (1960). "Oxidation-Reduction Potentials of Organic Systems." Waverly Press, Baltimore, Maryland.

Clayton, R. K., and Sistrom, W. R., eds. (1977). "The Photosynthetic Bacteria" Plenum, New York (in press).

Clayton, R. K., and Straley, S. C. (1970). *Biochem. Biophys. Res. Commun.* **39**, 1114–1119.

Clayton, R. K., and Wang, R. T. (1971). *In* "Methods in Enzymology" (A. San Pietro, ed.), Vol. 23A, pp. 696–704. Academic Press, New York.

Cline, J. D., and Richards, F. A. (1969). *Environ. Sci. Technol.* **3**, 838–843.

Cogdell, R. J., and Crofts, A. R. (1974). *Biochim. Biophys. Acta* **347**, 264–272.

Cogdell, R. J., Jackson, J. B., and Crofts, A. R. (1972). *Bioenergetics* **4**, 413–429.

Cohen, Y., Padan, E., and Shilo, M. (1975a). *J. Bacteriol.* **123**, 855–861.

Cohen, Y., Jørgensen, B. B., Padan, E., and Shilo, M. (1975b). *Nature (London)* **257**, 489–492.

Cox, R. P., and Delosme, R. (1976). *C. R. Hebd. Seances Acad. Sci., Ser. D* **282**, 775–778.

Cramer, W. A. (1977). *In* "Encyclopedia of Plant Physiology" (M. Auron and A. Trebst, eds.). Springer-Verlag, Heidelberg (in press).

Cramer, W. A., and Butler, W. L. (1967). *Biochim. Biophys. Acta* **143**, 332–339.

Cramer, W. A., and Butler, W. L. (1969). *Biochim. Biophys. Acta* **172**, 503–510.

Cramer, W. A., and Horton, P. (1975). *Photochem. Photobiol.* **22**, 304–308.

Cramer, W. A., Horton, P., and Donnell, J. J. (1974). *Biochim. Biophys. Acta* **368**, 361–370.

Crespi, H. L., Smith, U., Gajda, L., Tisue, T., and Ammeraal, R. M. (1972). *Biochim. Biophys. Acta* **256**, 611–618.

Crofts, A. R. (1974). *In* "Perspectives in Membrane Biology" (S. Estrada-O and C. Gitler, eds.), pp. 373–412. Academic Press, New York.

Crofts, A. R., Wraight, C. A., and Fleischmann, D. E. (1971). *FEBS Lett.* **15**, 89–100.

Crofts, A. R., Jackson, J. B., Evans, E. H., and Cogdell, R. J. (1972). *Photosynth., Two Centuries Its Discovery Joseph Priestley, Proc. Int. Congr. Photosynth. Res., 2nd, 1971* Vol. 2, pp. 873–902.

Crofts, A. R., Prince, R. C., Holmes, N. G., and Crowther, D. (1974a). *Proc. Int. Congr. Photosynth. Res., 3rd, 1974* pp. 1131–1146.

Crofts, A. R., Evans, E. H., and Cogdell, R. J. (1974b). *Ann. N.Y. Acad. Sci.* **227**, 227–243.

Crofts, A. R., Crowther, D., and Tierney, G. V. (1975a). *In* "Electron Transfer Chains and Oxidative Phosphorylation" (E. Quagliariello *et al.*, eds.), pp. 233–241. North-Holland Publ., Amsterdam.

Crofts, A. R., Holmes, N. G., and Crowther, D. (1975b). *Proc. 10th FEBS Meet.* **40**, 287–304.

Crofts, A. R., Crowther, D., Celis, H., Almanza de Celis, S., and Tierney, G. (1977). *Biochem. Soc. Trans.* **5**, 491–495.

Deamer, D. W., Prince, R. C., and Crofts, A. R. (1972). *Biochim. Biophys. Acta* **274**, 323–335.

De Grooth, B. G., and Amesz, J. (1976). *Abstr. Symp. Primary Electron Transp. Energy Transduction Photosynth. Bacteria*, p. WB1.

Delosme, R. (1972). *Photosynth., Two Centuries Its Discovery Joseph Priestley, Proc. Int. Congr. Photosynth. Res., 2nd, 1971* Vol. 1, pp. 187–196.

Delrieu, M. J. (1976). *Biochim. Biophys. Acta* **440**, 176–188.

Den Haan, G. A. (1976). Doctoral Thesis, University of Leiden.

Den Haan, G. A., Duysens, L. N. M., and Egberts, D. J. N. (1974). *Biochim. Biophys. Acta* **368**, 409–421.

Dickerson, R. E., and Timkovich, R. (1975). *In* "The Enzymes" (P. D. Boyer, ed.), 3rd ed., Vol. 11, pp. 397–547. Academic Press, New York.

Diner, B. (1974). *Biochim. Biophys. Acta* **368**, 371–385.

Diner, B., and Joliot, P. (1976). *Biochim. Biophys. Acta* **423**, 479–498.

Döring, G. (1976). *Z. Naturforsch., Teil C* **31**, 78–81.

Döring, G., Stiehl, H. H., and Witt, H. T. (1967). *Z. Naturforsch., Teil B* **22**, 639–644.

Droop, M. R. (1974). *In* "Algal Physiology and Biochemistry" (W. D. Stewart, ed.), pp. 530–559. Univ. of California Press, Berkeley.

Dutton, P. L. (1971). *Biochim. Biophys. Acta* **226**, 63–80.

Dutton, P. L., and Jackson, J. B. (1972). *Eur. J. Biochem.* **30**, 495–510.

Dutton, P. L., and Wilson, D. F. (1974). *Biochim. Biophys. Acta* **346**, 165–212.

Dutton, P. L., Leigh, J. S., and Reed, D. W. (1973). *Biochim. Biophys. Acta* **292**, 654–666.

Dutton, P. L., Petty, K. M., Bonner, H. S., and Morse, S. D. (1975). *Biochim. Biophys. Acta* **387**, 536–556.

Dutton, P. L., Petty, K. M., and Prince, R. C. (1976). *Fed. Proc., Fed. Am. Soc. Exp. Biol.* **35**, 1597.

Duysens, L. N. M., and Sweers, H. E. (1963). *In* "Studies of Microalgae and Photosynthetic Bacteria," pp. 353–372. Univ. of Tokyo Press, Tokyo.

Duysens, L. N. M., den Haan, G. A., and van Best, J. A. (1974). *Proc. Int. Congr. Photosynth. Res., 3rd, 1974* pp. 1–12.

Ellenson, J., and Sauer, K. (1976). *Photochem. Photobiol.* **23**, 113–123.

Elliott, P. B., and Bamforth, S. S. (1975). *J. Protozool.* **22**, 514–519.

Entsch, B., and Smillie, R. M. (1972). *Arch. Biochem. Biophys.* **151**, 378–386.

Epel, B. L., Butler, W. L., and Levine, R. P. (1972). *Biochim. Biophys. Acta* **275**, 395–400.

Eppley, R. W., and MaciasR, F. M. (1963). *Limnol. Oceanogr.* **8**, 411–416.

Etienne, A. L. (1974). *Biochim. Biophys. Acta* **333**, 497–508.

Evans, E. H., and Crofts, A. R. (1974a). *Biochim. Biophys. Acta* **357**, 78–88.

Evans, E. H., and Crofts, A. R. (1974b). *Biochim. Biophys. Acta* **357**, 89–102.

Evans, M. C. W. (1969). *Prog. Photosynth. Res., Proc. Int. Congr. [1st], 1968* pp. 1474–1475.

Evans, M. C. W., Reeves, S. G., and Cammack, R. (1974). *FEBS Lett.* **49**, 111–114.

Feher, G. (1971). *Photochem. Photobiol.* **14**, 373–387.

Fenchel, T. M., and Riedl, R. J. (1970). *Mar. Biol.* **7**, 255–268.

Fleischman, D. E. (1971). *Photochem. Photobiol.* **14**, 277–286.

Fogg, G. E. (1974). *In* "Algal Physiology and Biochemistry" (W. D. Stewart, ed.), pp. 560–582. Univ. of California Press, Berkeley.

Forti, G., Bertolè, M. L., and Zanetti, G. (1965). *Biochim. Biophys. Acta* **109**, 33–40.

Fowler, C. F. (1974). *Biochim. Biophys. Acta* **357**, 327–331.

Fowler, C. F. (1976). *Abstr. Symp. Primary Electron Transp. Energy Transduction Photosynth. Bacteria, 1976* p. WB9.

Fowler, C. F. (1977). *Biochim. Biophys. Acta* **459**, 351–363.

Fowler, C. F., and Kok, B. (1974a). *Biochim. Biophys. Acta* **357**, 299–307.

Fowler, C. F., and Kok, B. (1974b). *Biochim. Biophys. Acta* **357**, 308–318.

Fowler, C. F., and Kok, B. (1976). *Biochim. Biophys. Acta* **423**, 510–523.

Fredericks, W. W., and Gehl, J. M. (1976). *Arch. Biochem. Biophys.* **174**, 666–674.

Fujita, Y. (1974). *Plant Cell Physiol.* **15**, 861–874.
Garland, P. B., Clegg, R. A., Boxer, D., Downie, J. A., and Haddock, B. A. (1975). *In* "Electron Transfer Chains and Oxidative Phosphorylation" (E. Quagliariello *et al.*, eds.), pp. 351–358. North-Holland Publ., Amsterdam.
Gest, H. (1972). *Adv. Microb. Physiol.* **7**, 243–282.
Giaquinta, R. T., and Dilley, R. A. (1975). *Biochim. Biophys. Acta* **387**, 288–305.
Giaquinta, R. T., Dilley, R. A., Selman, B. R., and Anderson, B. J. (1974). *Arch. Biochem. Biophys.* **162**, 200–209.
Girault, G., and Galmiche, J. M. (1974). *Biochim. Biophys. Acta* **333**, 314–319.
Gitlitz, P. H., and Krasna, A. I. (1975). *Biochemistry* **14**, 2561–2568.
Glaser, M., Wolff, C., and Renger, G. (1976). *Z. Naturforsch., Teil C* **31**, 712–721.
Gorman, D. S., and Levine, R. P. (1966a). *Plant Physiol.* **41**, 1637–1642.
Gorman, D. S., and Levine, R. P. (1966b). *Plant Physiol.* **41**, 1643–1647.
Gorman, D. S., and Levine, R. P. (1966c). *Plant Physiol.* **41**, 1648–1656.
Govindjee, Pulles, M. P. J., Govindjee, R., Van Gorkom, H. J., and Duysens, L. N. M. (1976). *Biochim. Biophys. Acta* **449**, 602–605.
Graber, P., and Witt, H. T. (1975). *FEBS Lett.* **59**, 184–189.
Graber, P., and Witt, H. T. (1976). *Biochim. Biophys. Acta* **423**, 141–163.
Gray, B. H., Fowler, C. F., Nugent, N. A., Rigopoulos, N., and Fuller, R. C. (1973). *Int. J. Syst. Bacteriol.* **23**, 256–264.
Grimme, L. H., and Boardman, N. I. (1974). *Proc. Int. Congr. Photosynth. Res., 3rd, 1974* Vol. 3, pp. 2115–2124.
Haehnel, W. (1976). *Biochim. Biophys. Acta* **440**, 506–521.
Hauska, G. A., McCarty, R. E., Berzborn, R. J., and Racker, E. (1971). *J. Biol. Chem.* **246**, 3524–3531.
Hauska, G. A., Reimer, S., and Trebst, A. (1974). *Biochim. Biophys. Acta* **357**, 1–13.
Hauska, G. A., Oettmeier, W., Reimer, S., and Trebst, A. (1975). *Z. Naturforsch., Teil C* **30**, 37–45.
Haveman, J., and Lavorel, J. (1975). *Biochim. Biophys. Acta* **408**, 269–283.
Heber, U., Boardman, N. K., and Anderson, J. M. (1976). *Biochim. Biophys. Acta* **423**, 275–292.
Henningsen, K. W., and Boardman, N. K. (1973). *Plant Physiol.* **51**, 1117–1126.
Hochman, A., Fridberg, I., and Carmeli, C. (1975). *Eur. J. Biochem.* **58**, 65–72.
Holmes, N. G. (1976). Ph.D. Thesis, University of Bristol.
Holmes, N. G., and Crofts, A. R. (1977a). *Biochim. Biophys. Acta* **459**, 492–505.
Holmes, N. G., and Crofts, A. R. (1977b). *Biochim. Biophys. Acta* **461**, 141–150.
Holton, R. W., and Myers, J. (1967). *Biochim. Biophys. Acta* **131**, 375–384.
Honda, S. I., Baker, J. E., and Muenster, A.-M. (1961). *Plant Cell Physiol.* **2**, 151–163.
Horio, T., Higashi, T., Yamanaka, T., Matsubara, H., and Okunuki, K. (1961). *J. Biol. Chem.* **236**, 944–951.
Horton, P., and Cramer, W. A. (1974). *Biochim. Biophys. Acta* **368**, 348–360.
Husain, A., Hutson, K. G., Andrew, P. W., and Rogers, L. J. (1976). *Biochem. Soc. Trans.* **4**, 488.
Ikegami, I., and Katoh, S. (1973). *Plant Cell Physiol.* **14**, 829–836.
Ikegami, I., Katoh, S., and Takamiya, A. (1968). *Biochim. Biophys. Acta* **162**, 604–606.
Izawa, S., and Ort, D. R. (1974). *Biochim. Biophys. Acta* **357**, 127–143.
Jackson, J. B., and Crofts, A. R. (1968). *Biochem. Biophys. Res. Commun.* **32**, 908–915.
Jackson, J. B., and Crofts, A. R. (1969). *FEBS Lett.* **4**, 185–189.
Jackson, J. B., and Crofts, A. R. (1971). *Eur. J. Biochem.* **18**, 120–130.
Jackson, J. B., and Dutton, P. L. (1973). *Biochim. Biophys. Acta* **325**, 102–113.
Jackson, J. B., Crofts, A. R., and von Stedingk, L.-V. (1969). *Eur. J. Biochem.* **6**, 41–54.

Jagendorf, A. T. (1975). In "Bioenergetics of Photosynthesis" (Govindjee, ed.), pp. 413–492. Academic Press, New York.

Jennings, J. V., and Evans, M. C. W. (1977). FEBS Lett. 75, 33–36.

Jolchine, G., and Reiss-Husson, F. (1974). FEBS Lett. 40, 5–8.

Jolchine, G., and Reiss-Husson, F. (1975). FEBS Lett. 52, 33–36.

Joliot, P., and Delosme, R. (1974). Biochim. Biophys. Acta 357, 267–284.

Joliot, P., and Joliot, A. (1974). Proc. Int. Congr. Photosynth. Res., 3rd, 1974, Vol. 1, pp. 25–39.

Joliot, P., and Joliot, A. (1976). C.R. Hebd. Seances Acad. Sci., Ser. D 283, 393–396.

Joliot, P., and Kok, B. (1975). In "Bioenergetics of Photosynthesis" (Govindjee, ed.), pp. 387–412. Academic Press, New York.

Joliot, P., Barbieri, G., and Chabaud, R. (1969). Photochem. Photobiol. 10, 309–329.

Joliot, P., Joliot, A., Bouges-Bocquet, B., and Barbieri, G. (1971). Photochem. Photobiol. 14, 287–305.

Joliot, P., Delosme, R., and Joliot, A. (1977). Biochim. Biophys. Acta 459, 47–57.

Junge, W. (1975). Ber. Dtsch. Bot. Ges. 88, 283–301.

Junge, W., and Ausländer, W. (1974). Biochim. Biophys. Acta 333, 59–70.

Junge, W., and Ausländer, W. (1975). In "Electron Transfer Chains and Oxidative Phosphorylation" (E. Quagliariello et al., eds.), pp. 243–250. North-Holland Publ., Amsterdam.

Junge, W., and Witt, H. T. (1968). Z. Naturforsch., Teil B 23, 244–254.

Jursinic, P., Warden, J., and Govindjee. (1976). Biochim. Biophys. Acta 440, 322–330.

Kadenbach, B. (1971). In "Autonomy and Biogenesis of Mitochondria and Chloroplasts" (N. K. Boardman, A. W. Linnane, and R. S. Smillie, eds.), pp. 360–371. North-Holland Publ., Amsterdam.

Kagawa, Y., and Racker, E. (1971). J. Biol. Chem. 246, 5477–5487.

Kaiser, W., and Urbach, W. (1976). Biochim. Biophys. Acta 423, 91–102.

Katan, M. B., Pool, L., and Groot, G. S. P. (1976). Eur. J. Biochem. 65, 95–105.

Katoh, S. (1959). J. Biochem. (Tokyo) 46, 629–632.

Katoh, S. (1960). Nature (London) 186, 533–534.

Katoh, S., Suga, I., Shiratori, I., and Takamiya, A. (1961). Arch. Biochem. Biophys. 94, 136–141.

Kaufman, K. J., Dutton, P. L., Netzel, T. L., Leigh, J. S., and Rentzepis, P. M. (1975). Science 188, 1301–1304.

Ke, B., Chaney, T. H., and Reed, D. W. (1970). Biochim. Biophys. Acta 216, 373–383.

Ke, B., Hansen, R. E., and Beinert, H. (1973). Proc. Natl. Acad. Sci. U.S.A. 70, 2941–2945.

Ke, B., Hawkridge, F. M., and Sahu, S. (1976). Proc. Natl. Acad. Sci. U.S.A. 73, 2211–2215.

Keister, D. L., and Yike, N. J. (1967). Arch. Biochem. Biophys. 121, 415–422.

Kelly, J., and Ambler, R. P. (1974). Biochem. J. 143, 681–690.

Kenyon, C. N., and Gray, A. M. (1974). J. Bacteriol. 120, 131–138.

Kessler, E. (1974). In "Algal Physiology and Biochemistry" (W. D. Stewart, ed.), pp. 456–473. Univ. of California Press, Berkeley.

Kihara, T., and Chance, B. (1969). Biochim. Biophys. Acta 189, 116–124.

King, M. T., and Drews, G. (1975). Arch. Microbiol. 102, 219–231.

Knaff, D. B. (1975). FEBS Lett. 60, 331–335.

Knaff, D. B., and Buchanan, B. B. (1975). Biochim. Biophys. Acta 376, 549–560.

Knaff, D. B., and Malkin, R. (1973). Arch. Biochem. Biophys. 159, 555–562.

Knaff, D. B., and Malkin, R. (1976). Biochim. Biophys. Acta 430, 244–252.

Knaff, D. B., Buchanan, B. B., and Malkin, R. (1973). Biochim. Biophys. Acta 325, 94–101.

240 A. R. CROFTS AND P. M. WOOD

Kok, B., Forbush, B., and McGloin, M. (1970). *Photochem. Photobiol.* 11, 457–475.

Kok, B., Radmer, R., and Fowler, C. F. (1974). *Proc. Int. Congr. Photosynth. Res., 3rd, 1974* Vol. 1, pp. 485–496.

Kraan, G. P. B. (1971). Doctoral Thesis, University of Leiden.

Kunert, K.-J., and Böger, P. (1975). *Z. Naturforsch., Teil C* 30, 190–200.

Kunert, K.-J., Böhme, H., and Böger, P. (1976). *Biochim. Biophys. Acta* 449, 541–553.

Kusai, A., and Yamanaka, T. (1973a). *Biochim. Biophys. Acta* 292, 621–623.

Kusai, A., and Yamanaka, T. (1973b). *Biochim. Biophys. Acta* 325, 304–314.

La Monica, R. F., and Marrs, B. L. (1976). *Biochim. Biophys. Acta* 423, 431–439.

Lavorel, J. (1975). *In* "Bioenergetics of Photosynthesis" (Govindjee, ed.), pp. 223–317. Academic Press, New York.

Laycock, M. V., and Craigie, J. S. (1971). *Can. J. Biochem.* 49, 641–646.

Lemasson, C., and Etienne, A. L. (1975). *Biochim. Biophys. Acta* 408, 135–142.

Levine, R. P. (1971). *In* "Methods in Enzymology" (A. San Pietro, ed.), Vol. 23A, pp. 119–129.

Lien, S., and Racker, E. (1971). *J. Biol. Chem.* 246, 4298–4307.

Lightbody, J. J., and Krogmann, D. W. (1967). *Biochim. Biophys. Acta* 131, 508–515.

Liptay, W. (1969). *Angew. Chem., Int. Ed. Engl.* 8, 177–188.

Madigan, M. T., and Brock, T. D. (1975). *J. Bacteriol.* 122, 782–784.

Madigan, M. T., Petersen, S. R., and Brock, T. D. (1974). *Arch. Microbiol.* 100, 97–103.

Malkin, R., and Aparicio, P. J. (1975). *Biochem. Biophys. Res. Commun.* 63, 1157–1160.

Malkin, R., and Bearden, A. J. (1975). *Biochim. Biophys. Acta* 396, 250–259.

Manahan, S. E., and Smith, M. J. (1973). *Environ. Sci. Technol.* 7, 829–833.

Marrs, B. L. (1976). *Abstr. Symp. Primary Electron Transp. Energy Transduction Photosynth. Bacteria, 1976* p. ThA2 1-3.

Mathis, P. (1966). *C.R. Hebd. Seances Acad. Sci., Ser. D* 263, 1770–1772.

Mathis, P., and Galmiche, J. M. (1967). *C.R. Hebd. Seances Acad. Sci., Ser. D* 264, 1903–1906.

Mathis, P., and Vermeglio, A. (1975). *Biochim. Biophys. Acta* 396, 371–381.

Matsuzaki, E., and Kamimura, Y. (1972). *Plant Cell Physiol.* 13, 415–425.

Matsuzaki, E., Kamimura, Y., Yamasaki, T., and Yakushiji, E. (1975). *Plant Cell Physiol.* 16, 237–246.

Mayhew, S. G., and Ludwig, M. L. (1975). *In* "The Enzymes" (P. D. Boyer, ed.), 3rd ed., Vol. 12, pp. 57–118. Academic Press, New York.

Mayne, B. (1967). *Brookhaven Symp. Biol.* 19, 460–466.

Mayne, B. (1968). *Photochem. Photobiol.* 6, 189.

Menke, W., and Schmid, G. H. (1976). *Plant Physiol.* 57, 716–719.

Metzner, H. (1975). *J. Theor. Biol.* 51, 201–231.

Meyer, T. E., Bartsch, R. G., Cusanovich, M. A., and Mathewson, J. H. (1968). *Biochim. Biophys. Acta* 153, 854–861.

Mitchell, P. (1961). *Nature (London)* 191, 144–148.

Mitchell, P. (1966). "Chemiosmotic Coupling in Oxidative and Photosynthetic Phosphorylation." Glynn Res., Bodmin, Cornwall, England.

Mitchell, P. (1968). "Chemiosmotic Coupling and Energy Transduction." Glynn Res., Bodmin, Cornwall, England.

Mitchell, P. (1975). *In* "Electron Transfer Chains and Oxidative Phosphorylation" (E. Quagliariello *et al.*, eds.), pp. 305–316. North-Holland Publ., Amsterdam.

Mitchell, P. (1976). *J. Theor. Biol.* 62, 327–367.

Montal, M. (1974). *In* "Perspectives in Membrane Biology" (S. Estrada-O, and C. Gitler, eds.), pp. 591–622. Academic Press, New York.

Montal, M. (1976). *Annu. Rev. Biophys. Bioeng.* 35, 119–175.

Murano, F., and Fujita, Y. (1967). *Plant Cell Physiol.* 8, 673–682.

Murata, N., and Fork, D. C. (1971). *Biochim. Biophys. Acta* **245**, 356–364.
Nalbandyan, R. M. (1972). *Biokhimiya* **37**, 1161–1165.
Nelson, N., and Neumann, J. (1972). *J. Biol. Chem.* **247**, 1817–1824.
Nelson, N., and Racker, E. (1972). *J. Biol. Chem.* **247**, 3848–3853.
Nishimura, M. (1968). *Biochim. Biophys. Acta* **153**, 838–847.
Noel, H., van der Rest, M., and Gingras, G. (1972). *Biochim. Biophys. Acta* **275**, 219–230.
Norris, J. R., Crespi, H. L., and Katz, J. J. (1972). *Biochem. Biophys. Res. Commun.* **49**, 139–146.
Okada, M., Murata, N., and Takamiya, A. (1970). *Plant Cell Physiol.* **11**, 519–530.
Okamura, M. Y., Isaacson, R. A., and Feher, G. (1975). *Proc. Natl. Acad. Sci. U.S.A.* **72**, 3491–3495.
Okayama, S. (1976). *Biochim. Biophys. Acta* **440**, 331–336.
Okayama, S., and Butler, W. L. (1972). *Biochim. Biophys. Acta* **267**, 523–529.
Österberg, R. (1974). *Nature (London)* **249**, 382–383.
Palmer, G. (1975). In "The Enzymes" (P. D. Boyer, ed.), 3rd ed., Vol. 12, pp. 2–56. Academic Press, New York.
Parson, W. W., and Cogdell, R. J. (1975). *Biochim. Biophys. Acta* **416**, 105–149.
Penna, F. J., Reed, D. W., and Ke, B. (1974). *Proc. Int. Congr. Photosynth. Res., 3rd, 1974* Vol. 1. pp. 421–425.
Peroni, F., Schiff, J. A., and Kamen, M. D. (1964). *Biochim. Biophys. Acta* **88**, 74–90.
Pettigrew, G. W. (1974). *Biochem. J.* **139**, 449–459.
Pettigrew, G. W., Meyer, T. E., Bartsch, R. G., and Kamen, M. D. (1976). *Biochim. Biophys. Acta* **430**, 197–208.
Petty, K. M., and Dutton, P. L. (1976a). *Arch. Biochem. Biophys.* **172**, 335–345.
Petty, K. M., and Dutton, P. L. (1976b). *Arch. Biochem. Biophys.* **172**, 346–353.
Pfennig, N., and Trüper, H. G. (1974). In "Bergey's Manual of Determinative Bacteriology" (R. E. Buchanan and N. E. Gibbons, eds.), 8th ed., pp. 24–64. Williams & Wilkins, Baltimore, Maryland.
Pierson, B. K., and Castenholz, R. W. (1974). *Arch. Microbiol.* **100**, 5–24.
Powls, R., and Redfearn, E. R. (1969). *Biochim. Biophys. Acta* **172**, 429–437.
Powls, R., Wong, J., and Bishop, N. I. (1969). *Biochim. Biophys. Acta* **180**, 490–499.
Prince, R. C., and Crofts, A. R. (1973). *FEBS Lett.* **35**, 213–216.
Prince, R. C., and Dutton, P. L. (1975). *Biochim. Biophys. Acta* **387**, 609–613.
Prince, R. C., and Dutton, P. L. (1976a). *FEBS Lett.* **65**, 117–119.
Prince, R. C., and Dutton, P. L. (1976b). *Arch. Biochem. Biophys.* **172**, 329–334.
Prince, R. C., and Dutton, P. L. (1976c). *Abstr. Symp. Primary Electron Transp. Energy Transduction in Photosynth. Bacteria, 1976* p. TB4.
Prince, R. C., and Dutton, P. L. (1977). *Biochim. Biophys. Acta* **459**, 573–577.
Prince, R. C., and Olson, J. M. (1976). *Biochim. Biophys. Acta* **423**, 357–362.
Prince, R. C., Cogdell, R. J., and Crofts, A. R. (1974). *Biochim. Biophys. Acta* **305**, 597–609.
Prince, R. C., Lindsey, J. G., and Dutton, P. L. (1975a). *FEBS Lett.* **51**, 108–111.
Prince, R. C., Baccarini-Melandri, A., Hauska, G. A., Melandri, B. A., and Crofts, A. R. (1975b). *Biochim. Biophys. Acta* **387**, 212–227.
Prince, R. C., Leigh, J. S., and Dutton, P. L. (1976). *Biochim. Biophys. Acta* **440**, 622–636.
Pulles, M. P. J., Van Gorkom, H. J., and Verschoor, G. A. M. (1976a). *Biochim. Biophys. Acta* **440**, 98–106.
Pulles, M. P. J., Van Gorkom, H. J., and Willemsen, J. G. (1976b). *Biochim. Biophys. Acta* **449**, 536–540.
Radmer, R., and Kok, B. (1975). *Annu. Rev. Biochem.* **44**, 409–433.
Redfearn, E. R., and Powls, R. (1968). *Biochem. J.* **106**, 50P.

Reich, R., Scheerer, R., Sewe, K.-U., and Witt, H. T. (1976). *Biochim. Biophys. Acta* **449**, 285–294.

Renger, G. (1972a). *Biochim. Biophys. Acta* **256**, 428–439.

Renger, G. (1972b). *Eur. J. Biochem.* **27**, 259–269.

Renger, G. (1972c). *Physiol. Veg.* **10**, 329–345.

Renger, G. (1972d). *FEBS Lett.* **23**, 321–324.

Renger, G. (1973). *Biochim. Biophys. Acta* **314**, 390–402.

Renger, G. (1975). *FEBS Lett.* **52**, 30–32.

Renger, G. (1976). *Biochim. Biophys. Acta* **440**, 287–300.

Renger, G., Bouges-Bocquet, B., and Delosme, R. (1973). *Biochim. Biophys. Acta* **292**, 796–807.

Renger, G., Erixon, K., Döring, G., and Wolff, C. (1976). *Biochim. Biophys. Acta* **440**, 278–286.

Rieske, J. S. (1976). *Biochim. Biophys. Acta* **456**, 195–247.

Rockley, M. G., Windsor, M. W., Cogdell, R. J., and Parson, W. W. (1975). *Proc. Natl. Acad. Sci. U.S.A.* **72**, 2251–2255.

Romjin, J. C., and Amesz, J. (1976). *Biochim. Biophys. Acta* **423**, 164–173.

Ross, E., and Schatz, G. (1976). *J. Biol. Chem.* **251**, 1991–1996.

Ross, R. T., Anderson, R. J., and Hsiao, T. L. (1976). *Photochem. Photobiol.* **24**, 267–278.

Rottenberg, H., Grunwald, T., and Avron, M. (1972). *Eur. J. Biochem.* **25**, 54–63.

Rumberg, B., and Siggel, U. (1969). *Naturwissenschaften* **56**, 130–132.

Ryden, L., and Lundgren, J.-O. (1976). *Nature (London)* **261**, 344–346.

Saphon, S., and Crofts, A. R. (1977). *Z. Naturforsch., Teil C* **32**, 617–626.

Sauer, K. (1975). *In* "Bioenergetics of Photosynthesis (Govindjee, ed.), pp. 115–181. Academic Press, New York.

Saunders, V. A., and Jones, O. T. G. (1975). *Biochim. Biophys. Acta* **396**, 220–228.

Scawen, M. D., Ramshaw, J. A. M., and Boulter, D. (1975). *Biochem. J.* **147**, 343–349.

Schmidt, S., Reich, R., and Witt, H. T. (1971). *Naturwissenschaften* **58**, 414.

Schmidt, S., Reich, R., and Witt, H. T. (1972). *Photosynth., Two Centuries Its Discovery Joseph Priestley, Proc. Int. Congr. Photosynth. Res., 2nd, 1971* Vol. 2, pp. 1087–1095.

Schneeman, R., and Krogmann, D. W. (1975). *J. Biol. Chem.* **250**, 4965–4971.

Schwartz, M. (1968). *Nature (London)* **219**, 915–919.

Selman, B. R., Bannister, T. T., and Dilley, R. A. (1973). *Biochim. Biophys. Acta* **292**, 566–581.

Shipman, L. L., Cotton, T. M., Norris, J. R., and Katz, J. J. (1976). *Proc. Natl. Acad. Sci. U.S.A.* **73**, 1791–1794.

Shuvalov, V. A., Krakhmaleva, I. N., and Klimov, V. V. (1976). *Biochim. Biophys. Acta* **449**, 597–601.

Siedow, J. N., Curtis, V. A., and San Pietro, A. (1973a). *Arch. Biochem. Biophys.* **158**, 889–897.

Siedow, J. N., Yocum, C. F., and San Pietro, A. (1973b). *Curr. Top. Bioenerg.* **5**, 107–123.

Siggel, U. (1975). *Proc. Int. Congr. Phosynth. Res., 3rd, 1974* Vol. 1, pp. 645–654.

Siggel, U., Renger, G., Stiehl, H. H., and Rumberg, B. (1972). *Biochim. Biophys. Acta* **256**, 328–335.

Sillén, L. G., and Martell, A. E. (1964). *Chem. Soc., Spec. Publ.* **17**.

Simonis, W., and Urbach, W. (1973). *Annu. Rev. Plant Physiol.* **24**, 89–114.

Sinclair, J., and Arnason, T. (1974). *Biochim. Biophys. Acta* **368**, 393–400.

Skulachev, V. P. (1972). "Energy Transformations in Biomembranes." Nauka Press, Moscow.

Slooten, L. (1972). *Biochim. Biophys. Acta* **275**, 208–218.

Soininen, R., and Ellfolk, N. (1972). *Acta Chem. Scand.* **26**, 861–872.

Solomon, E. I., Hare, J. W., and Gray, H. B. (1976). *Proc. Natl. Acad. Sci. U.S.A.* **73,** 1389-1393.

Spencer, C. P. (1957). *J. Gen. Microbiol.* **16,** 282-285.

Stanier, R. Y., Kunisawa, R., Mandel, M., and Cohen-Bazire, G. (1971). *Bacteriol. Rev.* **35,** 171-205.

Stemler, A., and Radmer, R. (1975). *Science* **190,** 457-458.

Stewart, W. D. P., and Pearson, H. W. (1970). *Proc. R. Soc. London, Ser. B* **175,** 293-311.

Stiehl, H. H., and Witt, H. T. (1968). *Z. Naturforsch., Teil B* **23,** 220-224.

Straley, S. C., Parson, W. W., Mauzerall, D. C., and Clayton, R. K. (1973). *Biochim. Biophys. Acta* **305,** 597-609.

Stuart, A. L., and Wasserman, A. R. (1973). *Biochim. Biophys. Acta* **314,** 284-297.

Sugimura, Y., Toda, F., Murata, T., and Yakushiji, E. (1968). *In* "Structure and Function of Cytochromes" (K. Okunuki, M. D. Kamen, and I. Sekuzu, eds.), pp. 452-458. Univ. of Tokyo Press, Tokyo.

Takahashi, M., and Asada, K. (1975). *Plant Cell Physiol.* **16,** 191-194.

Tiede, D. M., Prince, R. C., Reed, G. H., and Dutton, P. L. (1976). *FEBS Lett.* **65,** 301-304.

Trebst, A. (1974). *Annu. Rev. Plant Physiol.* **25,** 423-458.

Trumpower, B. L., and Katki, A. (1975). *Biochemistry* **14,** 3635-3642.

Trüper, H. G. (1976). *Int. J. Syst. Bacteriol.* **26,** 74-75.

Tsuji, T., and Fujita, Y. (1972). *Plant Cell Physiol.* **13,** 93-99.

Van Beeumen, J., and Ambler, R. P. (1973). *Antonie van Leeuwenhoek* **39,** 355-356.

Van Gorkom, H. J. (1974). *Biochim. Biophys. Acta* **347,** 439-442.

Van Gorkom, H., Pulles, M. P. J., and Wessels, J. S. C. (1975). *Biochim. Biophys. Acta* **408,** 331-339.

Van Gorkom, H. J., Pulles, M. P. J., Haveman, J., and den Haan, G. A. (1976). *Biochim. Biophys. Acta* **423,** 217-226.

Van Grondelle, R., Duysens, L. N. M., and Van der Wal, H. N. (1976). *Biochim. Biophys. Acta* **449,** 169-187.

Velthuys, B. R. (1975). *Biochim. Biophys. Acta* **396,** 392-401.

Velthuys, B. R. (1976). Doctoral Thesis, University of Leiden.

Velthuys, B. R., and Amesz, J. (1974). *Biochim. Biophys. Acta* **333,** 85-94.

Vermeglio, A. (1977). *Biochim. Biophys. Acta* **459,** 516-524.

Vermeglio, A., and Clayton, R. K. (1976). *Biochim. Biophys. Acta* **449,** 500-515.

Visser, J. W. M., Amesz, J., and Van Gelder, B. F. (1974). *Biochim. Biophys. Acta* **333,** 279-287.

Vredenberg, W. J., and Amesz, J. (1967). *Brookhaven Symp. Biol.* **19,** 49.

Walker, D. A., and Crofts, A. R. (1970). *Annu. Rev. Biochem.* **39,** 389-428.

Warden, J. T., Blankenship, R. E., and Sauer, K. (1976). *Biochim. Biophys. Acta* **423,** 462-478.

Wharton, D. C., Gudat, J. C., and Gibson, Q. H. (1973). *Biochim. Biophys. Acta* **292,** 611-620.

Wikström, M. K. F. (1973). *Biochim. Biophys. Acta* **301,** 155-193.

Wildner, G. F. (1976). *Z. Naturforsch., Teil C* **31,** 157-162.

Wildner, G. F., and Hauska, G. (1974a). *Arch. Biochem. Biophys.* **164,** 127-135.

Wildner, G. F., and Hauska, G. (1974b). *Arch. Biochem. Biophys.* **164,** 136-144.

Witt, H. T. (1971). *Q. Rev. Biophys.* **4,** 365-477.

Witt, H. T. (1975). *In* "Bioenergetics of Photosynthesis" (Govindjee, ed.), pp. 493-554. Academic Press, New York.

Witt, H. T., and Zickler, A. (1973). *FEBS Lett.* **37,** 307-310.

Witt, H. T., and Zickler, A. (1974). *FEBS Lett.* **39,** 205-208.

Witt, K. (1973). *FEBS Lett.* **38,** 116-118.

Wolff, C., and Witt, H. T. (1969). Z. Naturforsch., Teil B 24, 1031–1037.
Wolff, C., Buchwald, H.-E., Ruppel, H., Witt, K., and Witt, H. T. (1969). Z. Naturforsch., Teil B 24, 1038–1041.
Wollman, F.-A., and Thorez, D. (1976). C.R. Hebd. Seances Acad. Sci. 283, 1345–1348.
Wood, P. M. (1974). Biochim. Biophys. Acta 357, 370–379.
Wood, P. M. (1976). FEBS Lett. 65, 111–116.
Wood, P. M. (1977). Eur. J. Biochem. 72, 605–612.
Wood, P. M., and Bendall, D. S. (1975). Biochim. Biophys. Acta 387, 115–128.
Wood, P. M., and Bendall, D. S. (1976). Eur. J. Biochem. 61, 337–344.
Wraight, C. A. (1972). Biochim. Biophys. Acta 283, 247–258.
Wraight, C. A. (1977). Biochim. Biophys. Acta 459, 525–531.
Wraight, C. A., and Crofts, A. R. (1971). Eur. J. Biochem. 19, 386–397.
Wraight, C. A., Kraan, G. P. B., and Gerrits, N. M. (1972). Biochim. Biophys. Acta 283, 259–267.
Wraight, C. A., Cogdell, R. J., and Clayton, R. K. (1975). Biochim. Biophys. Acta 396, 242–249.
Wraight, C. A., Cogdell, R. J., and Chance, B. (1977). In "The Photosynthetic Bacteria" (R. K. Clayton and W. R. Sistrom, eds.). Plenum, New York (in press).
Wydrzynski, T., Zumbulyachis, N., Schmidt, P. G., and Govindjee. (1975). Biochim. Biophys. Acta 408, 349–354.
Yamanaka, T., De Klerk, H., and Kamen, M. D. (1967). Biochim. Biophys. Acta 143, 416–424.
Zannoni, D., Melandri, B. A., and Baccarini-Melandri, A. (1976). Biochim. Biophys. Acta 423, 413–430.
Zumft, W. G., and Spiller, H. (1971). Biochem. Biophys. Res. Commun. 45, 112–118.

The ATPase Complex of Chloroplasts and Chromatophores

RICHARD E. McCARTY
Section of Biochemistry, Molecular and Cell Biology
Cornell University
Ithaca, New York

I. Introduction

Over 30 years ago, Lardy and Elvehjem (1945) suggested that the ATPase activity of isolated mitochondria might be related to oxidative phosphorylation. Since then, it has been firmly established that the ATPase activity is an expression of oxidative phosphorylation operating in reverse. Components of the ATPase complex of mitochondrial, chloroplast, and bacterial inner membranes have been isolated and purified. Pullman *et al.* (1960) reported that mechanical disruption of mitochondria causes the release of a soluble, Mg^{2+}-dependent ATPase from the membranes concomitant with an uncoupling of oxidative phosphorylation. Since purified preparations of the Mg^{2+}-ATPase partially restored phosphorylation to the ATPase-deficient membranes (Penefsky *et al.*, 1960) the ATPase and other components with similar activity were termed "coupling factors." The Mg^{2+}-ATPase was denoted F_1[1], for factor 1, since other factors that enhance oxidative phosphorylation were also found.[1]

In addition to the hydrophilic F_1, the ATPase system of mitochondria also contains a number of more hydrophobic components (see Pederson, 1975, for a review), which are likely to be more integral parts of the mitochondrial inner membrane. These components are essential for oxidative phosphorylation, and some are involved in the binding of F_1 to the membrane (Racker, 1975). Other proteins in the complex may form a transmembrane proton channel (Mitchell, 1966).

Although much of the early work was done with mitochondria, it is now realized that the energy-transducing membranes of chloroplasts and bacteria contain similar ATPase systems. This review is primarily concerned with the ATPase complexes of the chloroplast inner membrane of higher plants and of the membranes of some photosynthetic bacteria. The structure of the complex and suspected functions of the individual components in photosynthetic phosphorylation will be emphasized. The properties of the ATPases of chloroplasts (Jagendorf, 1975, 1977) and of bacterial chromatophores (Melandri and Baccarini-Melandri, 1976) have been reviewed. The reader may wish to consult those articles to gain more information about those aspects of these ATPase systems that I have not considered in detail.

II. Properties of the Soluble (Coupling Factor-ATPase) Component

The first report of the isolation of a coupling factor for photophosphorylation was made in 1963, nine years after the discovery of photophosphorylation. Jagendorf and Smith (1962) found that exposure of spinach chloroplast thylakoids to dilute solutions of EDTA at low ionic strengths uncoupled phosphorylation from electron flow. Although they were unsuccessful in their attempts to restore phosphorylation to the EDTA-extracted thylakoids, Avron (1963) had better luck. The EDTA extract of thylakoids when incubated with the EDTA-treated thylakoids in the presence of Mg^{2+} prior to the assay of photophosphorylation restored phosphorylation up to 35% of control, nonextracted thylakoids. Since the coupling factor activity was labile, the coupling

---

[1] Abbreviations: F_1, the coupling factor-ATPase from mitochondria; CF_1, the coupling factor-ATPase from chloroplast thylakoids; NBD-Cl, 7-chloro-4-nitrobenzo-2-oxa-1,3-diazole; etheno-ADP, 1-N^6-etheno-adenosine 5'-diphosphate.

factor was not characterized beyond the point of demonstrating that it was heat labile and nondialyzable.

Guided by a strong feeling that the mechanism of photophosphorylation resembles that of oxidative phosphorylation, Vambutas and Racker began an attempt to solubilize an ATPase-coupling factor from chloroplast thylakoids. In view of the fact that thylakoids catalyze only very low rates of ATP hydrolysis under conditions where the photophosphorylation rates are high, there was some cause to doubt whether this approach would be successful. However, Vambutas and Racker (1965) found that the treatment of spinach chloroplast thylakoids with trypsin resulted in a marked activation of a Ca^{2+}-dependent ATPase. They were able to extract a trypsin-activated, Ca^{2+}-ATPase from the membranes. Precipitation of thylakoids with cold acetone rendered the activity extractable with a buffer solution. Partially purified preparations of the latent ATPase stimulated photophosphorylation in sonicated thylakoids which had been washed with a medium of low ionic strength. Further work on the coupling factor-ATPase from thylakoids was facilitated by the stability of the preparation.

EDTA treatment also extracts a latent, Ca^{2+}-dependent ATPase from the thylakoids (McCarty and Racker, 1966). EDTA extracts of thylakoids stimulated phosphorylation in thylakoids depleted of the Vambutas and Racker coupling factor. Purified preparations of the latent, Ca^{2+}-ATPase enhanced phosphorylation in EDTA-treated chloroplasts as well. The coupling factor-ATPases extracted by both methods were purified to homogeneity, as judged by analytical polyacrylamide gel electrophoresis, and were found to have identical electrophoretic mobilities (McCarty and Racker, 1966). Thus, the coupling factor solubilized by Avron (1963) and by Vambutas and Racker (1965) are identical. Treatment of thylakoids with pyrophosphate also causes the solubilization of CF_1 (Strotmann et al., 1973). In analogy with the mitochondrial coupling factor ATPase, the chloroplast factor was called CF_1, for chloroplast factor one.

More recently, resolution of a coupling factor-ATPase from photosynthetic bacteria has been achieved. Chromatophores from *Rhodospseudomonas capsulata* and from *Rhodospirillum rubrum* contain a largely manifest Mg^{2+}-dependent ATPase. Most of the ATPase activity may be solubilized from the chromatophore membranes of these organisms by sonication of the membranes in the presence of EDTA (Baccarini-Melandri et al., 1970; Johansson, 1972). Exposure of the chromatophores to 2 M LiCl removes nearly all of the ATPase from the membranes (Konings and Guillory, 1973). Concomitant with the removal of the ATPase activity, photophosphorylation and $^{32}P_i$-ATP exchange activities were lost. These activities were restored to the depleted membranes by the ATPase preparations. The chromatophores of *Chromatium vinosum*, strain D resemble chloroplasts in that the ATPase activity is low and is markedly enhanced by trypsin treatment of the membranes (Hochman and Carmeli, 1971). A Ca^{2+}- and Mg^{2+}-activated ATPase may be solubilized from *Chromatium* chromatophores merely by suspending the chromatophores in buffers of low ionic strength.

Certain photosynthetic bacteria, including *Rhodospeudomonas capsulata* grow heterotrophically, deriving energy from oxidative phosphorylation. Since the photosynthetic and oxidative multienzyme complexes apparently are present in the same membrane, it was of interest to establish whether they use the same coupling factor-ATPase. The fact that an antiserum to the coupling factor isolated from photosynthetic membranes inhibits oxidative phosphorylation in membranes from heterotrophic bacteria (Melandri *et al.,* 1971) strongly suggests that oxidative and photosynthetic phosphorylation share a common coupling factor-ATPase.

Terminology can be a problem when ATPases of energy transducing organelles are discussed. It must be remembered that CF_1 and related proteins from other photosynthetic membranes are but a part of an ATPase system. Although CF_1 itself will catalyze rapid rates of ATP hydrolysis, the temptation to call CF_1 the chloroplast ATPase should be squelched. In my view, the chloroplast ATPase is CF_1 plus the other proteins required for the actual function of the ATPase in the membranes; all these together comprise the chloroplast ATPase. This aggregate of proteins I will call the ATPase complex. This complex probably functions as a reversible, proton translocating ATPase (Mitchell, 1966).

A. SIZE, COMPOSITION, AND STRUCTURE

CF_1 in a high state of purity may be obtained in relatively large amounts (Farron, 1970; Lien and Racker, 1971a). Farron (1970) accomplished the most extensive physical characterization of the protein. The sedimentation coefficient was essentially independent of protein concentration over the range of 0.7–7.1 mg/ml. The sedimentation coefficient, extrapolated to zero protein concentration and corrected for the viscosity of the solvent relative to that of water at 20°C, was 13.8 S. A molecular weight of 325,000 ($\pm$ 600) was derived from high speed centrifugation to equilibrium. In contrast, a molecular weight of 358,000 was calculated from data from low speed sedimentation to equilibrium runs. This discrepancy is likely a consequence of the formation of aggregates during the prolonged low speed runs.

The amino acid composition of CF_1 is not exceptional. It appears to be devoid of tryptophan (Farron, 1970), a fact that probably accounts for the relatively low extinction of CF_1 solutions at 280 nm. Approximately 2 SH groups per mole of CF_1 were accessible to dithiobisnitrobenzoate in the native enzyme whereas in the presence of 8 M urea, about 8 SH were titrated. If the protein was incubated in the presence of 8 M urea and 50 mM dithiothreitol, a total of 12 SH was found, indicating that CF_1 may contain 2 disulfides per mole (Farron and Racker, 1970).

Farron (1970) was unable to detect N-terminal amino acids in CF_1 using the dansyl-Cl technique. This finding suggests that the N termini in CF_1 are blocked. A minimal molecular weight of CF_1 of about 28,000 was calculated from the amino acid composition; furthermore, an apparent molecular weight of CF_1 in 6 M guanidine-HCl of 28,000 was found. These observations, coupled with the cysteine content of the enzyme, led to the tentative conclusion that CF_1 contains 12 similar subunits.

However, gel electrophoresis of dissociated CF_1 in the presence of sodium dodecyl sulfate revealed five components, labeled α, β, γ, δ, and ϵ in order of decreasing molecular weight from 59,000 to 13,000 (Racker et al., 1972). The nonidentity of the five CF_1 subunits was proved not only by their differing amino acid compositions, but also by their differing antigenic properties. Nelson et al. (1973) purified each of the CF_1 subunits and prepared antisera against them. Each antiserum interacted only with the subunit used to raise the antiserum. The coupling factor-ATPase from R. rubrum is composed of 5 subunits (Johnasson and Baltscheffsky, 1975) and has a molecular weight of 350,000 (Johansson et al., 1973). Coupling factor-ATPases from other photosynthetic bacteria have not been as extensively purified.

The stoichiometry of the subunits of CF_1 has not been established with certainty. The extensive cross-linking studies with soluble CF_1 by Baird and Hammes (1976) indicate a structure of $\alpha_2\beta_2\gamma\delta\epsilon_2$, but the evidence is not conclusive. The cysteic acid contents of the partially purified subunits of CF_1 after performic acid oxidation were reported to be: α, 2; β, 3; γ, 6 (Nelson et al., 1973), and ϵ, 1 (Nelson et al., 1972b). The δ subunit seems to be devoid of cysteine. Assuming a subunit structure of $\alpha_2\beta_2\gamma\delta\epsilon_2$, the CF_1 should contain 18 SH/mole. More recent determinations (A. Binder and A. T. Jagendorf, personal communication) confirm that CF_1 has 12 SH/mole and that ϵ has one SH/mole. However, they find the cysteic acid content of the γ subunit to be only 3/mole. At this writing, the cysteic acid contents of the α and β components had not been determined. By allowing purified, reduced, and denatured CF_1 to react with radioactive N-ethylmaleimide, followed by separation of the subunits by electrophoresis, M. A. Weiss and R. E. McCarty (unpublished) found: 7 to 8 maleimides per CF_1 in the α and β subunits together, 3 in the γ subunit, none in the δ, and 1 in ϵ. Although

TABLE I

DISTANCES BETWEEN SITES ON CF_1 DERIVED FROM RESONANCE ENERGY TRANSFER
EXPERIMENTS[a]

Donor	Acceptor	Distance calculated (Å)
Etheno-ADP on tight ADP sites (α?)[b]	NBD (β)	40
Etheno-ADP on tight ADP sites (α?)	Fluorescent maleimide (γ)	>47
Etheno-ADP on tight ADP sites (α?)	Quercetin (α or β)	40–47
Fluorescent maleimide (γ)	Quercetin (α or β)	37–41
Aminonaphthylsulfonate on quercetin sites (α or β)	NBD (β)	30

[a] From Cantley and Hammes (1975b, 1976a,b).

[b] Greek letters in parentheses refer to the subunits to which the various fluorescent probes are likely to be bound under the conditions of the experiments.

more data are needed, it would appear that the cysteic acid contents determined by Nelson *et al.* (1973) may be too high. For the present, however, the cysteine content of CF_1 seems consistent with the $\alpha_2\beta_2\gamma\epsilon$ stoichiometry.

Some information about the distances between various sites on CF_1 was obtained by Cantley and Hammes (1975a, 1976a,b). They employed a fluorescent ADP analog, etheno-ADP which binds to soluble CF_1, like ADP itself, at two, apparently identical sites characterized by dissociation constants of about 2 μM. 7-Chloro-4-nitrobenzo-2-oxa-1,3-diazole (NBD-Cl) was shown by Deters *et al.* (1975) to react stoichiometrically with one tyrosine in a β subunit of CF_1. Light absorbed by the etheno-ADP causes the bound NBD to fluoresce. From fluorescence energy transfer considerations, a distance of about 40 Å between the etheno-ADP sites and the NBD site was calculated. The nucleotide sites were

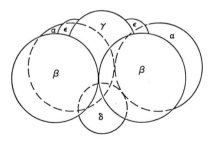

FIG. 1. An interpretation of the structure of CF_1 by Baird and Hammes (1976). Used with permission.

thus thought to reside on the α subunits. Other distances between various sites, as shown in Table I, were also established. From the resonance energy-transfer experiments and from the cross-linking studies, Baird and Hammes (1976) proposed a working model for the stoichiometry and arrangement of CF_1 subunits, which is shown in Fig. 1.

B. NUCLEOTIDE BINDING AND ATPASE ACTIVITY

Soluble CF_1 contains multiple nucleotide binding sites. Roy and Moudrianakis (1971a) demonstrated by equilibrium dialysis a rather slow binding of 2 moles of ADP per mole of CF_1. The dissociation constant for the binding was about 2 μM. Cantley and Hammes (1975b) showed that these tight ADP sites also bound etheno-ADP and adenylyl imidodiphosphate with dissociation constants similar to that of ADP binding. Nucleotides bound to these sites were readily lost when the enzyme was subjected to gel filtration. Multiple nucleotide sites were also found by Vandermeulen and Govindjee (1975) and Girault et al. (1973). Another site capable of binding adenylyl imidodiphosphate was uncovered when the CF_1 was heat-treated, which activates the ATPase (Vambutas and Racker, 1965; Farron and Racker, 1970). This site, characterized by a dissociation constant for adenylyl imidodiphosphate binding of 7.6 μM, was tentatively identified as the active site of the ATPase. The dissociation constant is very similar to the K_i for the ATP analog as a competitive inhibitor of the ATPase (Cantley and Hammes, 1975a). Much weaker nucleotide binding sites, which were not characterized, were also exposed by the heat treatment.

ATPase is the only enzymic activity found so far to be associated with soluble ATPase-coupling factors. CF_1 is devoid of ADP–ATP exchange or P_i–ATP exchange activities (Vambutas and Racker, 1965). The ATPase activity of CF_1 from chloroplast thylakoids and the coupling factor from *Chromatium* (Gepshtein and Carmeli, 1974) is latent whereas the ATPase of solubilized *R. rubrum* is manifest (Johannson et al., 1973). Pronounced activation of the ATPase of CF_1 occurs when the enzyme is exposed to heat, trypsin (Vambutas and Racker, 1965), or dithiothreitol (McCarty and Racker, 1968). Although trypsin treatment greatly enhanced the ATPase of a crude *Chromatium* coupling factor, heat and dithiothreitol treatments were ineffective (Gepshtein and Carmeli, 1974).

The differences in the latency of the ATPase of coupling factors from various sources are probably a consequence of differences in the ease with which an inhibitory subunit of the coupling factor may be dissociated or displaced. Nelson et al. (1972b) purified a trypsin-sensitive protein from CF_1 which inhibited the ATPase of CF_1 activated by heat treatment in the presence of digitonin. The ATPase inhibitor, which was quite hydrophobic, seems to be identical to the ϵ subunit. As was the case for F_1, the mitochondrial-ATPase coupling factor (Penefsky et al., 1960), the total activity of the ATPase of the coupling factor from R. rubrum increased during purification (Johansson et al., 1973). Dissociation of an ATPase inhibitor is probably the cause of the increased activity. Since 7 M urea or heating at 65°C in the presence of digitonin are required to dissociate the inhibitor from CF_1, it is clear that the inhibitor binds very tightly. Farron (1970) could not show dissociation of an inhibitor upon heat activation of CF_1 since the inhibitor, although probably dislocated by the heat treatment, remains bound unless digitonin is present during heating.

The metal specificities of the ATPase of coupling factors present a confusing picture. Although the light- and sufhydryl-activated ATPase (Petrack and Lipmann, 1961) of chloroplast thylakoids is much more active with Mg^{2+} than with Ca^{2+}, the ATPase of soluble CF_1 assayed under similar conditions is activated much more strongly by Ca^{2+} than by Mg^{2+} (Vambutas and Racker, 1965; McCarty and Racker, 1968). A similar situation exists for the membrane-bound and soluble ATPases of R. rubrum (Johansson et al., 1973), although a Mg^{2+}-dependent ATPase was solublized from R. rubrum chromatophores (Konings and Guillory, 1973). This difference, however, is not absolute. The presence of dicarboxylic acids or bicarbonate allows trypsin- or heat-activated CF_1 to catalyze Mg^{2+}-ATP hydrolysis at rapid rates (Nelson et al., 1972a). Moreover, CF_1 bound to thylakoids can be induced to catalyze a low rate of Ca^{2+}-dependent ATP hydrolysis under continuous illumination (Bennun and Avron, 1964). Ca^{2+} also seems to activate a light-dependent $^{32}P_i$-ATP exchange (Bakker-Grunwald, 1974). Mg^{2+}- or Ca^{2+}-ATP are effective substrates for the coupling factor-ATPase from Chromatium chromatophores (Gepshtein and Carmeli, 1974) or from Euglena chloroplasts (Chang and Kahn, 1966; Porat et al., 1976).

Trypsin-treated CF_1 catalyzes ATP hydrolysis at a more rapid rate than that of GTP or ITP. Pyrimidine nucleoside triphosphates, ADP, and pyrophosphate were not cleaved (Vambutas and Racker, 1965). ADP, but not IDP, inhibits ATPase activity. ADP inhibition of the ATPase of CF_1 was found by Nelson et al. (1972a) and Datta et al. (1974) to be apparently cooperative. That is, plots of rate of ATP hydrolysis against ATP concentration changed from hyperbolic in the absence of ADP to sigmoidal in its presence. Cantley and Hammes (1975b) proposed that the binding of ADP to one or both of the tight nucleotide sites on CF_1 inactivates the ATPase, whereas the binding of ATP to these sites enhances ATPase activity. The kinetic data seem consistent with this allosteric model. Adenylyl imidodiphosphate is a potent competitive inhibitor of the ATPase (Cantley and Hammes, 1975b) and excess ATP is also inhibitory.

Prolonged digestion of CF_1 with trypsin free of chymotryptic activity results in a preparation which is an active ATPase and which is devoid of the three smaller subunits. Thus, the ATP catalytic site is on the α and/or β subunits. Since NBD-Cl reacts predominantly with a component in the β subunit and inhibits the ATPase of either the two- or five-subunit enzyme a role for the β subunit in the ATPase reactions seems assured (Deters et al., 1975). Moreover, NBD-Cl inhibits the binding of adenylyl imidodiphosphate to the site exposed upon heat treatment of CF_1 (Cantley and Hammes, 1975a).

C. Reconstitution Studies

Homogeneous CF_1 partially restores phosphorylation to EDTA-treated thylakoids (McCarty and Racker, 1966). However, it is still unclear whether the added CF_1 is active in phosphorylation in a catalytic sense. EDTA or pyrophosphate extraction removes only 30–70% of the membrane-bound enzyme (McCarty and Racker, 1968; Strotmann et al., 1973) even though uncoupling may be complete. More CF_1 may be removed by repeated EDTA extraction, but the resulting thylakoids are inactive in phosphorylation even after reconstitution. In fact, the less CF_1 that is removed by the extractions, the better the reconstitution. Some irreversible changes in the thylakoid membranes probably take place during EDTA treatment. The marked swelling of the chloroplasts that accompanies EDTA uncoupling could lead to irreversible damage. Why a fraction of the CF_1 is resistant to extraction by EDTA is unknown, especially in view of the observations that CF_1 is on the outer surfaces of the thylakoids (Vambutas and Racker, 1965; Howell and Moudrianakis, 1967; Garber and Steponkis, 1974), and that it is probably not present in regions where the membranes are appressed (Miller and Staehelin, 1976). Moreover, antibodies to CF_1 cause a clumping rather than random distribution of CF_1 over the surface of the thylakoids, indicating that CF_1 can move laterally through the membrane (Berzborn et al., 1974).

The key to how partial extraction of CF_1 could result in complete uncoupling is the finding that the apparent proton permeability of thylakoid membranes is markedly enhanced by EDTA extraction (McCarty and Racker, 1966). CF_1 partially restores proton uptake to the extracted thylakoids. Moreover, in preparations of CF_1 exposed to the cold in the presence of high salt to inactivate its ATPase activity (McCarty and Racker, 1966), coupling factor activity may be retained (R. E. McCarty, unpublished). Dicyclohexyl carbodiimide, an energy transfer inhibitor of the oligomycin type (McCarty and Racker, 1967), also restores low proton permeability and some phosphorylation to EDTA-extracted thylakoids. Silicotungstate (Girault and Galmiche, 1972) and tri-n-butyltin chloride (Kahn, 1968) also will enhance proton uptake in EDTA-treated thylakoids. The phosphorylation restored by dicyclohexyl carbodiimide is quite sensitive to an antiserum to CF_1. Thus, extraction of CF_1 appears to enhance the permeability of thylakoid membranes to protons, resulting in a partial collapse of the proton electrochemical gradient, the driving force for phosphorylation. Added CF_1 could then merely act to prevent the proton leak (a "structural role") rather than to catalyze the terminal phosphorylation reaction. The fact that CF_1 restores low ion permeability to extracted thylakoids is confirmed by the observations of Schmid and Junge (1975) and of Girault et al. (1974). The rate of postillumination decay of the 518 nm absorbance change, an indicator of ion permeability of thylakoid membranes (Witt, 1971), was markedly enhanced by EDTA extraction and was slowed by reconstitution with CF_1.

It is tempting to speculate that removal of the CF_1 part of the ATPase complex merely allows communication of a transmembrane channel, formed by the membrane components of the complex, with the medium. However, dislocation of components could occur during EDTA extraction that could alter the proton permeability of the thylakoids. It is important to note, moreover, that proton permeability of chromatophores of *Rhodopseudomonas capsulata* is not enhanced upon extraction of the coupling factor (Melandri et al., 1970). This, however, is not characteristic of the coupling membranes of photosynthetic bacteria since the proton permeability of chromatophores from *Rhodopseudomonas spheroides* is enhanced by extraction of the coupling factor (Mitchell, 1967).

Silicotungstate treatment removes most of the CF_1 from thylakoids and abolishes photophosphorylation (Lien and Racker, 1971b). Unfortunately, only 2–5% of the phosphorylation of untreated thylakoids can be restored by CF_1. Clearly, more work on reconstitution is needed before it may be decided that added CF_1 functions catalytically in reconstituted photophosphorylation.

The coupling factor activity of CF_1 is sensitive to trypsin (Vambutas and Racker, 1965; Deters et al., 1975) and to heat (Farron, 1970). Nelson et al. (1972a) reported that CF_1 containing only the α, β, and γ subunits could be prepared by heating CF_1 in the presence of digitonin, followed by gel filtration. Although this preparation was reported to have coupling factor activity, it was subsequently found that digitonin stimulates phosphorylation in EDTA-extracted thylakoids (Deters et al., 1975). Coupling factor activity is not affected by activation of the ATPase by dithiothreitol (McCarty and Racker, 1968), suggesting that ATPase activation itself is not the cause of the inability of trypsin- or heat-treated CF_1 to enhance phosphorylation.

More recent work (Younis et al., 1977) suggests that the δ subunit is required for the binding of CF_1 to thylakoids. CF_1 deficient in the δ component can be prepared by a chloroform extraction method. Although this preparation is a latent ATPase, it does not serve as a coupling factor. Partially purified preparations of the δ component restored coupling factor activity to the four-subunit enzyme. Since heat can dissociate the δ subunit and since trypsin digests it, the inability of heat or trypsin-treated CF_1 to serve as a coupling factor is explained.

III. Properties of the More Hydrophobic Components of the ATPase Complex

A. Isolation and Partial Characterization

An approach to the isolation and identification of the non-CF_1 components of the ATPase complex has been to isolate the complex in its reconstitutively active form. Based on the pioneering work of Kagawa and Racker (1971), Carmeli and Racker (1973) fractionated thylakoid membranes, previously incubated with dithiothreitol in the dark to activate Mg^{2+}-ATPase and P_i-ATP exchange (McCarty and Racker, 1968), with cholate. After exposure of the thylakoids to 2% sodium cholate in the presence of $(NH_4)_2SO_4$, nearly all the ATPase activity remained in the supernatant even after centrifugation of the mixture at 165,000 g. Dialysis of the cholate extract, which was rich in lipids, but deficient in chlorophyll and cytochromes, resulted in the formation of vesicles. The vesicles that clearly contained CF_1 catalyzed a low rate of uncoupler-sensitive $^{32}P_i$-ATP exchange. An ammonium sulfate precipitate obtained from the cholate extract did not form vesicles upon removal of the cholate upon dialysis and eluted from a Bio-Gel A-1.5 M column when cholate was present in the eluting buffer in a position consistent with a particle weight of 800,000.

This work was extended by Winget et al. (1977). The cholate extract was carefully fractionated with ammonium sulfate. As revealed by sodium dodecyl sulfate gel electrophoresis, the ammonium sulfate precipitate contains approximately 13 components, 5 of which are the subunits of CF_1. The complex was virtually inactive in $^{32}P_i$-ATP exchange unless it was incorporated into liposomes made either from crude soybean phospholipids or from more purified phospholipids. With the purified lipids, maximal activity was obtained with mixtures of phosphatidylcholine and phosphatidylethanolamine. Jaynes et al. (1975) obtained some reconstitution of photophosphorylation when a Triton X-

100 extract of thylakoids, which was enriched in Photosystem I and must have contained the ATPase complex, was incorporated into phospholipid liposomes.

Winget *et al.* (1977) also prepared liposomes containing bacteriorhodopsin and the thylakoid ATPase complex. The bacteriorhodopsin purified from the outer membrane of *Halobacter halobium* functions as a light-dependent proton pump, and Racker and Stoeckenius (1974) previously showed that liposomes containing bacteriorhodopsin and the ATPase complex of mitochondria catalyzed photophosphorylation. Vesicles prepared with phospholipids, bacteriorhodopsin, and the chloroplast ATPase complex carried out rather active light-dependent ATP synthesis which was sensitive to proton ionophores (uncouplers) and to energy transfer inhibitors (Winget *et al.*, 1977). Thus, it is clear that the ATPase complex isolated by the cholate extraction procedure contains all the components necessary for ATP synthesis. Moreover, the ATPase complex used is free of chlorophyll and deficient in components of the electron transport chain. Interaction between the ATPase complex and electron transport components is therefore probably not obligatory for ATP synthesis.

B. Membrane Proteins That May Be Related to the ATPase Complex

The elucidation of the function of the non-CF_1 components of the ATPase complex awaits further purification and resolution of the complex. In view of the fact that the mitochondrial ATPase complex contains only 8–10 polypeptide chains (Serrano *et al.*, 1976), it seems likely that the chloroplast ATPase, which contains about 13 components, could be further purified. Preliminary work (G. D. Winget, personal communication) suggests that the complex contains a protein, probably a proteolipid, which binds dicyclohexyl carbodiimide. This protein, which may be analogous to the dicyclohexyl carbodiimide-binding protein from mitochondria (Cattell *et al.*, 1970; Stekhoven *et al.*, 1972), has the remarkable property of being soluble in ethanol.

Cattell *et al.* (1970) used chloroform–methanol to extract the dicyclohexyl carbodimide-binding protein from mitochondria. Water-soluble proteins precipitate from the chloroform–methanol extract after addition of ether. Proteins with this solubility behavior are termed "proteolipids." Racker (1975) briefly reported that proteolipid from either mitochondria or thylakoids appears to act as a proton ionophore. Proton uptake by the reconstituted bacteriorhodopsin proton pump was inhibited by the proteolipids. Moreover, proteolipid, like uncouplers, enhances the rate of oxygen reduction by cytochrome oxidase vesicles. Although it is tempting to conclude from this study that the proteolipid forms the transmembrane channel for protons within the part of the ATPase complex more intimately associated with the membrane, there is as yet no evidence that the proteolipid is associated with the ATPase complex of chloroplasts.

It is clear from the above description that a great deal of work remains to be done before the functions of the hydrophobic parts of the complex can be established. Very little work on this subject in photosynthetic bacteria seems to have been initiated. Melandri *et al.* (1974) reported, however, the isolation of a factor that would restore oligomycin sensitivity to the ATPase activity of chromatophores treated with NH_4OH. Resolution and reconstruction analysis, while tedious and often frustrating when applied to membranes, would appear to be the most promising approach to this problem.

IV. Functions of the ATPase Complex

A. ROLES IN PHOTOPHOSPHORYLATION AND RELATED MEMBRANE ACTIVITIES

Mitchell (1961, 1966) proposed that energy transducing membranes contain a reversible, proton translocating ATPase which serves to couple the synthesis of ATP to electron flow using a proton electrochemical gradient as the common intermediate. Electron transport in chloroplasts generates rather large transmembrane pH differentials and a relatively small membrane potential (for recent reviews, see McCarty, 1976; Jagendorf, 1977). Cyclic electron flow in *R. rubrum* chromatophores appears to generate a large membrane potential in addition to a pH differential (Casadio *et al.*, 1974). The outward translocation of protons through the ATPase was proposed to drive the synthesis of ATP. Moreover, the reverse reaction, ATP hydrolysis, should result in the inward translocation of protons. The evidence which has accumulated in the past few years clearly indicates that the ATPase complexes of chloroplasts and chromatophores has properties remarkably consistent with its being a reversible, proton translocating ATPase.

The dramatic experiments of Jagendorf and Uribe (1966), showed that artificially generated pH differentials across thylakoid membranes can serve as the driving force for ATP synthesis. The relationships between photophosphorylation in chloroplasts and the pH differential (ΔpH) generated by light-driven electron flow has been examined in some detail. If internal protons are consumed during phosphorylation, ATP synthesis should result in a decrease in the magnitude of ΔpH. Pick *et al.* (1973) and Portis and McCarty (1974) clearly demonstrated this to be the case. A decrease in ΔpH of about 0.4 unit was observed when phosphorylation occurred which was fully sensitive to energy-transfer inhibitors.

The reconstitution experiments of Winget *et al.* (1977) strongly suggest that the ATPase complex of thylakoids can function as a proton translocating ATPase. The bacteriorhodopsin clearly functions as a light-driven proton pump (Racker and Stoeckenius, 1974). It is difficult to escape the conclusion that the photophosphorylation observed in vesicles reconstituted with bacteriorhodopsin and the ATPase complex is a consequence of light-dependent proton uptake catalyzed by the bacteriorhodopsin and ATP synthesis linked to proton efflux catalyzed by the ATPase complex.

Moreover, ATP hydrolysis in the dark by chloroplasts (Carmeli, 1970) and chromatophores (Mitchell, 1967) drives the inward translocation of H^+, as predicted by Mitchell. The ATPase activity of chloroplasts, like that of CF_1 itself, is latent. To unmask this activity, chloroplasts can be illuminated in the presence of sulfhydryl compounds, such as dithiothreitol (Petrack and Lipmann, 1961; Hoch and Martin, 1963; Marchant and Packer, 1963). Brief trypsin treatment of illuminated thylakoids also activates the ATPase (Lynn and Staub, 1969). Once activated, chloroplasts will catalyze ATP hydrolysis in darkness. Light also enhances the Mg^{2+}-ATPase activity in chromatophores from *R. capsulata*, but in this instance, continuous illumination is required (Melandri *et al.*, 1972). Although it is not yet proved, energy-dependent activation of the ATPase of thylakoids and chromatophores probably is a consequence of conformational changes in the coupling factor-ATPase which result in a dislocation of the ϵ subunit. The function of the dithiothreitol in the induction of ATPase activity of chloroplasts could be to reduce a disulfide. In mitochondria, the ATPase inhibitor is apparently not as tightly bound to the coupling factor-ATPase as it is in chloroplasts. An energy-dependent dissociation of the ATPase inhibitor from mitochondrial vesicles has been observed (van de Stadt *et al.*, 1973).

Carmeli (1970) showed that activated chloroplasts catalyze an uncoupler sensitive, inward proton translocation associated with ATP hydrolysis. ATP hydrolysis in chloroplasts also drives the uptake of NH_4Cl (Crofts, 1966) and of amines (Gaensslen and McCarty, 1971) against a concentration gradient, suggesting that ATP hydrolysis generates a pH differential across thylakoid membranes. Substantial ΔpH's were found to be elicited by ATP hydrolysis in chloroplasts (Bakker-Grunwald and van Dam, 1973). The facts that activated chloroplasts catalyze $^{32}P_i$–ATP exchange (Carmeli and Avron, 1966; McCarty and Racker, 1968) and H_2O–P_i exchange (Skye *et al.*, 1967) in the dark, clearly indicate that the ATPase is acting in a reversible manner. ATP hydrolysis provides the energy for the synthesis of ATP from ADP and P_i. Uncouplers, which act as proton ionophores, collapse the proton gradient and enhance the rate of ATP hydrolysis in both chloroplasts (Hoch and Martin, 1963;

McCarty and Racker, 1968; Carmeli *et al.*, 1975) and chromatophores (Horio *et al.*, 1965). In view of the sensitivity of ATP hydrolysis and $^{32}P_i$-ATP exchange to an antiserum to CF_1 (McCarty and Racker, 1968) a participation of the ATPase complex in these activities seems assured. The number of protons translocated per ATP formed or hydrolyzed (the H^+/P ratio) has not yet been established with certainty. A precise stoichiometry of the ATPase reaction with respect to protons is predicted by Mitchell's chemiosmotic hypothesis. Schwartz (1968) calculated an H^+/P ratio of 2 by comparison of the rate of phosphorylation in the steady state to the initial rate of H^+ efflux after switching of the light. Schröder *et al.* (1972) criticized Schwartz's experiments because he did not have valinomycin and K^+ in his reaction mixtures. In the absence of valinomycin and K^+, Schröder *et al.* (1972) argued, rapid outward proton translocation in the dark could result in the build up of a diffusion potential (positive on the outside) which would slow the efflux rate. In agreement with this notion, Schröder *et al.* (1972) found that valinomycin and K^+ enhance the H^+ efflux rate by about 2-fold and they calculated an H^+/P ratio of approximately 3.

Correlations between the steady-state rate of phosphorylation and ΔpH have been obtained (Portis and McCarty, 1974, 1976), and it is very evident that the rate of phosphorylation is critically dependent on the magnitude of ΔpH. At constant external pH, the log of the phosphorylation rate was found to be a linear function of ΔpH. Assuming that internal proton concentration ($[H^+]_{in}$) is the rate-limiting factor for photophosphorylation, provided that ADP and P_i are present in saturating concentrations, it can readily be shown that log phosphorylation rate should be proportional to $-(H^+/P)pH_{in}$. The slope of plots of log phosphorylation rate vs pH_{in} (or ΔpH since pH_{out} was held constant) was always close to 3. Moreover, Hill (1910) plots of log phosphorylation rate vs $[H^+]_{in}$ also were linear with a slope close to 3 (McCarty and Portis, 1976).

The maximum phosphorylation efficiency (P/e) in chemiosmotic terms is equivalent to the ratio of protons translocated inward per electron transferred down the chain (H^+/e ratio) divided by the H^+/P ratio. The H^+/e ratio in thylakoids appears to be 2 (Junge and Ausländer, 1974), and, thus, the maximum P/e ratio would be 0.66 if the H^+/P ratio is 3, and 1.0 if the H^+/P ratio is 2. Portis and McCarty (1976) and McCarty and Portis (1976) have shown that the observed P/e ratio varies in a predictable manner with respect to $[H^+]_{in}$ when either light intensity or various concentrations of an energy-transfer inhibitor was used to vary these parameters. A P/e max close to 0.66 was found in agreement with the notion that the H^+/P ratio is 3. The observation that the observed P/e ratio in specially prepared intact chloroplasts lysed prior to assay can exceed 0.66, is troublesome. Neither H^+/e nor, H^+/P ratios, however, have been determined in these efficient preparations (Reeves and Hall, 1973).

Energetics also dictates that the H^+/P ratio of the chloroplast ATPase complex should be greater than 2. The phosphorylation system of chloroplasts comes into equilibrium at a potential of about 14.5 kcal/mole (Kraayenhof, 1969; A. R. Portis and R. E. McCarty, unpublished). The maximum ΔpH, determined under similar conditions, is 3.5 units, which is equivalent to about 4.8 kcal/mole. Clearly, $3H^+$ would be required to drive ATP synthesis against its maximum potential. In chromatophores of *R. capsulata*, only $2H^+$ may be required per ATP synthesized (Casadio *et al.*, 1974).

Estimates of the number of H^+ translocated inward per ATP hydrolyzed by activated chloroplast thylakoids have been made (Carmeli, 1970; Carmeli *et al.*, 1975). H^+/P ratios approaching 2 were calculated from the initial rates of H^+ uptake and the steady-state rate of ATP hydrolysis, assayed at pH 8. At pH 7, however, H^+/P ratios approaching 4 were estimated. The ordinary glass electrode responds rather slowly to changes in pH, and it is difficult to obtain true initial rates of H^+ uptake with conventional electrodes (Izawa and Hind, 1967). The fact that higher apparent H^+/P ratios are obtained at pH 7 than at pH 8 may be a reflection of the fact that the rate of H^+ efflux is considerably slower at the lower pH (Rumberg *et al.*, 1969). A reinvestigation of the ratio of protons translocated per ATP hydrolyzed, using a more rapidly responding method to determine the rate of proton uptake, is clearly needed.

The properties of the ATPase complex are consistent with the complex being a reversible, proton-translocating ATPase system. It is somewhat unsettling that the H^+/P ratio for the ATPase complex in chloroplasts appears to be 3, whereas the ratio is probably 2 in mitochondria (Mitchell, 1967) and in chromatophores (Casadio *et al.*, 1974). If the higher H^+/P ratio for the chloroplast ATPase complex stands the tests of further experimentation, it would suggest that the mechanism of the ATPase reaction in chloroplasts differs from that in other energy-transducing organelles. The many similarities between the ATPase complexes from the various sources makes the possibility that they differ mechanistically somewhat unpalatable.

The functions of the individual subunits of CF_1 in ATP synthesis or hydrolysis are just beginning to be elucidated. The ϵ subunit is an ATPase inhibitor (Nelson *et al.*, 1972b) and regulates the expression of the ATPase. The δ component is required for binding of CF_1 to depleted membranes (Younis *et al.*, 1977). An antiserum to the γ subunit strongly inhibits both photophosphorylation and the light- and sulfhydryl-induced ATPase of chloroplast thylakoids (Nelson *et al.*, 1973). Moreover, the reaction of a group on the γ component with N-ethylmaleimide inhibits photophosphorylation (McCarty *et al.*, 1972; McCarty and Fagan, 1973) and the light-induced ATPase of chloroplasts (R. E. McCarty, unpublished). Since the γ subunit is not required for ATP hydrolysis by isolated CF_1 (Deters *et al.*, 1975), the γ component may not contain catalytic activity in itself. The active site(s) for ATP hydrolysis are probably contained within the β subunits (Deters *et al.*, 1975). Moreover, illumination of thylakoids in the presence of radioactive ADP results in the incorporation (likely an exchange) of ADP into CF_1 (Harris and Slater, 1975). This ADP is bound to the α and/or β subunit (Magnusson and McCarty, 1976a). Since the bound ADP may also be phosphorylated (Magnusson and McCarty, 1976b), the active site(s) for ATP synthesis are possibly also within the α and/or β subunits. The γ subunit may be involved in the transmission of the electrochemical potential to the α or β subunits.

B. ROLE IN MODULATION OF ELECTRON TRANSPORT AND OF ΔpH

Coupling factor-ATPases are not likely to play a direct role in the inward proton translocation driven by light-dependent electron flow. Antibodies to CF_1 have much less effect on proton uptake than on phosphorylation (McCarty and Racker, 1966; Nelson *et al.*, 1973). Moreover, energy-transfer inhibitors, such as Dio-9 (McCarty *et al.*, 1965) or phlorizin (Winget *et al.*, 1969), do not inhibit light-dependent proton uptake (McCarty and Racker, 1966). The lack of a direct involvement of coupling factor-ATPases in electron transport-driven proton uptake is also supported by the finding that proton uptake in *R. capsulata* is largely unaffected by removal of the coupling factor (Melandri *et al.*, 1970). ATP-driven proton translocation would be abolished by energy-transfer inhibitors and by antibodies to CF_1.

Rather low concentrations of ADP or ATP enhance the extent of light-dependent H^+ uptake in thylakoids (McCarty *et al.*, 1971; Telfer and Evans, 1972). This stimulation, which occurs maximally at pH values where photophosphorylation occurs at its highest rates, is not due to an ATP-dependent proton translocation superimposed upon that resulting from electron flow. Instead, ADP or ATP seems to slow the rate of proton efflux and, in effect, decreases the proton permeability of thylakoid membranes. Since the magnitude of ΔpH is also augmented by ATP or ADP (Portis and McCarty, 1974), the stimulation of proton uptake by these nucleotides cannot be a consequence of increased internal buffering or of increased proton binding to the membrane. An antiserum to CF_1 abolished the ATP stimulation of proton uptake (McCarty *et al.*, 1971), suggesting that CF_1 plays some role in this process.

Under steady-state nonphosphorylating conditions, the rate of electron transport (called "basal" electron transport) in thylakoids should be proportional to the rate of H^+ efflux (Rumberg and Siggel, 1969). The rate of proton influx is simply (H^+/e) times the rate of electron flow, whereas the rate of H^+ efflux may be described as $k_{H^+}[H^+]_{in}$, where k_{H^+} is a constant related to the permeability constant of the thylakoid membrane for protons. Thus, reagents that decrease k_{H^+} could also inhibit electron flow. Indeed, in their early investigations of the relationships between photophosphorylation and electron flow, Avron et al. (1958) found that low concentrations of ATP or ADP inhibit basal electron flow.

Portis et al. (1975) measured basal electron flow and $[H^+]_{in}$ as a function of light intensity. Assuming that k_{H^+} is independent of light intensity, the rate of electron flow should be strictly proportional to $[H^+]_{in}$. This was found to hold true only at relatively low light intensities when ATP was not present. At ΔpH values of about 2.8 or above the rate of electron flow increased markedly with light intensity whereas the internal proton concentration increased only slightly. Thus, k_{H^+} apparently increases at light intensities sufficient to elicit high ΔpH values. In the presence of ATP, however, proportionality between the rate of electron flow and internal proton concentration was maintained at all light intensities even though the ΔpH was larger at high light intensities in the presence of ATP than in its absence. Clearly, then, ATP prevents the increase of the proton permeability that is brought on by high ΔpH values. We proposed (McCarty et al., 1971; Portis et al., 1975) that CF_1 undergoes conformational changes, the extent of which is sharply dependent on ΔpH. At high ΔpH values, CF_1 assumes a modified conformation that allows a more rapid rate of efflux of protons and, consequently, a faster rate of electron flow. ATP (or ADP) by binding to CF_1 prevents these conformational changes (see Section VI,A) and the induced proton leak. In a sense, CF_1 acts as a gated proton translocator. The induced proton leak is likely to occur through the hydrophobic components of the ATPase complex. Either CF_1 itself or some other component of the complex which is influenced by the conformational state of CF_1 could be the proton gate.

Antibodies against the α subunit of CF_1 abolished the stimulation of proton uptake by ATP whereas anti-β and anti-δ had no effect (Nelson et al., 1973). This result could mean that the ATP binding is to the α subunit or that the α subunit is involved in the conformational transformation. Anti-γ serum stimulated proton uptake about 1.3-fold, and ATP gave little enhancement in addition. Since ATP stimulated H^+ uptake in the control thylakoids by over 2-fold, anti-γ serum at least partially prevents the ATP stimulation of proton uptake.

A role for the γ subunit of CF_1 in the enhancement of the extent of proton uptake by ATP is also indicated by experiments in which membrane bound CF_1 is modified with maleimides. Although the reaction of a group within the γ subunit with N-ethylmaleimide has no effect on either the extent of proton uptake (McCarty *et al.*, 1972) or of ΔpH (Portis *et al.*, 1975), it partially blocks the ATP stimulation of proton uptake. Moreover, reaction of CF_1 with a bifunctional maleimide, which probably cross-links groups within the γ subunit, causes an increase in the proton permeability of thylakoid membranes without causing detachment of CF_1 from the membranes (R. E. McCarty and M. A. Weiss, unpublished). As is the case in EDTA-extracted thylakoids, dicyclohexylcarbodiimide restores low proton permeability to the maleimide-treated thylakoids.

The physiological significance of the CF_1-mediated change in proton permeability of the membrane is likely to be minor. It is unlikely that the adenine nucleotide content of the stroma compartment of chloroplasts *in vivo* would ever reach the low values necessary for the effect to set in. Moreover, when chloroplasts are actively fixing CO_2, ΔpH values are probably not as high as those that can be reached with isolated thylakoids. Nonetheless, the changes in the permeability of the membrane to protons give interesting clues as to the possible functions of CF_1 and its subunits in energy transduction.

V. Suggested Mechanisms for the ATPase Complex

As Mitchell's concept that electron transport and phosphorylation were coupled to each other via a proton electrochemical gradient became more accepted, more attention was turned to how the ATPase complex might accomplish the terminal steps of ATP synthesis. Mitchell's early mechanisms for the ATPase system of coupling membranes invoked the existence of an intermediate which was thought to contain a "high energy" anhydride bond, such as a thioester linkage. More recently, however, Mitchell has advanced a molecular mechanism for the ATPase complex in which proton translocation is more directly involved (Mitchell, 1974; Mitchell and Moyle, 1974). Mitchell abandoned his earlier concepts of the ATPase mechanism in part because no evidence for covalent intermediates has been found.

Mitchell suggested that ADP and P_i (both probably as Mg^{2+} complexes) enter the coupling factor-ATPase part of the complex in specific states of protonation. The binding site for P_i or Mg^{2+}–P_i was proposed to be arranged so that one side of the P_i is exposed to a more acid environment than the other. ADP (as $ADPO^-$) is thought to approach the P_i from the more basic side so that protonation of $ADPO^-$, which would reduce the nucleophilicity of the $-O^-$, would not occur. Nucleophilic attack by the O^- in ADP, coupled with the protonation of an $-O^-$ in the P_i, produces a transition state which contains an oxonium group ($-OH_2^+$). The oxonium group then leaves after electronic rearrangements, and H_2O and ATP are formed. The driving force for ATP synthesis is essentially a proton gradient within the coupling factor-ATPase. The protons which protonate the bound P_i are thought to approach the binding site from the acid side of the coupling membrane via a proton channel. In contrast, the ADP site is proposed to have a more basic environment because it has more contact with the basic side of the membrane. Mitchell's proposals are illustrated in Fig. 2.

Note that the mechanism as written in Fig. 2 involves only two protons for each ATP formed. As pointed out in Section IV,A, however, the chloroplast ATPase complex is likely to be characterized by an H^+/P ratio of three. This is not necessarily inconsistent with Mitchell's proposals. In fact in his original communication Mitchell (1974) wrote: "If the passage of inorganic phosphate into and out of the F_1 complex were specific for a more deprotonized species . . . a H^+/P quotient of 3 or 4 instead of 2 might be accounted for." Although Mitchell and Moyle (1974) attempted to elucidate the specific states of protonation of the phosphorylation substrates through analysis of the pH dependence of steady-state kinetic parameters associated with the ATPase of inverted

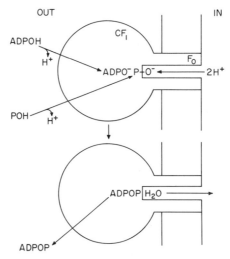

FIG. 2. A representation of Mitchell's (1974) mechanism for proton-translocating ATPases. ADPOH and POH represent ADP and P_i in specific states of protonation. ADPOP is ATP. Note the direct attack of protons on phosphate oxygen bound to a region in CF_1 in contact with a proton channel formed by F_0, defined here as the components of the ATPase complex that were integral parts of the membrane.

mitochondrial vesicles, this has not been attempted with the ATPase of photosynthetic membranes.

Mitchell's (1974) mechanism for the proton translocating ATPase was criticized by Boyer (1975a) and by Williams (1975). Boyer (1975a) considered the mechanism to be unlikely because it did not appear to him that the scheme was consistent with the chemistry of phosphate ester cleavage and formation. Mitchell (1975) pointed out, however, that many of Boyer's criticisms were based on misconceptions on Boyer's part of Mitchell's proposed mechanism. In a later article, Boyer (1975b) pointed out in a footnote that Mitchell's (1975) rebuttals failed in his opinion to answer his criticisms.

Boyer (1965) proposed that conformational changes elicited by electron transport were transmitted to the ATP synthesizing enzymes and provide the driving force for ATP synthesis. More recently, he has modified his proposals to explain how proton translocation may be coupled to ATP synthesis by the ATPase complex (Boyer, 1974, 1975b). It should be pointed out, however, that Boyer is not convinced that the proton electrochemical gradient is the sole driving force for ATP synthesis (Boyer, 1977). However, he concedes that any explanation for ATP synthesis must include a mechanism for how proton translocation may be coupled to ATP synthesis.

Boyer et al. (1973) and Harris et al. (1973) suggested that the release of the newly synthesized ATP from coupling factor-ATPases may be a major energy-dependent step in phosphorylation. This suggestion was prompted by the finding by the California group that exchange of ^{18}O-labeled water into P_i catalyzed by mitochondria is rather insensitive to uncouplers. Assuming that this exchange is caused by reversible hydrolytic cleavage of bound ATP, its uncoupler insensitivity would indicate that actual synthesis of ATP from ADP and P_i could occur in deenergized mitochondria. Interestingly, the exchange of ^{18}O-labeled P_i into water is uncoupler sensitive, suggesting the possibility that P_i binding to F_1 is energy-dependent (Rosing et al., 1977). The Amsterdam group found firmly bound adenine nucleotides in both F_1 (Harris et al., 1973) and CF_1 (Harris and Slater, 1975). Although nucleotide exchange into CF_1 in thylakoids in the dark does not readily occur, an uncoupler sensitive, light-dependent exchange was detected (Harris and Slater, 1975). Thus, CF_1 appears to undergo energy-dependent changes in its affinity for adenine nucleotides.

Boyer and Slater suggested that the dissociation of bound ATP is a major energy-requiring step in phosphorylation. This dissociation is proposed to be elicited by conformational changes in coupling factor-ATPases which could be brought about by the proton electrochemical gradient. A detailed "alternating site" mechanism has been advanced by

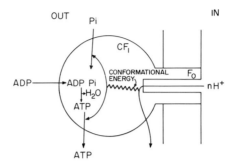

FIG. 3. A representation of an indirect mechanism for the ATPase patterned after Boyer (1975b). Proton efflux is shown as causing conformational changes in CF_1, which in turn provide the energy for P_i binding and the release of newly synthesized ATP from CF_1.

Boyer (1977). In this case, protons are involved in the ATPase mechanism in a rather indirect way. Mitchell (1975) points out that proton translocation and reversible ATP hydrolysis are not, in the Boyer–Slater scheme, linked by a common intermediate. Moreover, changes in the affinity of coupling factor-ATPases in membranes upon formation of the proton gradient may be rationalized in terms of Mitchell's ideas. A representation of the Boyer–Slater hypothesis is given in Fig. 3.

Based on the notion that proton-translocating ATPases may resemble other cation-translocating ATPases, Racker (1977) proposed that Mg^{2+} and P_i bind tightly to the coupling factor ATPases. The function of protons in the reaction is suggested to be displacement of the bound Mg^{2+} which provides the driving force for ATP formation. Racker's scheme suggests that coupling factor-ATPases should undergo energy-dependent changes in affinity for Mg^{2+}. This should be experimentally approachable.

At present, there is no compelling reason to accept or reject any of the proposed mechanisms for the protein-translocating ATPase. Other mechanisms can and probably will be advanced and existing ones will likely be modified as new data become available. Nevertheless, these models provide a framework for experimentation and a forum for interesting discussion.

VI. Observations Possibly Relevant to Mechanism of the ATPase Complex

A. CONFORMATIONAL CHANGES OF COUPLING FACTOR-ATPASES

The first evidence that coupling factor-ATPases undergo energy-dependent conformational changes was obtained by Ryrie and Jagendorf

(1971b, 1972). Chloroplast thylakoids were either illuminated or kept in the dark in the presence of highly labeled tritiated water. CF_1 was detached from the thylakoids by EDTA extraction and was partially purified. An uncoupler sensitive, light-dependent incorporation of tritium into CF_1 was observed. This incorporation, which is probably an exchange, could also be brought about by subjecting thylakoids to an acid-to-base transition in the presence of ^3H_2O. It seems clear, therefore, that sites bearing exchangeable hydrogens on CF_1 become exposed to the suspending medium upon the formation of a proton gradient across the thylakoid membranes. Once the thylakoids are returned to the dark, these sites become inaccessible.

About 100 moles of ^3H per mole of CF_1 were incorporated in the light (Ryrie and Jagendorf, 1972). The actual amount of exchange enhanced by formation of the proton gradient could be somewhat higher than that observed. Since several hours were required to isolate the CF_1, only the most slowly exchangeable hydrogens would have been detected.

The fluorescence of fluorescamine is sensitive to the environment in which it finds itself. This fluorophore may be covalently linked to CF_1 and the modified protein added to EDTA-treated thylakoids for reconstitution. A rapid, light-induced shift in the fluorescent emission maximum of the fluorophore was detected (Kraayenhof and Slater, 1974), indicating a change in the conformation of CF_1.

In a search for ways to investigate the interactions between CF_1 bound to thylakoids and adenine nucleotides, McCarty et al. (1972) found that the alkylating reagent N-ethylmaleimide would inhibit phosphorylation only when the chloroplasts were illuminated in the presence of this reagent prior to the assay of photophosphorylation. Although incubation of chloroplast thylakoids with N-ethylmaleimide in the dark caused incorporation of about 2 moles (1 in γ and 1 in ϵ) of N-ethylmaleimide per mole of CF_1, no effect on phosphorylation or related processes was observed. Illumination of thylakoids in the presence of N-ethylmaleimide allows the extra incorporation of somewhat less than 1 mole of the reagent per mole of CF_1 (McCarty and Fagan, 1973). Nearly all of the incorporated N-ethylmaleimide was present in the γ subunit of CF_1 and incorporation was correlated to inhibition of photophosphorylation (McCarty and Fagan, 1973; Magnusson and McCarty, 1975). Since the incorporation of N-ethylmaleimide and tritium into CF_1 show similar properties, it was concluded that an energy-dependent change in CF_1 conformation results in the exposure of a group to reaction with N-ethylmaleimide. Baccarini-Melandri et al. (1975) showed that N-ethylmaleimide also inhibits phosphorylation and ATPase activity of R. capsulata chromatophores only when the chromatophores were illuminated in the presence of the reagent prior to the assay of activity. Thus, the coupling factor-ATPase of a photosynthetic bacterium is also likely to undergo energy-dependent changes in its conformation.

The reagents 2,2'-dithiobis-(5-nitropyridine) (Andreo and Vallejos, 1976) and o-iodosobenzoate (Vallejos and Andreo, 1976) also inhibit photophosphorylation in thylakoids in a manner similar to that of N-ethylmaleimide. The inhibition was attributed to an oxidation of 2 thiols in CF_1, exposed upon illumination, to a disulfide, but definitive proof for this has not yet been published. The Ca^{2+}-ATPase of heat-activated CF_1 was not affected by the SH reagents unless the reagents were present during heat activation (Vallejos et $al.$, 1977), indicating that groups are exposed upon heat treatment which can react with the reagents.

Amino groups in membrane-bound CF_1 are also exposed upon illumination. Oliver and Jagendorf (1976) reacted accessible amino groups in thylakoids with methylacetamidate in the dark. Subsequent exposure of the modified thylakoids to trinitrobenzene sulfonate resulted in the incorporation of 4 moles of trinitrobenzene sulfonate per mole of CF_1 as well as inhibition of Ca^{2+}-dependent ATPase activity. Two amino groups in the γ subunit and one in α and β reacted with trinitrobenzene sulfonate in the light. Thus, although the SH of the α and β subunits of CF_1 in illuminated thylakoids are not accessible to SH reagents, at least some lysine ϵ-amino groups are. An indication of the extent of the light-induced conformational change as well as its subunit location may be obtained by more extensive modification studies.

Light accelerates the inhibition of the Ca^{2+}-dependent ATPase activity of CF_1 by permanganate (Datta et $al.$, 1974) and is required for the irreversible inhibition of photophosphorylation by sulfate (Ryrie and Jagendorf, 1971a). Interestingly, ADP at low concentrations is absolutely required for these anion-induced alterations. ADP binding probably alters the conformation of CF_1 in such a way as to render it receptive to interactions with anions. Since P_i counteracted the effects of either permanganate or sulfate, P_i binding to a site in CF_1 may require a particular conformation of the enzyme.

B. Nucleotide Binding to Membrane-Bound Coupling Factor-ATPases

The interactions between nucleotides and coupling factor-ATPases bound to coupling membranes have been studied in several ways. Roy and Moudrianakis (1971a,b) found that illumination of chloroplast thylakoids in the presence of AMP and P_i resulted in the incorporation of ADP into CF_1. This bound ADP remained associated with the enzyme during isolation and purification of the CF_1 and was, therefore, quite firmly bound. Harris and Slater (1975) detected the presence of 2.5

nmoles ATP and 1.3 nmoles ADP per milligram of chlorophyll in well washed thylakoids. Most of the bound nucleotide was present in CF_1. Isolated CF_1 contained slightly less than 1 mole of ATP and ADP per mole of enzyme. Bound nucleotides were also found in chromatophores of *R. rubrum* (Yamamoto *et al.*, 1972; Lutz *et al.*, 1974). Although the exchange of ATP or ADP bound to CF_1 in darkened thylakoids for that in the medium was very slow, complete exchange occurred if the thylakoids were illuminated in the presence of these nucleotides. Nucleotide exchange into bound CF_1 is sensitive to uncouplers, but not to the energy-transfer inhibitor, Dio-9 (Harris and Slater, 1975). When thylakoids were illuminated in the presence of dithioerythritol, ADP was lost from the bound CF_1. Although it is tempting to correlate the dissociation of ADP from CF_1 with the activation of the ATPase, ADP is not lost from the enzyme when Mg^{2+} is present during illumination. Mg^{2+}, however, does not prevent the activation of the ATPase.

The properties of the light-induced exchange of nucleotides into CF_1 in thylakoids were examined in more detail by Magnusson and McCarty (1976c), who used a silicone fluid microcentrifugation procedure to determine nucleotide binding. This method for measuring binding, previously used to estimate amine uptake for the determination of ΔpH (Gaensslen and McCarty, 1971, 1972; Portis and McCarty, 1973, 1974), is much less laborious than the multiple washing procedure used by Harris and Slater (1975). The light-dependent portion of the binding of ATP or ADP was found to be rapid and to saturate at relatively low concentrations of these nucleotides. Half-maximal exchange was achieved at about 5 μM ADP or ATP, and the half-time for the exchange was less than 0.1 second. Neither GTP nor adenylyl imidodiphosphate at concentrations 10-fold in excess of that of radioactive ATP markedly affected the exchange. These nucleotides also bound much more poorly than either ADP or ATP at equivalent concentrations. The exchange thus appears to be at least somewhat specific for ADP and ATP. Mg^{2+} was required for the binding of ATP, but not for ADP binding. These results conflict with those of Harris and Slater (1975). However, the Amsterdam group used 200 μM nucleotide and 5 minutes of illumination. It seems likely that sufficient ADP could have been present as a contaminant of the ATP or could have been slowly formed during the prolonged incubations, to have saturated the binding sites with ADP.

Thylakoids previously illuminated in the absence of nucleotides retain their capacity to bind ADP or ATP in the dark long after the proton electrochemical gradient had decayed (Magnusson and McCarty, 1976c). This result indicates that the energy-dependent step in the nucleotide exchange is probably the dissociation of the bound nucleotide from CF_1. The nucleotide exchange may result from the exposure of nucleotide binding sites by light-dependent changes in the conformation of CF_1. If this is so, it may be concluded that the conformation of CF_1 in thylakoids after their illumination in the absence of added ADP or ATP differs from that in thylakoids either kept in the dark or illuminated in the presence of ADP or ATP. Strotmann et al. (1976) showed that either illumination or an acid to base transition causes the dissociation of adenylates from thylakoids.

CF_1 purified from thylakoids briefly illuminated in the presence of a low concentration of radioactive ADP or ATP was found to contain largely labeled ADP (Magnusson and McCarty, 1976a). Although about 0.6 moles of radioactive ADP was bound per mole of CF_1, only traces of AMP and ATP were detected. Remarkably, much of the ADP remained bound even after prolonged digestion of CF_1 with trypsin, which removes the smaller subunits. Strotmann et al. (1976), however, found that less than 50% of the radioactive nucleotide bound to thylakoids as a result of illumination in the presence of labeled ADP was present as ADP. AMP and ATP comprised about 25% each. Harris and Slater (1975) failed to detect much AMP bound to thylakoids. It seems plausible that the AMP and ATP found by Strotmann et al. (1976) arose from ADP via adenylate kinase activity. The significance of the ATP found by Harris and Slater (1975) in CF_1 is not clear. The exchange between the bound ATP and ATP in the medium is apparently quite slow, since ADP, not ATP, is found bound to CF_1 only briefly illuminated in the presence of radioactive ATP.

The interactions between thylakoid-bound CF_1 and adenylates have also been studied by indirect means. The enhancement of proton uptake in illuminated thylakoids (McCarty et al., 1971; Telfer and Evans, 1972) by ADP or ATP causes an inhibition of electron flow (Avron et al., 1958; Mukohata and Yagi, 1974) and a stimulation of delayed light emission (Vambutas and Bertsch, 1975; Felker et al., 1974). Since these effects are elicited only by ADP or ATP even though other nucleoside diphosphates may be phosphorylated by thylakoids, it is clear that bound CF_1 contains multiple nucleotide sites.

Adenine nucleotides also partially protect photophosphorylation from inhibition by maleimides (McCarty et al., 1972; Magnusson and Mc-Carty, 1975; Baccarini-Melandri et al., 1975), presumably by preventing the conformational change in CF_1 which exposes a group in the γ subunit to reaction with the maleimide (McCarty and Fagan, 1973; Magnusson and McCarty, 1975). This protective effect was specific for adenine nucleotides. The characteristics of adenine nucleotide protection from maleimide inhibition of phosphorylation are similar to those of adenine nucleotide exchange (Magnusson and McCarty, 1976b) with the exception that inorganic phosphate enhances the protective action of ADP and ATP, but does not affect their exchange.

Although ADP and P_i alone had no effect on the light-induced exchange of tritium from ^3H_2O into CF_1, their presence together reduced the extent of exchange by about half (Ryrie and Jagendorf, 1972). This effect of ADP and P_i cannot be ascribed to photophosphorylation since phlorizin at concentrations which completely blocked ATP synthesis did not prevent it. The simultaneous presence of ADP and P_i thus alters the conformation of CF_1. Whether ATP and P_i would have a similar effect was not tested. P_i also markedly enhances the ability of low concentrations of either ADP or ATP to prevent the inhibition of photophosphorylation by N-ethylmaleimide (Magnusson and McCarty, 1975), although P_i alone is ineffective. Since GDP and P_i do not protect, even though GDP may be phosphorylated rapidly, the protection afforded by ADP and P_i is not a consequence of photophosphorylation. In contrast, Oliver and Jagendorf (1976) showed that P_i by itself partially prevents the inhibition of the Ca^{2+}-ATPase activity of CF_1 which is brought on by illumination of thylakoids in the presence of trinitrobenzene sulfonate. ATP (or ADP) also gave partial protection, and full protection was observed when ATP and P_i were both present. Moreover, ADP and ADP only is required for the development of the irreversible inhibition of photophosphorylation by sulfate (Ryrie and Jagendorf, 1971a; Grebanier and Jagendorf, 1977). It is quite difficult to fit all these observations into a simple scheme for nucleotide and phosphate binding to CF_1. Part of the difficulty must be attributable to the probability that different regions of the CF_1 molecule are affected by the different inhibitory treatments. P_i binding by itself, for example, could affect the conformation of CF_1 in the region that contains a lysine that can react with trinitrobenzene sulfonate, but not in the region of the γ subunit that contains the maleimide-reactive group. More work on the direct interactions between CF_1 and P_i (as well as other anions) is clearly needed. Multiple kinds of P_i binding sites may be present in CF_1 and P_i binding to one of these kinds of sites may require the presence of ADP.

The role of bound nucleotides in photophosphorylation is an area of quite active investigation. Roy and Moudrianakis (1971a,b) demonstrated that illumination of thylakoids in the presence of AMP and P_i results in ADP being tightly bound to CF_1. Some β-labeled ATP was detected, and it was proposed that bound ADP, formed from AMP and P_i, served as the phosphoryl donor to ADP from the medium. Using rapid mixing techniques to estimate the initial rate of ATP formation by thylakoids following an acid-to-base transition, Smith *et al.* (1976) and Yamamoto and Tonomura (1975) found that the rate of formation of β-labeled ADP was much slower than the rate of formation of γ-labeled ATP. These experiments suggest that the AMP pathway for ATP synthesis is not kinetically competent to be on the main path. The situation is, however, not entirely clear since a small amount of β-

labeled ADP is rapidly formed during illumination of the thylakoids (Smith et al., 1976). The acid-to-base transition may modify the properties of CF_1.

Bound nucleotides in chromatophores (Yamamoto et al., 1972; Lutz et al., 1974) and in thylakoids (Pflugshaupt and Bachofen, 1975; Boyer et al., 1975; Aflafo and Shavit, 1976) can be phosphorylated in the absence of added nucleotides by P_i either in the light or, in the case of thylakoids also as a result of an acid-to-base transition. Although Aflafo and Shavit (1976) found no labeling of bound nucleotides in thylakoids by $^{32}P_i$ in the dark, Pflugshaupt and Bachofen (1975) reported an extensive dark labeling. Similar findings were also published for chromatophores (Yamamoto et al., 1972; Lutz et al., 1974). The only obvious difference between the experiments of Aflafo and Shavit and Pflugshaupt and Bachofen is that the former group used lettuce chloroplasts and the latter used spinach chloroplasts. Bound inorganic phosphate serves as the phosphoryl donor in the phosphorylation of bound ADP (Magnusson and McCarty, 1976b). It could be that the exchange between bound P_i and free P_i in spinach chloroplast thylakoids will take place in the dark more readily than in lettuce chloroplasts.

Smith et al. (1976) point out, however, that the rate of phosphorylation of bound nucleotide is slower than the initial rate of phosphorylation when ADP is present in the medium. This finding is not conclusive evidence against the role of bound ADP in phosphorylation. In the absence of added nucleotides, newly formed ATP may not dissociate from the CF_1, and it may be simply converted back to ADP and P_i before quenching can occur. Moreover, the proton permeability of thylakoids is decreased by exogenous ATP or ADP (McCarty et al., 1971; Telfer and Evans, 1972).

Magnusson and McCarty (1976b) showed that the ADP bound to the nucleotide site in CF_1 which undergoes light-dependent exchange can be phosphorylated in the light. Thylakoids were briefly illuminated in the presence of either $[^3H]$ATP or $[^{14}C]$ATP (or radioactive ADP). Unbound nucleotides were removed by washing in the dark. Nearly all the bound radioactivity was present in ADP. Exposure of the washed, radioactive thylakoids to a 2-second light flash caused the conversion of 20–25% of the bound ADP to ATP. Bound P_i serves as the phosphate donor for this synthesis. The extent of incorporation of bound $^{32}P_i$ into ATP was not influenced by the presence of 2 mM nonradioactive P_i in the medium, indicating that bound and free phosphate do not readily exchange. Preliminary experiments (S. Ketcham and R. E. McCarty, unpublished) indicate that there is a light-dependent P_i binding to thylakoids.

Incredible as it may seem, rapid acidification of thylakoids to pH 2 or below seems to cause synthesis of ATP from bound ADP and bound P_i (Magnusson and McCarty, 1976c). When heat, urea, or mildly acidic conditions are used to denature thylakoids, ADP is by far the major nucleotide which dissociates. If, however, strong acids, including sulfuric, perchloric, and trichloroacetic acids, or buffers at pH values of 2 or below are used to dissociate nucleotides, up to 30% of the nucleotide released is ATP. The possibility that denaturing treatments other than strong acids allow expression of ATPase activity and cause the hydrolysis of ATP bound to the CF_1 cannot be totally excluded. Little, if any, light-dependent phosphorylation of bound ADP may be detected if strong acids are used to dissociate the nucleotides. Although acid-induced phosphorylation is totally insensitive to uncouplers, acid-induced phosphorylation did not occur in preparations of isolated CF_1. Only radioactive ADP is dissociated by strong acids from CF_1 isolated from thylakoids illuminated in the presence of ATP (Magnusson and McCarty, 1976a). However, bound P_i is lost from the enzyme during its isolation.

Although the significance of acid-induced phosphorylation to the mechanism of phosphorylation is not clear, the phenomenon makes the interpretation of the results of previous studies on the phosphorylation of bound nucleotides open to question. Most investigators have used either perchloric or trichloroacetic acid to dissociate nucleotides, and this should result in the phosphorylation of bound ADP. For example, the high labeling of ATP by $^{32}P_i$ in thylakoids in the dark could be a result of acid-induced phosphorylation.

A somewhat fuzzy picture, which, it is hoped, will be brought into sharper focus with time and more data, has emerged from the nucleotide exchange and phosphorylation experiments. CF_1 in thylakoids contains a specific nucleotide binding site (or sites), which is exposed to exchange upon formation of the proton electrochemical gradient across thylakoid membranes. The exposure of this site may be the consequence of a conformational change in the membrane-bound CF_1. ADP is bound to this exchangeable site, and P_i is probably bound to a site close by. Illumination causes ATP to be synthesized from the bound precursors. Since thylakoids illuminated in the presence of ATP bind ADP and P_i, the newly synthesized ATP would probably be reconverted back to bound $ADP + P_i$ in the dark unless it dissociates from the enzyme. Since the dissociation of ADP is energy dependent, it is not illogical to conclude that the dissociation of ATP might also require energy. According to Aflafo and Shavit (1976), uncouplers inhibit the phosphorylation of bound nucleotides with *medium* P_i. This sensitivity could reflect either a requirement for energy in the actual synthesis of the phosphate ester bond of ATP or for binding of P_i, or both. It should be kept in mind that there is certainly no guarantee that any of these phenomena are directly related to phosphorylation. The nucleotides bound to CF_1 could well serve a regulatory, rather than catalytic, function.

VII. Conclusions

Considerable progress has been made in the past decade toward an understanding of the broad outlines of energy conversion in photosynthetic and nonphotosynthetic systems. The detailed mechanisms are, however, almost entirely unknown. We know that the ATPase complex of chloroplasts and chromatophores operates as a reversible, proton-translocating ATPase. We know that the coupling factor-ATPases of the complexes undergo energy-dependent changes in their affinity for nucleotides and in their conformation. Even though these changes could be coincidental to the function of these proteins in phosphorylation, they will have to be studied in more detail before a complete picture of the roles of CF_1 and its bacterial analog can be drawn. Further purification of the ATPase complex and identification of its components is also sorely needed. It will probably be some years before the mechanism of the proton translocating ATPase of chloroplasts and chromatophores will be elucidated. This mechanism might include elements of many different postulated mechanisms. Perhaps we will be able to draw upon the experiences of the researchers who toil with other cation translocating ATPases, such as the Ca^{2+} "pump" of sarcoplasmic reticulum or the Na^+K^+ "pump" of certain plasma membranes which have certain similarities to the proton "pump." It is hoped that studies on the ATPase complex of photosynthetic membranes will continue to contribute to an understanding of the mechanism of the proton-translocating ATPases.

ACKNOWLEDGMENTS

I thank Drs. Baccarini-Melandri, Melandri, Hammes, Racker, Winget, Jagendorf, Boyer, Vallejos, and Andreo for providing reprints and preprints of their work. I am also indebted to my introductory biochemistry class of 390 students who patiently suffered through my lectures for which I was not as well prepared as I should have been because of the time spent in writing this article. Mrs. Joyce Broadhead's prompt and efficient typing is very much appreciated.

REFERENCES

Aflafo, C., and Shavit, N. (1976). *Biochim. Biophys. Acta* **440**, 1522.
Andreo, C. S., and Vallejos, R. H. (1976). *Biochim. Biophys. Acta* **423**, 590.
Avron, M. (1963). *Biochim. Biophys. Acta* **77**, 699.

Avron, M., Krogmann, D. W., and Jagendorf, A. T. (1958). *Biochim. Biophys. Acta* **30**, 144.

Baccarini-Melandri, A., Melandri, B. A., Gest, H., and San Pietro, A. (1970). *J. Biol. Chem.* **245**, 1224.

Baccarini-Melandri, A., Fabbri, E., Fistater, E., and Melandri, B. A. (1975). *Biochim. Biophys. Acta* **376**, 72.

Baird, B. A., and Hammes, G. G. (1976). *J. Biol. Chem.* **251**, 6953.

Bakker-Grundwald, T. (1974). *Biochim. Biophys. Acta* **347**, 141.

Bakker-Grunwald, T., and van Dam, K. (1973). *Biochim. Biophys. Acta* **292**, 808.

Bennun, A., and Avron, M. (1964). *Biochim. Biophys. Acta* **79**, 646.

Berzborn, R. J., Kopp, F., and Mühlenthaler, K. (1974). *Z. Naturforsch., Teil C* **29**, 694.

Boyer, P. D. (1965). *In* "Oxidases and Related Redox Systems" (T. E. King, H. S. Mason, and M. Morrison, eds.), p. 994. Wiley, New York.

Boyer, P. D. (1974). *In* "Dynamics of Energy Transducing Membranes"(L. Ernster *et al.*, eds.), p. 289. Elsevier, Amsterdam.

Boyer, P. D. (1975a). *FEBS Lett.* **50**, 91.

Boyer, P. D. (1975b). *FEBS Lett.* **58**, 1.

Boyer, P. D. (1977). *Annu. Rev. Biochem.* **46**, 957.

Boyer, P. D., Cross, R. L., and Momsen, W. (1973). *Proc. Natl. Acad. Sci. U.S.A.* **70**, 2837.

Boyer, P. D., Smith, D. J., Rosing, J., and Kayalar, C. (1975). *In* "Electron Transport Chains and Oxidative Phosphorylation" (E. Quagliarielo *et al.*, eds.), p. 361. North-Holland Publ., Amsterdam.

Cantley, L. C., Jr., and Hammes, G. G. (1975a). *Biochemistry* **14**, 2968.

Cantley, L. C., Jr., and Hammes, G. G. (1975b). *Biochemistry* **14**, 2976.

Cantley, L. C., Jr., and Hammes, G. G. (1976a). *Biochemistry* **15**, 1.

Cantley, L. C., Jr., and Hammes, G. G. (1976b). *Biochemistry* **15**, 9.

Carmeli, C. (1970). *FEBS Lett.* **7**, 297.

Carmeli, C., and Avron, M. (1966). *Biochem. Biophys. Res. Commun.* **24**, 93.

Carmeli, C., and Racker, E. (1973). *J. Biol. Chem.* **248**, 8281.

Carmeli, C., Lifshitz, Y., and Gepshtein, A. (1975). *Biochim. Biophys. Acta* **376**, 249.

Casadio, R., Baccarini-Melandri, A., Zannoni, D., and Melandri, B. A. (1974). *FEBS Lett.* **49**, 203.

Cattell, K. J., Knight, I. G., Lindop, C. R., and Beechey, R. B. (1970). *Biochem. J.* **125**, 169.

Chang, I. C., and Kahn, J. S. (1966). *Arch. Biochem. Biophys.* **117**, 282.

Crofts, A. R. (1966). *Biochem. Biophys. Res. Commun.* **24**, 725.

Datta, D. B., Ryrie, I. J., and Jagendorf, A. T. (1974). *J. Biol. Chem.* **249**, 4404.

Deters, D. W., Racker, E., Nelson, N., and Nelson, H. (1975). *J. Biol. Chem.* **250**, 1041.

Farron, F. (1970). *Biochemistry* **9**, 3823.

Farron, F., and Racker, E. (1970). *Biochemistry* **9**, 3829.

Felker, P., Izawa, S., Good, N. E., and Houg, A. (1974). *Arch. Biochem. Biophys.* **162**, 345.

Gaensslen, R. E., and McCarty, R. E. (1971). *Arch. Biochem. Biophys.* **147**, 55.

Gaensslen, R. E., and McCarty, R. E. (1972). *Anal. Biochem.* **48**, 504.

Garber, M. P., and Steponkis, P. L. (1974). *J. Cell Biol.* **63**, 24.

Gepshtein, A., and Carmeli, C. (1974). *Eur. J. Biochem.* **44**, 593.

Girault, G., and Galmiche, J. M. (1972). *FEBS Lett.* **19**, 315.

Girault, G., Galmiche, J.-M., Michel-Villaz, M., and Thiery, J. (1973). *Eur. J. Biochem.* **38**, 473.

Girault, G., Galmiche, J.-M., and Vermeglio, A. (1974). *Proc. Int. Congr. Photosynth. 3rd, 1974* p. 839.

Grebanier, A., and Jagendorf, A. J. (1977). *Plant Cell Physiol.* (in press).

Harris, D. A., and Slater, E. C. (1975). *Biochim. Biophys. Acta* **387**, 335.

Harris, D. A., Rosing, J., van de Stadt, R. J., and Slater, E. C. (1973). *Biochim. Biophys. Acta* **314**, 149.

Hill, A. V. (1910). *J. Physiol. (London)* **40**, 4.

Hoch, G., and Martin, I. (1963). *Biochem. Biophys. Res. Commun.* **12**, 223.

Hochman, A., and Carmeli, C. (1971). *FEBS Lett.* **13**, 36.

Horio, T., Nishikawa, K., Katsumata, M., and Yamashita, J. (1965). *Biochim. Biophys. Acta* **64**, 577.

Izawa, S., and Hind, G. (1967). *Biochim. Biophys. Acta* **143**, 377.

Jagendorf, A. T. (1975). In "Bioenergetics of Photosynthesis" (Govindjee, ed.), p. 414. Academic Press, New York.

Howell, S. H., and Moudrianakis, E. N. (1967). *Proc. Natl. Acad. Sci. U.S.A.* **58**, 1261.

Jagendorf, A. T. (1977). In "Encyclopedia of Plant Physiology, New Series" (A. Trebst and M. Avron, eds.), Vol. 5, p. 307. Springer-Verlag, Berlin and New York.

Jagendorf, A. T., and Smith, M. (1962). *Plant Physiol.* **37**, 135.

Jagendorf, A. T., and Uribe, E. (1966). *Proc. Natl. Acad. Sci. U.S.A.* **55**, 170.

Jaynes, J. M., Vernon, L. P., and Klein, S. M. (1975). *Biochim. Biophys. Acta* **408**, 240.

Johansson, B. C. (1972). *FEBS Lett.* **20**, 339.

Johansson, B. C., and Baltscheffsky, M. (1975). *FEBS Lett.* **53**, 221.

Johansson, B. C., Baltscheffsky, M., Baltscheffsky, H., Baccarini-Melandri, A., and Melandri, A. (1973). *Eur. J. Biochem.* **40**, 109.

Junge, W., and Ausländer, W. (1974). *Biochim. Biophys. Acta* **333**, 59.

Kagawa, Y., and Racker, E. (1971). *J. Biol. Chem.* **246**, 5477.

Kahn, J. S. (1968). *Biochim. Biophys. Acta* **153**, 203.

Konings, A. W. T., and Guillory, R. J. (1973). *J. Biol. Chem.* **248**, 2522.

Kraayenhof, R. (1969). *Biochim. Biophys. Acta* **180**, 213.

Kraayenhof, R., and Slater, E. C. (1974). *Proc. Int. Congr. Photosynth. Res., 3rd, 1974* Vol. 2, p. 985.

Lardy, H. A., and Elvehjem, C. A. (1945). *Annu. Rev. Biochem.* **14**, 1.

Lien, S., and Racker, E. (1971a). In "Methods in Enzymology" (A. San Pietro, ed.), Vol. 23, p. 547. Academic Press, New York.

Lien, S., and Racker, E. (1971b). *J. Biol. Chem.* **246**, 4298.

Lutz, H. V., Dahl, J. S., and Bachofen, R. (1974). *Biochim. Biophys. Acta* **347**, 359.

Lynn, W. S., and Staub, K. D. (1969). *Proc. Natl. Acad. Sci. U.S.A.* **63**, 540.

McCarty, R. E. (1976). In "Encyclopedia of Plant Physiology, New Series" (U. Heber and C. R. Stocking, eds.), Vol. 3, p. 347. Springer-Verlag, Berlin and New York.

McCarty, R. E., and Fagan, J. (1973). *Biochemistry* **12**, 1503.

McCarty, R. E., and Portis, A. R., Jr. (1976). *Biochemistry* **15**, 5110.

McCarty, R. E., and Racker, E. (1966). *Brookhaven Symp. Biol.* **19**, 202.

McCarty, R. E., and Racker, E. (1967). *J. Biol. Chem.* **242**, 3435.

McCarty, R. E., and Racker, E. (1968). *J. Biol. Chem.* **243**, 129.

McCarty, R. E., Guillory, R. J., and Racker, E. (1965). *J. Biol. Chem.* **240**, 4822.

McCarty, R. E., Fuhrman, J. ., and Tsuchiya, Y. (1971). *Proc. Natl. Acad. Sci. U.S.A.* **68**, 2522.

McCarty, R. E., Pittman, P. R., and Tsuchiya, Y. (1972). *J. Biol. Chem.* **247**, 3048.

Magnusson, R. P., and McCarty, R. E. (1975). *J. Biol. Chem.* **250**, 2593.

Magnusson, R. P., and McCarty, R. E. (1976a). *Biochem. Biophys. Res. Commun.* **70**, 1283.

Magnusson, R. P., and McCarty, R. E. (1976b). *J. Biol. Chem.* **251**, 6874.

Magnusson, R. P., and McCarty, R. E. (1976c). *J. Biol. Chem.* **251**, 7417.

Marchant, R. H., and Packer, L. (1963). *Biochim. Biophys. Acta* **75**, 458.

Melandri, B. A., and Baccarini-Melandri, A. (1976). *J. Bioenerget.* **8**, 109.

Melandri, B. A., Baccarini-Melandri, A., San Pietro, A., and Gest, H. (1970). *Proc. Natl. Acad. Sci. U.S.A.* **67**, 477.

Melandri, B. A., Baccarini-Melandri, A., San Pietro, A., and Gest, H. (1971). *Science* **174**, 514.

Melandri, B. A., Baccarini-Melandri, A., and Fabbri, E. (1972). *Biochim. Biophys. Acta* **275**, 383.

Melandri, B. A., Fabbri, E., Firstater, E., and Baccarini-Melandri, A. (1974). *In* "Membrane Proteins in Transport and Phosphorylation" (G. F. Azzone *et al.*, eds.), p.55. Elsevier, North-Holland.

Miller, K. R., and Staehelin, L. A. (1976). *J. Cell Biol.* **68**, 30.

Mitchell, P. (1961). *Nature (London)* **191**, 144.

Mitchell, P. (1966). *Biol. Rev. Cambridge Philos. Soc.* **41**, 445.

Mitchell, P. (1967). *Fed. Proc., Fed. Am. Soc. Exp. Biol.* **26**, 1370.

Mitchell, P. (1974). *FEBS Lett.* **43**, 189.

Mitchell, P. (1975). *FEBS Lett.* **50**, 95.

Mitchell, P., and Moyle, J. (1974). *Biochem. Soc. Spec. Pub.* **4**, 91.

Mukohata, Y., and Yagi, T. (1974). *Bioenergetics* **7**, 111.

Nelson, N., Nelson, H., and Racker, E. (1972a). *J. Biol. Chem.* **247**, 6506.

Nelson, N., Nelson, H., and Racker, E. (1972b). *J. Biol. Chem.* **247**, 7657.

Nelson, N., Deters, D. W., Nelson, H., and Racker, E. (1973). *J. Biol. Chem.* **248**, 2049.

Oliver, D. J., and Jagendorf, A. T. (1976). *J. Biol. Chem.* **251**, 7168.

Pederson, P. L. (1975). *Bioenergetics* **6**, 243.

Penefsky, H. S., Pullman, M. E., Datta, A., and Racker, E. (1960). *J. Biol. Chem.* **235**, 3330.

Petrack, B., and Lipmann, F. (1961). *In* "Light and Life" (W. D. McElroy and H. B. Glass, eds.), p. 621. Johns Hopkins Press, Baltimore, Maryland.

Pflugshaupt, C., and Bachofen, R. (1975). *Bioenergetics* **7**, 49.

Pick, U., Rottenberg, H., and Avron, M. (1973). *FEBS Lett.* **32**, 91.

Porat, N., Ben-Shaul, Y., and Friedberg, I. (1976). *Biochim. Biophys. Acta* **440**, 365.

Portis, A. R., Jr., and McCarty, R. E. (1973). *Arch. Biochem. Biophys.* **156**, 621.

Portis, A. R., Jr., and McCarty, R. E. (1974). *J. Biol. Chem.* **249**, 6250.

Portis, A. R., Jr., and McCarty, R. E. (1976). *J. Biol. Chem.* **251**, 1610.

Portis, A. R., Jr., Magnusson, R. P., and McCarty, R. E. (1975). *Biochem. Biophys. Res. Commun.* **64**, 877.

Pullman, M. E., Penefsky, H. S., Datta, A., and Racker, E. (1960). *J. Biol. Chem.* **235**, 3322.

Racker, E. (1975). *In* "Electron Transport Chains and Oxidative Phosphorylation" (E. Quagliarello *et al.*, eds.), p. 407. North-Holland Publ., Amsterdam.

Racker, E. (1977). *Annu. Rev. Biochem.* **46**, 1006.

Racker, E., and Stoeckenius, W. (1974). *J. Biol. Chem.* **249**, 662.

Racker, E., Hauska, G. A., Lein, S., Berzborn, R. J., and Nelson, N. (1972). *Photosynth., Two Centuries Its Discovery Joseph Priestley, Proc. Int. Congr. Photosynth. Res., 2nd, 1971* Vol. III, p. 1097.

Reeves, S. G., and Hall, D. O. (1973). *Biochim. Biophys. Acta* **314**, 66.

Rosing, J., Kayalar, C., Smith, D. J., and Boyer, P. D. (1977). *J. Biol. Chem.* **252**, 2487.

Roy, H., and Moudrianakis, E. N. (1971a). *Proc. Natl. Acad. Sci. U.S.A.* **68**, 461.

Roy, H., and Moudrianakis, E. N. (1971b). *Proc. Natl. Acad. Sci. U.S.A.* **68**, 2720.

Rumberg, B., and Siggel, U. (1969). *Naturwissenshaften* **56**, 130.

Rumberg, B., Reinwald, E., Schröder, H., and Siggel, U. (1969). *Prog. Photosynth. Res., Proc. Int. Congr. [1st], 1968* Vol. III, p. 1374.

Ryrie, I. J., and Jagendorf, A. T. (1971a). *J. Biol. Chem.* **246**, 582.

Ryrie, I. J., and Jagendorf, A. T. (1971b). *J. Biol. Chem.* **246,** 3771.
Ryrie, I. J., and Jagendorf, A. T. (1972). *J. Biol. Chem.* **247,** 4453.
Schmid, R., and Junge, W. (1975). *Biochim. Biophys. Acta* **394,** 76.
Schröder, H., Muhle, H., and Rumberg, B. (1972). *Photosynth., Two Centuries Its Discovery Joseph Priestley, Proc. Int. Congr. Photosynth. Res., 2nd, 1971* Vol. III, p. 1571.
Schwartz, M. (1968). *Nature (London)* **219,** 915.
Serrano, R., Kanner, B. E., and Racker, E. (1976). *J. Biol. Chem.* **251,** 2453.
Skye, G. E., Shavit, N., and Boyer, P. D. (1967). *Biochem. Biophys. Res. Commun.* **28,** 724.
Smith, D. J., Stokes, B. O., and Boyer, P. D. (1976). *J. Biol. Chem.* **251,** 4165.
Stekhoven, F. S., Waittrus, R. F., and van Moerkerk, H. T. B. (1972). *Biochemistry* **11,** 1144.
Strotmann, H., Hesse, H., and Edelmann, K. (1973). *Biochim. Biophys. Acta* **314,** 202.
Strotmann, H., Bickel, S., and Huchzermeyer, B. (1976). *FEBS Lett.* **61,** 194.
Telfer, A., and Evans, M. C. W. (1972). *Biochim. Biophys. Acta* **256,** 625.
Vallejos, R. H., and Andreo, C. S. (1976). *FEBS Lett.* **61,** 95.
Vallejos, R. H., Ravizzini, R. A., and Andreo, C. S. (1977). *Biochim. Biophys. Acta* **459,** 20.
Vambutas, V., and Bertsch, W. (1975). *Biochim. Biophys. Acta* **376,** 169.
Vambutas, V. K., and Racker, E. (1965). *J. Biol. Chem.* **240,** 2660.
Vandermeulen, D. L., and Govindjee. (1975). *FEBS Lett.* **57,** 272.
van de Stadt, R. J., de Boer, B. L., and van Dam, K. (1973). *Biochim. Biophys. Acta* **292,** 338.
Williams, R. J. P. (1975). *In* "Electron Teansport Chains and Oxidative Phosphorylation" (E. Quagliariello *et al.*, eds.), p. 417. North-Holland Publ., Amsterdam.
Winget, G. D., Izawa, S., and Good, N. E. (1969). *Biochemistry* **8,** 2067.
Winget, G. D., Kanner, N., and Racker, E. (1977). *Biochim. Biophys. Acta* **460,** 490.
Witt, H. T. (1971). *Q. Rev. Biophys.* **4,** 365.
Yamamoto, N., Yoshimura, S., Higuti, T., Wishikawa, K., and Horio, T. (1972). *J. Biochem. (Tokyo)* **72,** 1397.
Yamamoto, T., and Tonomura, Y. (1975). *J. Biochem. (Tokyo)* **77,** 137.
Younis, H., Winget, G. D., and Racker, E. (1977). *J. Biol. Chem.* **252,** 1814.

Volume 8 will contain the third article in this section, *Electron Transport and Photophosphorylation,* as well as the articles comprising the sections on *The Photosynthetic Membrane* and *Genetic Control of the Photosynthetic Membrane.*

Subject Index

A

Absorbance-change measurements, in picosecond range, 48–51
ADRY reagents, role in electron transport chains, 210
Algae
  electron transport in, 183–191
  picosecond fluorescence measurements of photosynthesis in, 55–56
ATPase complex
  of chloroplasts and chromatophores, 245–278
  conformational changes in, 266–268
  functions of, 257–263
    in electron transport, 261–263
    in photophosphorylation, 257–261
  hydrophobic components, 255–257
  isolation, 255–256
  membrane proteins, 256–257
  mechanisms for, 263–266
  soluble component of, 246–255
    nucleotide binding, 251–253
    reconsititution studies, 253–255
    size and structure, 248–251

B

Bacteria
  electron-transport chains of, 175–244
  photosynthetic type
    electron acceptors in, 111–131
    electron transport in, 191–197
    picosecond studies on, 59
    proton pumps of, 217–234
Bacteriopheophytin, photosynthetic properties of, 125–126

C

C-550, primary electron acceptor and, 150–153

Carotenoid change, as indicator of membrane potential, 225–228
Carotenoid triple states, in bacterial photosynthesis, 64–65
Chlorophyll
  electron spin resonance studies on, 21–25
  electron transfer reactions of, 10–13
  energy transfer in, 6
  fluorescence quenching of, 6–8
  ion pair formation of, 12–17
    dissociation of, 13–17
  light absorption by, 4–5
  oxidatant and reductant effects on, 6–7
  oxidation of solvent by, 28–31
  oxidized, visible spectra of, 25–27
  photochemistry of, 1–37
  photopotentials of, 27–28
  quinone reaction with, 17–34
  self-quenching of, 7
  singlet excited state, reactivity of, 31–34
  singlet state decay of, 5–6
  triplet excited state of, 8–10
Chloroplasts
  ATPase complex of, 245–278
  electron-transport chain of, 140
  photosystem II of, 139–172
  picosecond fluorescence measurements of photosynthesis in, 53–55, 66
Chromatium, photosynthetic reaction center of, properties, 133
Cytochrome *f*
  in electron transport in algae, 186–188
Chromatophores
  acceptor pools in, 222–223
  ATPase complex of, 245–278
Coupling factor-ATPase, properties of, 246–255
Cytochromes, electron flow between redox potential effects, 228–229
Cytochrome b_{559}, photooxidation of, 161–162
Cytochrome c_2, electron transport to, 223–225

279